FUNDAMENTALS OF ALGEBRAIC MICROLOCAL ANALYSIS

PURE AND APPLIED MATHEMATICS

A Program of Monographs, Textbooks, and Lecture Notes

MONOGRAPHS AND TEXTBOOKS IN
PURE AND APPLIED MATHEMATICS

FUNDAMENTALS OF ALGEBRAIC MICROLOCAL ANALYSIS

Goro Kato

California Polytechnic State University
San Luis Obispo, California

Daniele C. Struppa

George Mason University
Fairfax, Virginia

CRC Press
Taylor & Francis Group
Boca Raton London New York

CRC Press is an imprint of the
Taylor & Francis Group, an **informa** business

CRC Press
Taylor & Francis Group
6000 Broken Sound Parkway NW, Suite 300
Boca Raton, FL 33487-2742

First issued in paperback 2019

© 1999 by Taylor & Francis Group, LLC
CRC Press is an imprint of Taylor & Francis Group, an Informa business

No claim to original U.S. Government works

ISBN-13: 978-0-8247-9327-2 (hbk)
ISBN-13: 978-0-367-40000-2 (pbk)

Visit the Taylor & Francis Web site at
http://www.taylorandfrancis.com

and the CRC Press Web site at
http://www.crcpress.com

This book is dedicated to our sons Alexander Benkei and Alessandro.

Human life is a long arduous journey to be undertaken patiently. Convince yourself insufficiency is The Way, and then you will miss nothing. If unattainable desire presents itself to you, recall your time of most difficult struggle. If you know only victories without experiencing defeats, self-absorption becomes your nature. Remember having quite not enough is superior to having in excess. Persistence is the foundation for your secure and enduring life.

Shōgun Iyeyasu Tokugawa

Preface

This book is an introduction to the algebraic theory of systems of differential equations, as developed by the Japanese school of M. Sato and his co-workers, mainly T. Kawai and M. Kashiwara. The book may be used for an advanced graduate course, or as a reference for some of the fundamental aspects of this theory. On the other hand, we also hope that our work will be of some use to scholars in partial differential equations who want to acquaint themselves with the algebraic methods of microlocal analysis. Finally, algebraic geometers will recognize that the flavor of our work is very much in tune with their own taste, and may be attracted to this beautiful subject, which has not yet received the attention it deserves.

The expression "microlocal analysis" refers, from a general point of view, to that approach to the study of partial differential equations which moves the problem of singularities to the cotangent bundle of the variety on which the differential equations are defined. From this point of view, it is a very well developed subject, which probably needs no introduction (since the interested student should then study the fundamental work of Hörmander [89], or, at a more elementary level, the excellent work of Treves [224]); on the other hand, what we are trying to describe in this book is the algebro-geometric approach, which can be best summarized under the heading of algebraic microlocal analysis. In their classical work [123], Kashiwara, Kawai, and Kimura pointedly observe that the Japanese algebraic analysis is really the algebraic analysis in the tradition of Euler; we may add that what we mean by algebraic microlocal analysis is the successful attempt to adapt the methods of abstract algebraic geometry to the non-commutative setting in which the base ring is now the ring of variable coefficients partial differential operators.

The origin of the theory goes back to the early papers of Sato on hyperfunctions in the late 1950s, but it was only in the early 1970s that the full potential of the theory of hyperfunctions and microfunctions became evident.

When Sato introduced hyperfunctions, he was mainly guided by the belief that the natural setting for a theory of differential equations should be the analytic setting, rather than the differentiable one as in Schwartz's theory of distributions. At the same time, his background in theoretical physics naturally led him to the development of the study of boundary values of holomorphic functions, and from there to hyperfunctions. As we will show in Chapter 1, hyperfunctions are indeed (sums of) boundary values of holomorphic functions, but, more precisely, they can be defined by the use of sheaf cohomology, and in this way, they present an obvious advantage over other theories of generalized functions, as they allow the use of algebraic tools for their study, and especially of sheaf theory. In a fundamental, but relatively little known, paper of Ehrenpreis [47], it was shown how sheaf theory could play a crucial role in the algebraic treatment of differential equations; the paper (maybe not surprisingly) had little repercussion, since the sheaf of distributions was not well suited for an algebraic treatment (in particular, as we know, it is not a flabby sheaf, and therefore distribution resolutions of solution sheaves of differential equations do not carry enough information). The advent of hyperfunctions was finally necessary for Ehrenpreis' ideas to be implemented.

We should point out, in this regard, that the local, sheaf-theoretic definition of hyperfunctions and of microfunctions allows us to study these objects on arbitrary real analytic manifolds (in fact, we don't even need to require orientability); however, in this book we have decided to stick to the euclidean case (i.e. we study only differential equations on $I\!\!R^n$), since this allows us to avoid requiring even more prerequisites from the theory of real analytic manifolds. Readers who are familiar with this theory will have no trouble in extending the results of this book to the more general setting.

The purpose of this book is therefore to introduce the reader to the developments that in the last forty years have led from the creation of hyperfunctions to the modern algebraic treatment of microlocal analysis. Our choice of topics has been oriented to give the reader a sound foundation in the necessary tools and to ensure that all the fundamental results are included. This has implied that many interesting and even important features have not been included, because they might have led us away from the main objective, or because they would have required more prerequisites than we were willing to accept.

An interesting feature of our work has been the attempt to provide a historical perspective for the development of these ideas. This does not mean at all that we are giving precise and appropriate credit to all those who have worked in the field. We have striven to be as complete as possible, but we are aware of our limitations in this respect. On the other hand, we have tried to provide the reader with a sense of the development of ideas, and of the motivation underlying these ideas.

We now outline the architecture of the book, and highlight the links among different chapters.

After an introductory chapter in which we try to give a self-contained description of hyperfunctions, we define and study the sheaf of singularities of hyperfunctions, which is usually known as the sheaf of microfunctions. The notion of microfunction is rather delicate, and to motivate it, we preface its introduction with a treatment of the differentiable case. In simple terms, a microfunction is a singularity of a hyperfunction (more precisely, the sheaf of microfunctions is the sheaf of singularities of hyperfunctions, pulled back to the cotangent bundle of the manifold on which the sheaf of hyperfunctions is defined; in our case this last sheaf is over $I\!\!R^n$, as we mentioned earlier).

We then proceed to give a rather extensive treatment of microfunctions, and we prove most of the fundamental results of Sato, Kawai, and Kashiwara which provide the foundations for this topic.

The algebraic treatment of systems of differential equations is approached in Chapters 3, 4, and 5, which can be read almost independently of the rest of the book, as long as the fundamental definitions from Chapters 1 and 2 are understood. The algebraic treatment of systems of differential equations is a well honored subject, which goes back to the early 1960s (see the references in Palamodov's fundamental book [178]), but its first application to variable coefficients differential equation is probably due to Kashiwara [108]. The point here is that, as in algebraic geometry, where ideals replace specific representations for algebraic varieties, in the theory of partial differential equations it is possible to replace specific representations of systems by what we call $\mathcal{D}$-modules, i.e., sheaf of modules over the sheaf $\mathcal{D}$ of variable coefficients differential operators. We regret to say that the treatment in these chapters is rather complicated, and some expertise in homological algebra is required. We do try to provide the necessary tools from the theory of derived categories, but these chapters may present some difficulties for the inexperienced reader. We have tried to compensate for the inherent difficulties by giving as many concrete examples as we could.

We finally come back to a more analytic point of view in the last chapter of the book, where the fundamental Sato's structure theorem for systems of differential equations is established. This is the crowning result of the pioneering period of the creation of algebraic microlocal analysis, and is definitely the starting point for any serious study of this subject.

We have said something in passing, but let us now be more specific about the prerequisites necessary for an understanding of this book. We have tried to provide the reader with the basic notions that will allow him or her to progress to the more advanced books in the field [103], [123], [206]. In view of the complexity of the topic, we have given as much background as possible, while limiting the size of the volume. We have not, however, been able to provide all the necessary prerequisites, and in fact we will assume that the reader is familiar with the content of a first course in complex analysis (at the undergraduate level), as well as with the rudiments of the theory of several complex variables as given, e.g., in [71], and [192]. These topics are necessary even for the first chapter in which hyperfunctions are discussed. This chapter, as well as the second one, does not require much algebraic know-how, but on the other hand they require some familiarity with the fundamental notions of sheaves and of sheaf cohomology.

As we proceed to the following chapters, the reader will need some background in algebra, as can usually be obtained from the first two algebra courses in graduate school. More precisely, we will require some fundamental notions from commutative algebra, as well as some notions from homological algebra.

We want to express our gratitude to Professors M. Sato, M. Kashiwara, and T. Kawai from whom we learned most of this subject. In fact, we want to offer this work as a modest and unworthy tribute to the great contribution of the Japanese school to the creation and development of modern algebraic analysis. We further want to express our indebtedness to the great masterpieces of algebraic analysis, namely, [123] and [206]. In the course of our studies, we have been lucky enough to be in contact with many of the people who, in one way or another, have contributed to the development of algebraic analysis. In particular, we would like to express our gratitude to Professors C. A. Berenstein, L. Ehrenpreis, V. P. Palamodov, and P. Schapira, for many enlightening discussions. We are further grateful to the members and staffs at the Research Institute for Mathematical Sciences, Kyoto, and the Institute for Advanced Study, Princeton, for their assistance during our visits. We are indebted to Dr. Irene Sabadini and Mr. Domenico Napoletani for reading several preliminary versions. The first author also wishes to thank his Chairman, Professor S. Weinstein, for providing him with some release time. We also thank Ms. Lynn Hanson for her impeccable typesetting.

On the nonmathematical side, the authors wish to express our gratitude to our spouses, Christine and Carmen, for their support while the book was written, and to the staff of Marcel Dekker, Inc. for their incredible patience.

Goro Kato
Daniele C. Struppa

Contents

FUNDAMENTALS OF
ALGEBRAIC
MICROLOCAL ANALYSIS

Chapter 1

Hyperfunctions

1.1   Introduction

In this chapter we will introduce the first object necessary for the study of
Microlocal Analysis, as developed by the Japanese school, namely the sheaf of
hyperfunctions. As it is known since Schwartz's introduction of the notion of
distribution [210], there is no hope of dealing with the subtle issues posed by
the theory of partial differential equations unless one resorts to some kind of
generalized functions. In the case of Schwartz, the notion which was developed
is that of distribution, i.e. of a continuous linear functional on the topological
space of infinitely differentiable functions with compact support. Other different
spaces have been studied and developed for a variety of different reasons, see e.g.
[14], [17], [37], but the choice of the Japanese school led by M. Sato has been to
employ a space of functions which can be defined on any analytic manifold and
which somehow generalizes the space of distributions itself.

As we will see, the definition of this space is quite natural in the case of a
single real variable (where this notion also has a well established history, as we
shall see in the historical notes to this chapter), but, on the other hand, is quite
complicated and not too intuitive in the case of several variables. For this reason
we will begin this chapter with the case of one variable, where all the details can
be easily explained, and only after this introductory treatment we will devote
ourselves to the more complex issues which stem from the generalization to
several variables.

Providing an introduction to the theory of hyperfunctions is not a straightforward task, since several different approaches could be followed; in the first
volume of his monumental work [89], Hörmander chooses to define hyperfunctions following an approach similar to the one used by Schwartz to introduce
distributions; this approach is quite interesting, and we will come back to it as
we develop the microlocalization technique in Chapter II, but it does not seem
to convey the spirit of the Japanese school, and we have therefore decided to
follow the approach originally used by Sato in his first groundbreaking papers

[195], [196], [197]. This choice, akin to the one followed by Kaneko in his beautiful book [103], seems more apt to the study of the generalizations which will follow in the subsequent chapters.

With respect to this last remark, it should be noted that this chapter is only introductory to the rest of our work, and that its scope is somehow limited, in view also of Kaneko's [103], which provides a complete, thorough and highly readable introduction to the study of the theory of hyperfunctions.

As for the structure of this chapter, we provide, in section 1.2, the first definitions of single variable hyperfunctions, while the first important theorems are given in section 1.3. The extension of the definitions and of the results to the case of several variables will be dealt with in sections 1.4 and 1.5. In keeping with the rest of the book, we confine our treatment to the case of hyperfunctions defined on the Euclidean spaces $\mathbb{R}^n$, but we provide, at the end of section 1.5, a brief description on how the notion of hyperfunction can be extended to the case of real analytic manifolds; the chapter ends with a rather extensive set of historical notes on the birth of the theory of hyperfunctions.

For the reader interested in a more thorough treatment of the concepts which will be introduced and developed in this chapter, we wish to point out three basic references which should be kept in consideration throughout this book: the first treatment by Sato and his coworkers Kashiwara and Kawai, [206], the treatise of Kashiwara, Kawai and Kimura, [123], and finally the work of Kaneko, [103]. One comment may be necessary on these three references; the first one (usually referred to as SKK) is a fundamental work, whose readability, however, is not optimal, also in view of the fact that it is the first introduction to a difficult subject, written as the subject itself was being developed; the second treatise, to which we will refer to as KKK, is a very comprehensive discussion of Algebraic Analysis, in which the Japanese algebraic methods to study partial differential equations are described; finally, Kaneko's book is the most readable, but his analysis stops at the notion of hyperfunction (with a treatment of the notion of microfunction) and no attempt is made to deal with the most interesting topics concerning the algebraic treatment of systems of differential equations (namely $\mathcal{E}$-modules and $\mathcal{D}$-modules).

1.2 Hyperfunctions of One Variable: Basic Definitions

Let Ω be an open set in $\mathbb{R}$, and let V be a **complex neighborhood** of Ω, i.e. V is an open set in the complex plane $\mathbb{C}$ such that Ω is relatively closed in V; by this term we mean that Ω is compact in V and, for simplicity, the reader may fix his/her attention on Ω being an open interval (a, b) in the real line and V being an open set in $\mathbb{C}$ whose intersection with $\mathbb{R}$ is exactly (a, b), as indicated

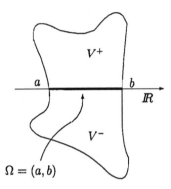

Figure 1.2.1

in figure 1.2.1. We can then define the open sets

$$V^+ := \{\, z \in V : Imz > 0 \,\}, V^- := \{\, z \in V : Imz < 0 \,\},$$

and the set $\mathcal{B}(\Omega)$ of hyperfunctions as the quotient

$$(1.2.1) \qquad \qquad \mathcal{B}(\Omega) = \mathcal{B}(\Omega; V) = \frac{\mathcal{O}(V \setminus \Omega)}{\mathcal{O}(V)}$$

where $\mathcal{O}$ denotes the sheaf of germs of holomorphic functions and, therefore, for any given open set U in $\mathbb{C}$, $\mathcal{O}(U)$ denotes the set of holomorphic functions on U.

Before we can make any remarks on the set $\mathcal{B}(\Omega)$, we must point out an apparent oddity in its definition, which seems to depend on the choice of the complex neighborhood V. This is not the case in view of a deep result in the theory of holomorphic functions of one complex variable, namely the so-called Mittag-Leffler Theorem. In classical textbooks, the Mittag-Leffler Theorem is also known as the theorem on the existence of a meromorphic function with assigned polar singularities and assigned principal parts. In cohomological terms, this theorem sounds as follows:

Theorem 1.2.1 *(Mittag-Leffler: cohomological formulation). For any open set U in the complex plane, the first cohomology group of U with coefficients in the sheaf of germs of holomorphic functions vanishes, i.e.*

$$H^1(U, \mathcal{O}) = 0.$$

Proof. This is an immediate consequence, almost a rephrasing, of Cartan's Theorem A. More directly, it can be shown, when in the form described in Theorem 1.2.1′ below, by a simple Cauchy integral formula argument, which is described in detail in [103].   □

Of course, such a statement has a more direct and less abstract formulation, which is what we will need to prove that hyperfunctions are a well defined object:

Theorem 1.2.1′. *(Mittag-Leffler; explicit formulation).* Let U, V be two open sets in $\mathbb{C}$. For any function F in $\mathcal{O}(U \cap V)$ there are functions F_1 in $\mathcal{O}(U)$ and F_2 in $\mathcal{O}(V)$ such that, on the intersection $U \cap V$, it is

$$F_1(z) - F_2(z) = F(z).$$

Let us now show how this result can be used to prove that the quotient (1.2.1) does not depend on the choice of V.

Proposition 1.2.1 *Let U and V be two complex neighborhood of the same open set Ω. Then there is a vector space isomorphism between the quotient spaces*

$$\mathcal{B}(\Omega; U) = \frac{\mathcal{O}(U \setminus \Omega)}{\mathcal{O}(U)} \text{ and } \mathcal{B}(\Omega; V) = \frac{\mathcal{O}(V \setminus \Omega)}{\mathcal{O}(V)}.$$

Proof. To begin with, we may assume U to contain V, since we can always prove the isomorphism between $\mathcal{B}(\Omega; V)$ and $\mathcal{B}(\Omega; U \cap V)$ and the isomorphism between $\mathcal{B}(\Omega; U)$ and $\mathcal{B}(\Omega; U \cap V)$, which would prove the isomorphism between $\mathcal{B}(\Omega; U)$ and $\mathcal{B}(\Omega; V)$. Consider now the natural map ρ, induced by the restriction mapping, and the inclusion of V into U,

$$(1.2.2) \qquad \rho : \frac{\mathcal{O}(U \setminus \Omega)}{\mathcal{O}(U)} \longrightarrow \frac{\mathcal{O}(V \setminus \Omega)}{\mathcal{O}(V)}.$$

It is immediately seen that the map in (1.2.2) is injective, since the restriction of a holomorphic function F from $U \setminus \Omega$ to $V \setminus \Omega$ is holomorphic on all of V only if F itself was already holomorphic on all of U. We now only have to prove that ρ is surjective. To do so, take F in $\mathcal{O}(V \setminus \Omega)$, and apply Theorem 1.2.1′ to the pair of open sets $U \setminus \Omega$ and V. We get a function F_1 in $\mathcal{O}(U \setminus \Omega)$ and a function F_2 in $\mathcal{O}(V)$ such that, on the intersection $(U \setminus \Omega) \cap V = V \setminus \Omega$, it is $F(z) = F_1(z) - F_2(z)$. Then the inverse image under ρ of the equivalence class $[F]$ of F is exactly the equivalence class $[F_1]$ in $\mathcal{B}(\Omega; U)$.   □

This result shows that the definition of the space of hyperfunctions does not depend on the choice of the open set in the defining equation (1.2.1), i.e.

$$\mathcal{B}(\Omega) = \varinjlim_{V \supseteq \Omega} \frac{\mathcal{O}(V \setminus \Omega)}{\mathcal{O}(V)};$$

throughout this book, we will denote by $[F]$ the hyperfunction defined, in the quotient (1.2.1), by the equivalence class of the holomorphic function F or, equivalently, by the pair (F^+, F^-), where $F^+ = F_{|V^+}$ and $F^- = F_{|V^-}$.

Note that we can interpret what we have just described by looking at the short exact sequence of global section functors

$$0 \longrightarrow \Gamma(V, V \setminus \Omega; -) \longrightarrow \Gamma(V, -) \overset{\varrho}{\longrightarrow} \Gamma(V \setminus \Omega, -)$$

where $\Gamma(V, V \setminus \Omega; -)$, also denoted by $\Gamma_\Omega(V, -)$, is the kernel of the restriction

$$\varrho : \Gamma(V, -) \longrightarrow \Gamma(V \setminus \Omega, -).$$

The sequence above then becomes the following exact sequence

$$0 \longrightarrow \Gamma(V, V \setminus \Omega; \mathcal{L}) \longrightarrow \Gamma(V, \mathcal{L}) \overset{\varrho}{\longrightarrow} \Gamma(V \setminus \Omega; \mathcal{L}) \longrightarrow 0$$

for $\mathcal{L}$ any flabby sheaf (we will come back later to these notions). But by the definition of flabbiness, the restriction becomes a surjective map and then a standard homological algebra method provides the long exact sequence

$$0 \longrightarrow H^0(V, V \setminus \Omega; \mathcal{O}) \longrightarrow H^0(V, \mathcal{O}) \longrightarrow H^0(V \setminus \Omega; \mathcal{O})$$

$$\longrightarrow H^1(V, V \setminus \Omega; \mathcal{O}) \longrightarrow H^1(V, \mathcal{O}) \longrightarrow H^1(V \setminus \Omega, \mathcal{O}) \longrightarrow \cdots$$

where $H^j(V, \mathcal{O})$ vanishes for all $j = 1, 2, \ldots$. Now the analytic continuation property for an analytic function implies that

$$H^0(V, V \setminus \Omega; \mathcal{O}) = 0$$

i.e. if $f_{|V \setminus \Omega} = 0$ for $f \in H^0(V, \mathcal{O}) = \Gamma(V, \mathcal{O})$, then f is identically zero in V. Then we obtain an isomorphism

$$\mathcal{B}(\Omega) = \frac{H^0(V \setminus \Omega; \mathcal{O})}{H^0(V, \mathcal{O})} \overset{\approx}{\longrightarrow} H^1(V, V \setminus \Omega; \mathcal{O}) = H^1_\Omega(V; \mathcal{O})$$

to which we will refer frequently later on.

Now that we have a properly defined notion of hyperfunction on an open subset of the real line it may be natural to look for an interpretation of such a concept, especially if we want to show that we are really defining a sort of generalized function. To begin with, we note that because of its definition, the set $\mathcal{B}(\Omega)$ actually inherits a structure of vector space over $\mathbb{C}$, even though, at least at this moment, we are unable to provide it with any topological structure. To understand what $\mathcal{B}(\Omega)$ stands for, we may recall a famous result by Painleve' [175], actually a formulation of a special case of Schwarz reflection principle, [192]:

Theorem 1.2.2 *(Painleve's Theorem) Let Ω be an open set in the real line, and let V be a complex neighborhood of Ω. If a function f is holomorphic in $V \setminus \Omega$, and continuous on V, then f is holomorphic on V.*

Proof.   Once again this is a consequence of Cauchy's integral formula and of Morera Theorem. For further details the reader is referred to any classical text in one complex variable.   □

For later purposes, it may be useful to point out a stronger version of this same result, in which the notion of continuity is replaced by a weaker form (actually a distributional form). To do so, note that the continuity hypothesis on F only meant that the limit of $F(x+iy)$, $x \in \Omega$, as y converges to 0 from the right, and the limit of $F(x+iy)$, $x \in \Omega$, as y converges to 0 from the left, coincide (in symbols, $\lim_{y \to 0+} F(x+iy) = \lim_{y \to 0-} F(x+iy)$). These limits, instead of being taken in the strong sense, may be taken in the sense of distributions, to obtain the following result:

Theorem 1.2.3 *Let F^+ and F^- be two functions holomorphic, respectively, in V^+ and in V^-. If, for any compactly supported infinitely differentiable function φ in Ω, i.e. for any φ in $\mathcal{D}(\Omega)$, it is*

$$\lim_{\alpha \to 0} < F^+(x+i\alpha), \varphi(x) > = \lim_{\alpha \to 0} < F^-(x-i\alpha), \varphi(x) >,$$

where $<,>$ denotes the duality bracket relative to the pair $(\mathcal{D}', \mathcal{D})$, then F^+ and F^- actually glue together to yield a function F holomorphic on V.

Proof.   It is sufficient to reduce the problem to Theorem 1.2.2 by repeated integration.   □

We now readily see that, in view of this result, a hyperfunction should be seen as a holomorphic function on $V \setminus \Omega$, and therefore as a pair of functions (F^+, F^-), with F^+ in $\mathcal{O}(V^+)$ and F^- in $\mathcal{O}(V^-)$, and with the proviso that (F^+, F^-) will denote the zero hyperfunction if there exists a function F, holomorphic on all of V, such that its restriction to $V^\pm$ coincides with $F^\pm$; by Painleve's Theorem this actually identifies a hyperfunction with the "difference" of the boundary values of F^+ and F^-, where the notion of boundary values has of course to be suitably defined. As we will mention in the historical notes, such a concept has been in the minds of mathematicians and physicists for quite a long time, and it was only with the introduction of hyperfunctions that a full and satisfactory treatment of such notion was given.

We now would like to show why and how it is possible to think of hyperfunctions as generalized functions, and which operations can be defined on the vector spaces $\mathcal{B}(\Omega)$. To begin with, we note that all real analytic functions can be naturally thought of as hyperfunctions, i.e. there is, for every open set in $\mathbb{R}$, a natural embedding of the vector space $\mathcal{A}(\Omega)$ into the vector space $\mathcal{B}(\Omega)$. As a matter of fact, if f is any real analytic function on Ω, then, by the definition itself of the space of real analytic functions as the inductive limit, taken on the inductive family of all complex neighborhoods of Ω, of spaces of holomorphic functions,

$$\mathcal{A}(\Omega) = \text{ind} \lim \mathcal{O}(V),$$

we have that f extends to a holomorphic function $\tilde{f}$ defined in some complex neighborhood V of Ω. If we now choose F^+ to be $\tilde{f}$ restricted to V^+, and F^- to be the function identically zero, then the pair (F^+, F^-) defines a hyperfunction which exactly coincides (if the boundary value interpretation is used) with the analytic function f.

We can actually say something more, since we can define a natural product between real analytic functions and hyperfunctions, so that $\mathcal{B}(\Omega)$ turns out to be an $\mathcal{A}(\Omega)$−module. To do so, we simply define the product of the hyperfunction $[F]$ by the real analytic function f to be the hyperfunction defined, in the quotient (1.2.1), by the holomorphic function $\tilde{f}F$, i.e.

$$(1.2.3) \qquad\qquad f[F] = [\tilde{f}F].$$

It can be easily shown that the definition (1.2.3) is well posed, i.e. it does not depend on the choice of the representative F of the hyperfunction F, nor on the choice of the domain of the extension $\tilde{f}$ of the real analytic function f. Such an independence, together with the independence of (1.2.2) from the choice of the complex neighborhoods chosen, can be established with a simple argument based on the Cauchy theorem. Unfortunately, it must be emphasized that hyperfunctions may not be, on the other hand, multiplied, so that they do not enjoy an algebra structure. This is a most unfortunate fact, since one of the problems which physicists have been posing to mathematicians has been the creation of a theory of generalized functions which would allow multiplications; this problem has been at the heart of several different attempts to modify Schwartz's theory of distributions (where the kernel theorem [210] shows the impossibility of imposing any multiplicative structure); among the most interesting attempts, we would like to point the reader's attention to the non-standard analysis approach by Colombeau in which multiplication between generalized functions is indeed possible. In our case, it can be easily seen why a product cannot be defined in any reasonable way; suppose, indeed, that two hyperfunctions $f = [F]$ and $g = [G]$ are defined on some open set Ω of $\mathbb{R}$. Then the first spontaneous attempt to define multiplication (and the only one which would be consistent with the structure of $\mathcal{A}(\Omega)$-module which we have given $\mathcal{B}(\Omega)$ in the previous few pages) would be to define

$$f \cdot g := [F \cdot G].$$

It is however immediately seen that such a definition would not be invariant under the choice of the holomorphic representative of the hyperfunctions. Indeed, since $z - 1$ is holomorphic everywhere, we have that the hyperfunction $[z]$ and the hyperfunction $[1]$ are one and the same hyperfunction (both represent the identically zero hyperfunction); still, the hyperfunctions

$$\left[\frac{1}{z} \cdot 1\right] = \left[\frac{1}{z}\right]$$

and

$$\left[\frac{1}{z} \cdot z\right] = [1]$$

are two different hyperfunctions since $1/z - 1$ is not holomorphic in any neighborhood V of the origin.

We will see in the next chapter, when dealing with microlocalization, that, in some cases, hyperfunctions may be multiplied; it turns out that these cases are exactly the ones of physical interest, so that we have at least a partial solution to the needs of physicists.

Generalized functions are usually introduced as a device towards constructing solutions to equations for which no solutions would otherwise exist. This is certainly the case with hyperfunctions as well; as a matter of fact, the treatment of real analytic coefficients differential equations becomes particularly transparent when hyperfunctions are employed. The deep reason for this is one of the greatest early contributions of Mikio Sato, and will not be completely clear until the last chapter of this book; we can at least begin to define how linear differential operators with real analytic coefficients act on the space of hyperfunctions.

To this purpose, let

$$P(x, \frac{d}{dx}) = \sum_{i=0}^{m} a_i(x) \frac{d^i}{dx^i}$$

be a linear differential operator with real analytic coefficients on an open set Ω of the real line, and let $f = [F]$ be a hyperfunction defined on the same open set; as we noticed before, each coefficient a_i of $P(x, d/dx)$ can be extended to a holomorphic function α_i on some complex neighborhood U of Ω; by possibly taking an intersection, we can assume that F is actually holomorphic on the open set $U \setminus \Omega$. We can therefore define the action of $P(x, d/dx)$ on f as follows:

$$P(x, \frac{d}{dx})[f] := [P(z, \frac{d}{dz})F],$$

where the operator $P(z, d/dz)$ is defined by

$$P(z, \frac{d}{dz}) = \sum_{i=0}^{m} \alpha_i(z) \frac{d^i}{dz^i}.$$

It is not difficult to check that this definition does not depend on the several choices which have been made (the choice of the representative of f, the choice of the open set U and the choice of the extensions α_i).

Before we proceed to the study of some elementary examples of hyperfunctions, we want to illustrate one more important concept. We say that a hyperfunction $f = [F]$ defined on an open set Ω vanishes on an open subset $\Omega' \subseteq \Omega$ if, on Ω', f coincides with the zero hyperfunction (note that, in general, given a hyperfunction f, it is not possible to speak about its value at a point, and therefore we cannot just say $f(x) = 0$). This is of course equivalent to saying that the function F which defines f is holomorphic through the real axis at every point of Ω'. It is not difficult to see that, for every hyperfunction f on Ω, there always exists a largest open subset Ω' of Ω on which f vanishes; it is indeed sufficient to take for Ω' the union of all opens subsets on which f vanishes. We can now prove that f vanishes also on such a union (this may seem trivial if we forget that f cannot be pointwise evaluated). This is however a consequence of the fact that the function F which defines f must be holomorphic across the real line in all points of the open sets whose union is Ω', and therefore in all points of Ω'. This shows that f vanishes on all of Ω'.

For a hyperfunction f in $\mathcal{B}(\Omega)$ we define its support to be the complement in Ω of the largest open subset on which f vanishes. This notion will prove to be very valuable, and is strictly linked with some deeper result which we will obtain in the next section.

For K a compact subset of $\mathbb{R}$, we will denote by $\mathcal{B}_K$ the space of hyperfunctions whose support is contained in K. It is possible to define a notion of integration for such functions in a rather natural way.

Let f be a hyperfunction with compact support K and let F be a defining function for f which, of course, can be chosen to be holomorphic in $U \setminus K$ where U is some complex open set properly containing K. Let now τ be a closed simple piecewise smooth curve contained in U and surrounding K once; we will assume τ to circle around K clockwise. We then define (see figure 1.2.2)

$$\int_{\mathbb{R}} f(x)dx = \int_{\tau} F(z)dz$$

and it is easily seen that the definition is consistent with the choices of F, U and of τ. This independence is, of course, once more a consequence of Cauchy's Theorem.

We have, in particular, all the elements to prove a simple yet fundamental result which, as it will be seen in the historical notes to this chapter, was actually already known to Fantappie' (it was actually the beginning of his theory of analytic functionals).

Theorem 1.2.4 *Let $\mathcal{B}_K$ be the space of hyperfunctions supported by the compact set K. For any complex neighborhood U of K, the following isomorphism holds:*

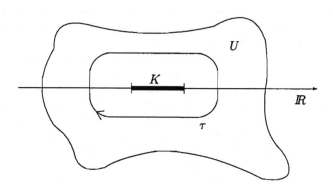

Figure 1.2.2

(1.2.4)
$$\mathcal{B}_K \cong \frac{\mathcal{O}(U \setminus K)}{\mathcal{O}(U)}.$$

Proof. Let f be in $\mathcal{B}_K$; for any open set Ω containing K, we can think of f as an element of $\mathcal{B}(\Omega)$, by simply continuing f to zero in $\Omega \setminus K$. In particular, we can choose Ω to be $U \cap \mathbb{R}$, which shows that any element of $\mathcal{B}_K$ can be expressed as in the quotient in (1.2.4). On the other hand, if we now choose a complex neighborhood U of Ω and we represent $\mathcal{B}(\Omega)$ as the quotient $\mathcal{O}(U \setminus \Omega)/\mathcal{O}(U)$, we immediately see that the quotient in (1.2.4) corresponds to the space of hyperfunctions having compact support in K. Note that, in particular, the embedding of $\mathcal{B}_K$ into $B(\Omega)$ is well defined for any open set Ω containing K. $\square$

The definition of integral for a compactly supported hyperfunction is a special case of a more general definition which applies to all hyperfunctions which are, at least in two points, real analytic. Let indeed f be a hyperfunction defined on a neighborhood of a closed interval $[a, b]$, and suppose that f be real analytic at the points a and b (this means that the defining functions F^+ and F^- are both holomorphic, or can be holomorphically continued, in a neighborhood of a and of b); let now τ_+ and τ_- be piecewise smooth arcs connecting a to b in such a way that τ_+ lies in the open set in which F^+ is defined and τ_- lies in the open set in which F^- is defined. Then, see figure 1.2.3, the integral of f from a to b is defined by

$$\int_a^b f(x)dx = \int_{\tau_+} F^+(z)dz - \int_{\tau_-} F^-(z)dz.$$

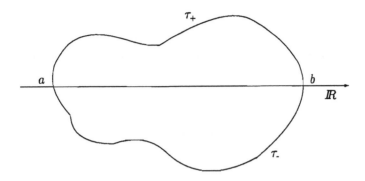

Figure 1.2.3

It is easily seen (by using again Cauchy's theorem), that this definition is independent of the many arbitrary choices which have been made.

We can now look at some of the simplest examples of hyperfunctions, which will lead us directly to the consideration of some deep results in the theory of hyperfunctions.

Example 1.2.1 The first example of a (non trivial) hyperfunction is given by the Dirac delta function which can be defined by

$$\delta(x) = \left[\frac{-1}{2\pi i} \cdot \frac{1}{z}\right].$$

It is clear from its definition that $\delta(x)$ is zero in $I\!\!R \setminus \{0\}$, since the function $\frac{1}{z}$ is holomorphic in $I\!\!R \setminus \{0\}$; in other words, the support of δ is the origin; it is also clear that, at the origin, the hyperfunction δ is "singular" (we have not yet defined this notion in any precise way) since it should represent the boundary value of a function which has a singularity at the origin. We have, however, given δ a rather famous name, since we claim that it is actually the well known delta function introduced by Dirac in 1936, [40]. If we look back at Dirac's original definition, we notice that his delta was defined as a "function" which was zero everywhere except at the origin, and such that, for any infinitely differentiable function $a(x)$ with compact support, it would give

(1.2.5) $$\int_{I\!\!R} a(x)\delta(x)dx = a(0).$$

Now, we have already easily checked that the delta hyperfunction vanishes outside the origin; in view of the definition of product between a real analytic function and a hyperfunction, as well as in view of the definition of the integral of a compactly supported hyperfunction, we immediately get the validity of (1.2.5), at least when $a(x)$ is real analytic. The delta hyperfunction has therefore all the characteristics to justify its name. It has still to be remarked that it has to be allowed for the test functions to be real analytic instead of simply infinitely differentiable, so that the analogy is not, at least at this moment, complete.

Example 1.2.2 Let us take a few derivatives of the delta hyperfunction, according to the rules of differentiation for hyperfunctions; we easily get

$$\delta'(x) = \left[\frac{1}{2\pi i} \cdot \frac{1}{z^2} \right],$$

$$\delta''(x) = \left[\frac{-1}{\pi i} \cdot \frac{1}{z^3} \right],$$

and, more generally,

$$\delta^{(n)}(x) = \left[\frac{(-1)^{n+1} n!}{2\pi i} \cdot \frac{1}{z^{(n+1)}} \right].$$

It is easily seen that all of these hyperfunctions are still compactly supported, and that they behave as the derivatives of the classical Dirac delta does; namely, for any real analytic function $a(x)$, it is

$$\int_{\mathbb{R}} a(x) \delta^{(n)}(x) dx = a^{(n)}(0).$$

Example 1.2.3 Classically, the Dirac delta function is seen as the derivative of the Heaviside function $H(x)$, this being defined as identically zero for negative values of x, and identically one for positive values of x. This function, which was originally used to describe electrical switches, is not differentiable in the classical sense, but, if we allow for weak derivation (i.e. integration by parts) to take place, it has the Dirac delta as its derivative. In our case, we define the Heaviside hyperfunction to be

$$H(x) = \left[\frac{-1}{2\pi i} \log(-z) \right]$$

where the logarithmic function is taken in its principal value; it is then clear that $H(x) = 1$ for any positive value of x, while $H(x) = 0$ for any negative value of x; on the other hand $H(x)$ is not defined at $x = 0$, if it is considered as a function. By applying the rules for differentiation which we have set up for hyperfunctions one immediately obtains that $H'(x) = \delta(x)$. One sees here an application of an important phenomenon, namely that differential operators do not enlarge the set where hyperfunctions are "singular" (we will make this concept much more precise in the next section). This could also be seen, of course, in Example 1.2.2.

Example 1.2.4 We now define two very simple and innocuous looking hyperfunctions, which will be needed in the immediate sequel; consider the following functions which are holomorphic on $\mathbb{C} \setminus \mathbb{R}$:

$$\epsilon(z) = \begin{cases} 1, & \text{if } Imz > 0 \\ 0, & \text{if } Imz < 0 \end{cases} \quad \text{and} \quad \bar{\epsilon}(z) = \begin{cases} 0, & \text{if } Imz > 0 \\ -1, & \text{if } Imz < 0 \end{cases}.$$

Then it is clear that $[\epsilon] = [\bar{\epsilon}]$, since the difference $\epsilon - \bar{\epsilon}$ is entire. The hyperfunction $[\epsilon]$, clearly defined on all of the real line, can be thought as somehow the unit hyperfunction, so that we will often denote it by 1. As a matter of fact, if we think of hyperfunctions (in the naive sense) as differences of boundary values of holomorphic functions, then it is quite natural to set $[\epsilon] = 1$. This hyperfunction, together with the structure of $\mathcal{A}(\Omega)$-module which we have given to $\mathcal{B}(\Omega)$, also allows us to show that every real analytic function can be thought of as a hyperfunction, or, equivalently, that, for any open set Ω, one has a natural injection of $\mathcal{A}(\Omega)$ into $\mathcal{B}(\Omega)$. Such an injection is obtained by multiplying a real analytic function $a(x)$ by the unit hyperfunction 1. The resulting hyperfunction will be identified with $a(x)$ itself.

Example 1.2.5 Let F be a function holomorphic in $U \setminus \Omega$, where we maintain the usual notations. We want to give a precise meaning to the expression "boundary value of F from above and from below"; we will show now that such a boundary value can be used to define precisely the value of a hyperfunction defined by $[F]$. In this new framework, the difference of the boundary values of F from above and from below is the value of the hyperfunction at that point. Define, to this purpose,

$$F(x + i0) = [F \cdot \epsilon]$$

and

$$F(x - i0) = [F \cdot \bar{\epsilon}].$$

With this notation, which the physicists have been employing for quite some time to denote boundary values in various weak senses (mainly in the sense of distributions), we now have the intuitive equality

$$[F(z)](x) = F(x + i0) - F(x - i0),$$

which justifies our use of the term "boundary value".

It may be worthwhile observing that Example 1.2.5 is the first example to provide us with objects which were not necessarily known before; indeed, Examples 1.2.1-1.2.3 gave us the Dirac distribution and the Heaviside function, while Example 1.2.4 only gave us the well known real analytic functions; now we finally get in touch with objects which are not known. As a matter of fact (the reader is referred to section 3 in this chapter), it is well known, [210] that all distributions on the real line can be obtained as boundary values (in the sense of distributions) of functions which are holomorphic in the upper half-plane; however it was shown by G. Köthe in [138] that not all boundary values of holomorphic functions yield distributions (this characterization of distributions is, interestingly enough, to be found in one of the papers which can be characterized as part of the prehistory of hyperfunctions, as we shall discuss in the historical notes). The following important result holds true:

Theorem 1.2.5 *Let Ω be an open set in $\mathbb{R}$, and let U be a complex neighborhood of Ω. Let F be a function holomorphic in U^+. Then $F(x + i0)$ is a distribution in Ω if and only if, for every compact K in Ω, there are positive constants C, N such that*

$$sup|F(x + iy)| < C|y|^{-N}.$$

Proof.   See [138], [142].   □

Example 1.2.6 As a consequence of this result, we see that all singularities of polar nature only produce distributions, while in order to obtain boundary values which cannot be treated by distribution theory, one needs to look for more singular holomorphic functions, such as

$$F(z) = \exp\left(\frac{1}{z}\right).$$

Then $[F(z)]$ is indeed a hyperfunction, supported at the origin, and which is certainly not a distribution. Other examples of hyperfunctions which are not distributions will be more evident as we progress with our theory.

It may be interesting to point out the relationships which link distributions with polar singularities, and hyperfunctions with essential singularities; Theorem 1.2.5 somehow points this out. As we shall see, the connection is even stronger than it may appear; it is, indeed, one of the important points about hyperfunctions.

Example 1.2.7 Let f be a hyperfunction supported at the origin, so that one can write $f = [F]$, with F holomorphic in $U \setminus \{0\}$, U being a suitably small open set in $\mathbb{C}$, and containing the origin. We have seen before that all finite order derivatives of f are well defined hyperfunctions; one can, however, do something more, and consider what we shall call, in later pages, an infinite differential operator. Namely, we consider an operator which, at least formally, is defined by

$$P(z, \frac{d}{dz}) = \sum_{i=0}^{+\infty} a_i(z)\frac{d^i}{dz^i};$$

one can then try to apply $P(z, d/dz)$ to the holomorphic function $F(z)$; as long as

$$P(z, \frac{d}{dz})F = \sum_{i=0}^{+\infty} a_i(z) \frac{d^i F(z)}{dz^i}$$

remains a holomorphic function, we can claim that a suitably defined operator $P(x, d/dx)$ has acted on f to yield a new hyperfunction. The reader is invited to compare this situation with what happens in the distributional case, where it is well known, [210], that every distribution supported at the origin is a finite linear combination of derivatives of Dirac's delta. Once again, as we have discussed before, we have this striking parallel between hyperfunctions as essential singularities and distributions as polar singularities.

Our example provokes, however, a natural question: when is the operator $P(z, d/dz)$ such that it actually maps holomorphic functions into holomorphic functions; more simply, which growth conditions should we impose on the series of coefficients $a_i(z)$ which defines $P(z, d/dz)$ in order that the formal series $P(z, d/dz)F$ converges in some punctured neighborhood of the origin?

In order to answer this natural question, let us first note the following characterization of hyperfunctions supported in the origin:

Lemma 1.2.1 *Let f be a hyperfunction supported at the origin; then one can express f as*

$$f(x) = \sum_{i=0}^{+\infty} b_i \frac{d^i \delta}{dx^i}(x),$$

where the sequence $\{b_i\}$ satisfies the following condition:

$$\lim_{i \to \infty} \sqrt[i]{|b_i| i!} = 0.$$

Proof. To begin with, let us recall that, in view of Theorem 1.2.4, the space of hyperfunctions with support at the origin can be seen as $\mathcal{O}(\mathbb{C} \setminus 0)/\mathcal{O}(\mathbb{C})$; henceforth, the hyperfunction $f(x)$ has a representative $F(z)$ which is holomorphic on $\mathbb{C} \setminus 0$. We now consider the Laurent expansion of F at the origin,

$$F(z) = \sum_{i=-\infty}^{+\infty} a_i z^i;$$

of course, when $i \geq 0$, the corresponding part of F is entire, and therefore (by the way hyperfunctions are defined) we can say that a different representative of $f(x)$ is given by the part of F composed only of negative powers, i.e. there is a one-to-one correspondence between hyperfunctions supported at the origin and Laurent expansions such as

$$\sum_{i=0}^{+\infty} \frac{a_i}{z^{i+1}},$$

with infinite radius of convergence (remember that F was to be holomorphic everywhere except at the origin). This condition on the radius of convergence of the Laurent expansion translates into the fact that, by the Cauchy-Hadamard theorem, one must have $\lim_{i \to \infty} \sqrt[i]{|a_i|} = 0$.

We now note that $1/z^{i+1}$ is nothing but the i-th derivative (up to a coefficient) of $1/z$, and therefore, by putting

$$b_i = 2\pi i (-1)^{i+1} \frac{a_i}{i!}$$

one sees that the defining function of $f(x)$ can be written as

$$\sum_{i=0}^{+\infty} b_i \frac{d^i}{dz^i} (-(2\pi i z)^{-1}).$$

If we now recall the definition of the delta hyperfunction, we realize that we have just proved that

$$f(x) = \sum_{i=0}^{+\infty} b_i \frac{d^i \delta}{dx^i}(x),$$

with the required growth conditions on the coefficients b_i. $\square$

We can now prove a necessary and sufficient condition for an infinite order differential operator to act on hyperfunctions as we have described before. We have:

Proposition 1.2.2 *Let $P(z, d/dz)$ be defined as above with coefficients $a_i(z)$ holomorphic in an open set U; then $P(z, d/dz)F(z)$ is a germ of a holomorphic function at z_0 for any germ F of holomorphic function at z_0, if and only if, for any compact K contained in U,*

$$(1.2.6) \qquad\qquad \lim_{i \to \infty} \sqrt[i]{sup_K |a_i(z)| i!} = 0.$$

Proof. We begin by proving the sufficiency of the limit condition (1.2.6). Let F be a germ of a holomorphic function at some point z_0. This means that F is holomorphic in some closed disk D of radius δ around z_0, and we can call M the maximum of $|F(z)|$ in that disk. By Cauchy's inequality we then obtain that, on D,

$$\left| \frac{d^i F(z)}{dz^i} \right| \leq \frac{i! M}{\delta^i}.$$

We can now couple this information with the hypothesis (1.2.6); since D is itself a compact set, the hypothesis can be rephrased by saying that for any $\epsilon > 0$ we can find a positive constant C_ϵ such that, for any i,

$$|a_i(z) i!| \leq C_\epsilon \epsilon^i.$$

By now choosing $\epsilon = \delta/2$, we immediately deduce from the previous inequalities that the series

$$\sum_{i=0}^{+\infty} a_i(z) \frac{d^i F(z)}{dz^i}$$

converges uniformly in D to a holomorphic function. We now proceed to prove the converse, i.e. the necessity of condition (1.2.6) for P to act on the space of germs of holomorphic functions.

By contradiction suppose that condition (1.2.6) is not satisfied; then, for some $\epsilon > 0$ there exists a sequence z_j, which we can assume to be convergent to some z_0, such that

$$\sqrt[j]{|a_j(z)j!|} \geq 2\epsilon, j = 1, 2, \ldots.$$

We now apply the operator $P(z, d/dz)$ to the function

$$F(z) = \frac{1}{(z - z_0 - \epsilon)},$$

which is clearly holomorphic in the open disk D centered in z_0 and of radius ϵ. It is immediate to see that the series

$$P(z, \frac{d}{dz})F(z) = \sum_{i=0}^{+\infty} \frac{a_i(z)(-1)^i i!}{(z - z_0 - \epsilon)^{i+1}} = \sum_{i=0}^{+\infty} F_i(z)$$

does not converge locally uniformly in D, because of the estimates which we have given on the F_i, and therefore the theorem is completely proved. □

We shall come back to this result later on, to show how it actually implies that condition (1.2.6) is a necessary and sufficient condition for $P(z, d/dz)$ to be a sheaf homomorphism on the sheaf of hyperfunctions.

In the sequel, we will refer to operators satisfying condition (1.2.6) as local operators, the reason for this terminology being that they do not move the support of the functions they act on. Local operators are, therefore, a special case of infinite order differential operators. It is interesting to note how other operators of infinite order do appear and have some important applications in theory; the most important, probably, is the translation operator defined by

$$\exp(\frac{d}{dz})F(z) = \sum_{i=0}^{+\infty} \frac{D^i F(z)}{i!} = F(z + 1),$$

where D denotes the differentiation operator d/dz. Such operators are clearly unsuited to act on sheaves, since if F is only a germ of holomorphic function at the origin, then $F(z + 1)$ may not even be defined. From a historical point of view, such operators were studied since the end of last century by the Italian mathematician Pincherle. Some more details on this will be given in section five of the historical appendix to this chapter.

Example 1.2.8 Infinite order local operators can be used to construct new examples of hyperfunctions; in connection with Example 1.2.6 (and to provide an immediate application of Proposition 1.2.2), we can easily see that

$$[\exp(-1/z)] = \sum_{k=0}^{+\infty} \frac{2\pi i}{k!(k+1)!} D^k \delta(x).$$

This equality is immediately established by expanding in its Laurent series around the origin the function $\exp{-1/z}$.

Example 1.2.9 We can also use the notion of boundary value to define other important examples of hyperfunctions. To begin with, let F be a meromorphic function on a complex neighborhood U of an open set Ω in $I\!R$; we may assume that the poles of F all lie in Ω. Following the classical example of Hadamard, we can define the so called "partie finie" or finite part of $F(x)$ as the hyperfunction defined by

$$\text{f.p.}(F(x)) = \frac{1}{2}\{F(x+i0) + F(x-i0)\}.$$

This hyperfunction (which, of course, is actually a distribution, as the reader will immediately see) can be obviously defined by the pair $(1/2F(z), -1/2F(z))$. A case of historical interest, and certainly well known, is the finite part of $1/x$, for which the following identity (known by the physicists as the Lippmann-Schwinger formula) holds:

$$\frac{1}{(x+i0)} = \text{f.p.}(1/x) - \pi i \delta(x).$$

The expression on the right hand side of the Lippmann-Schwinger formula is also known, in the literature on distributions, as Cauchy's principal value of $1/x$, denoted by p.v. $\left(\frac{1}{x}\right)$. The reasons for these terminologies are rather well known, and can be found in detail in Schwartz's treatise [210].

Example 1.2.10 Other interesting examples of hyperfunctions can be defined by taking boundary values of multivalued functions, such as the power function of a complex variable. To begin with, one may consider

$$(x+i0)^\mu, \text{whose defining function is the pair}(z^\mu, 0)$$

and

$$(x-i0)^\mu, \text{whose defining function is the pair}(0, -z^\mu).$$

Example 1.2.11 Along the same lines, we have the hyperfunctions x_+^μ, which are defined in different ways depending on whether μ is an integer or not. To be precise we have, for μ not an integer

$$x_+^\mu = \left[\frac{-(-z)^\mu}{2i\sin\pi\mu}\right] = \frac{1}{2i\sin\pi\mu}\{(-x+i0)^\mu - (-x-i0)^\mu\},$$

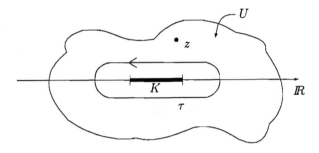

Figure 1.2.4

and

$$x_-^\mu = \left[\frac{(z)^\mu}{2i\sin\pi\mu}\right] = \frac{1}{2i\sin\pi\mu}\{(x+i0)^\mu - (-x-i0)^\mu\},$$

while, in the case in which μ is an integer, we define

$$x_\pm^\mu = \left[\mp\frac{(\pm z)^\mu}{2\pi i}\log\mp z\right].$$

To conclude this introductory section on single variable hyperfunctions, we show (by essentially quoting Cauchy's integral formula) how to choose a specific defining function for a given compactly supported hyperfunction; the choice which we will make is interesting from a particular point of view, and the reader may, after reading this result, directly jump to our historical appendix, to compare this result with some old Fantappie's theories.

Proposition 1.2.3 *Let f be a hyperfunction with compact support K. Define*

(1.2.7) $$G(z) = \frac{1}{2\pi i}\int_{-\infty}^{+\infty}\frac{f(x)}{x-z}dx.$$

Then $[G] = f$, G is holomorphic in $\mathbb{C}\,\mathbb{P}^1\setminus K$, and, finally, $G(\infty) = 0$. Moreover, there is only one defining function for f which satisfies these two properties; we will call such a defining function the standard defining function for f.

Proof. To begin with, let us note that the function $1/(x-z)$ is a real analytic function of x on K, for each z chosen in $\mathbb{C}\setminus K$; therefore, by the previous definitions, it can be multiplied by the hyperfunction $f(x)$; the result is a hyperfunction for which it makes sense to take the integral as in (1.2.7). Take now U a complex neighborhood of K, and F a representative of f which is holomorphic in $U\setminus K$. If we choose a closed path τ contained in U and which leaves z on the outside (as in fig. 1.2.4),

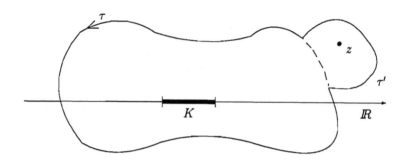

Figure 1.2.5

we see that the definition of definite integral for a hyperfunction yields

$$G(z) = \frac{-1}{2\pi i} \oint_\tau \frac{F(\varrho)}{\varrho - \tau} d\varrho.$$

We then deduce immediately that G is holomorphic in $\mathbb{C} \setminus K$, so that $[G]$ is a hyperfunction with support in K. We now want to prove that $[G] = [F]$ near K, which would show that $[G] = f$. By definition of hyperfunctions, it will suffice to show that $G - F$ can be holomorphically extended near K. To do so, choose τ' to be a closed curve encircling z once in the positive direction, as shown in figure 1.2.5.

Cauchy's integral formula immediately yields the following series of equalities

$$G(z) - F(z) = \frac{-1}{2\pi i} \oint_\tau \frac{F(\varrho)}{\varrho - z} d\varrho - \frac{1}{2\pi i} \oint_{\tau'} \frac{F(\varrho)}{\varrho - z} d\varrho = \frac{-1}{2\pi i} \oint_{\tau + \tau'} \frac{F(\varrho)}{\varrho - z} d\varrho,$$

which shows that $G - F$ actually does extend to a holomorphic function to the interior of the region bounded by $\tau + \tau'$. This region contains K, and therefore we have proved that G is a representing function for f. The fact that G is holomorphic on all of $\mathbb{C}\mathbb{P}^1 \setminus K$ is a consequence of its integral representation, and so is its vanishing at infinity. Finally, if H were another representative of f with these two properties, we would have that $G - F$ would be holomorphic everywhere on $\mathbb{C}\mathbb{P}^1$, and vanishing at infinity. By Liouville's theorem, $G - F$ would be a constant and hence the zero constant. This concludes the proof. $\square$

1.3 Hyperfunctions of One Variable: Main Results

Section 1.2 has been essentially devoted to a rather cursory description of the notion of hyperfunction, together with some fundamental examples. We have also described how to operate on hyperfunctions, so that they can be concretely considered as generalized functions. This section, on the other hand, will dwell more deeply into the properties of hyperfunctions.

We will show that hyperfunctions actually constitute a sheaf of generalized functions on $I\!R$ of which the sheaves of real analytic functions and of distributions are proper subsheaves. This is one of the fundamental results in the theory of hyperfunctions. We will also prove that the sheaf $\mathcal{B}$ of hyperfunctions is a flabby sheaf, a very useful result when dealing with systems of differential equations. Some simple applications of these notions to the study of differential equations are included to provide the flavor and show the power of the method.

After providing the reader with the statement and proof of the celebrated Köthe duality theorem (which, more appropriately, could be called Fantappie'-Köthe-Martineau-Sato duality theorem, as we shall show in the historical appendix to the chapter), this section will be concluded by a brief discussion on the possible topologies which the spaces of hyperfunctions can be endowed with; this is, as we shall see, one of the weak points of the theory, and it will be useful to have a preliminary look at the situation in the single variable case.

In what follows we have decided not to provide sheaf theory per se, since many adequate texts can be found to fill the gap, [64], [123]. In particular, Kaneko's monograph on the theory of hyperfunctions [103] contains a rather extensive section devoted to the theory of sheaves and to the theory of cohomology, with an eye towards the necessities of hyperfunction theory. The reader unfamiliar with sheaves is therefore invited to consult one of these texts.

The first result we wish to prove is the flabbiness of the sheaf of hyperfunctions, which marks the first important difference with the theory of distributions.

Theorem 1.3.1 *The assignment* $\mathcal{B}(\Omega)$, *with* Ω *any open set in* $I\!R$, *is a flabby sheaf of vector spaces.*

Proof. To prove that a presheaf, i.e. a collection of vector spaces indexed by the family of open sets of a topological space, is a sheaf one needs to prove two completeness properties; that local vanishing implies global vanishing, and that local belonging implies global belonging. To be more precise, the following properties have to be established to prove that hyperfunctions are a sheaf on $I\!R$; let Ω be an open set in $I\!R$, and let $\{\Omega_i\}$ be an open covering of Ω:

(i) if an element f in $\mathcal{B}(\Omega)$ vanishes when restricted to each Ω_i, then f is the zero in $\mathcal{B}(\Omega)$;

(ii) if a family $\{f_i\}$ of hyperfunctions in $\mathcal{B}(\Omega_i)$ satisfies the compatibility conditions $f_i = f_j$ on $\Omega_i \cap \Omega_j$, then there exists f in $\mathcal{B}(\Omega)$ such that its restriction to each Ω_i is exactly f_i.

The reader will remember that we have already proved (i) when establishing the existence of a support for hyperfunctions, so that we only need to establish (ii). The proof of (ii) is slightly deeper, as it requires the use of Mittag-Leffler Theorem. Let therefore Ω, Ω_i and f_i be as above, and let F_i be holomorphic functions in $U_i \setminus \Omega_i$, which define the hyperfunctions f_i (the U_i are just complex neighborhoods of Ω_i). We set U to be the union of the U_i, and, by the paracompactness of both $\mathbb{C}$ and $\mathbb{R}$, we can assume that both $\{\Omega_i\}$ and $\{U_i\}$ are locally finite (i.e. every point only belongs to a finite number of open sets). The compatibility condition immediately translates into the fact that the function F_{ij} defined by $F_j - F_i$ is holomorphic in $U_i \cap U_j$, even though not necessarily zero there. It is however immediate to see that the family $\{F_{ij}\}$ defines a 1-cocycle on U with values in the sheaf of holomorphic functions. Mittag-Leffler Theorem (Theorem 1.2.1 and 1.2.1'), then shows that $\{F_{ij}\}$ is actually a 1-coboundary, and therefore there are functions G_i, holomorphic on U_i such that

$$F_{ij} = G_j - G_i \text{ on } U_i \cap U_j.$$

The conclusion of the proof of part (ii) now follows by observing that the function $F_i - G_i = F_j - G_j$ is holomorphic on $U \setminus \Omega$, and since $[F_i - G_i] = [F_i]$, the hyperfunction $[F]$ is the desired global hyperfunction.

To conclude the proof of the theorem, we only need to establish the flabbiness of the sheaf of hyperfunctions. We recall that a sheaf $\mathcal{F}$ is said to be flabby if, given any two open sets Ω and Ω', with $\Omega \supseteq \Omega'$, the natural restriction map $\mu : \mathcal{F}(\Omega) \longrightarrow \mathcal{F}(\Omega')$ is always surjective. In particular, to prove the flabbiness of a sheaf $\mathcal{F}$ on a topological space X it is sufficient to prove the surjectivity of the restriction map $\mathcal{F}(X) \longrightarrow \mathcal{F}(\Omega)$, for any open set Ω in X. In our case, let Ω be an open set in $\mathbb{R}$. We know, by Proposition 1.2.1, that $\mathcal{B}(\Omega)$ can be represented as $\mathcal{O}(U \setminus \Omega)/\mathcal{O}(U)$, for U any complex neighborhood of Ω. To prove the flabbiness, it is sufficient to take $U = \mathbb{C} \setminus \partial\Omega$, where $\partial\Omega$ denotes the boundary of Ω in $\mathbb{R}$. Then every representative F in $\mathcal{O}(U \setminus \Omega)$ of f gives the required extension to all of $\mathbb{R}$, since $U \setminus \Omega$ is now nothing but $\mathbb{C} \setminus \bar{\Omega}$. More explicitly, one has that

$$\mathcal{B}(\Omega) = \frac{\mathcal{O}(\mathbb{C} \setminus \bar{\Omega})}{\mathcal{O}(\mathbb{C} \setminus \partial\Omega)}.$$

Then the restriction

$$\mathcal{O}(\mathbb{C} \setminus \bar{\Omega}) \longrightarrow \mathcal{O}(\mathbb{C} \setminus \mathbb{R})$$

induces a surjective map

$$\frac{\mathcal{O}(\mathbb{C} \setminus \Omega)}{\mathcal{O}(\mathbb{C})} \longrightarrow \frac{\mathcal{O}(\mathbb{C} \setminus \mathbb{R})}{\mathcal{O}(\mathbb{C})}$$

and therefore from the commutative diagram

$$0 \longrightarrow \frac{\mathcal{O}(\mathbb{C} \setminus \bar{\Omega})}{\mathcal{O}(\mathbb{C})} \longrightarrow \frac{\mathcal{O}(\mathbb{C} \setminus \mathbb{R})}{\mathcal{O}(\mathbb{C})} = \mathcal{B}(\mathbb{R})$$

$$\frac{\mathcal{O}(\mathbb{C} \setminus \bar{\Omega})}{\mathcal{O}(\mathbb{C} \setminus \partial \Omega)} = \mathcal{B}(\Omega)$$

$$0$$

we obtain the surjectiveness of $\mathcal{B}(\mathbb{R}) \longrightarrow \mathcal{B}(\Omega)$. The flabbiness of the sheaf of hyperfunctions is thus proved. □

The reader will note how the fact that $\mathcal{B}$ is a sheaf allows us to speak of a support for hyperfunctions (indeed, in our previous section we had already proved half of the sheafness of $\mathcal{B}$, in order to introduce the notion of support); this is not different from what happens with distributions (which, too, form a sheaf), but it contrasts with the situation for analytic functionals. Indeed, see e.g. [83], an analytic functional is defined as an element μ of $\mathcal{H}'(\mathbb{C})$, where the symbol $\mathcal{H}'(\mathbb{C})$ denotes the dual of the Frechet space of entire functions (i.e. the space of linear continuous complex valued functionals on the space of entire functions); analogously, if K is a compact set in $\mathbb{C}$, then one can define the space of germs of holomorphic functions on K as the inductive limit

$$\mathcal{H}(K) = \operatorname{ind} \lim \mathcal{H}(U).$$

It is possible to show that $\mathcal{H}(K)$, endowed with the inductive limit topology, is not anymore a Frechet space, but rather what is called a DFS space (see [223] for more details on topological vector spaces), and one can easily see that for μ an analytic functional, there exists a compact set K in $\mathbb{C}$ such that μ belongs to the dual $\mathcal{H}'(K)$ of $\mathcal{H}(K)$. This is equivalent to the following estimate on its values:

for every open neighborhood U of K, there exists a positive constant C_U such that, for every φ in $\mathcal{H}(U)$, it is

$$| < \mu, \varphi > | \leq C_U \sup_K |\varphi(z)|.$$

When such a condition is satisfied, we say that f is carried by the compact set K; it occurs naturally the question of whether a minimal such K exists for every analytic functional. It is by now well known that this is not always the case (see e.g. [83]), and that, therefore, a notion of support cannot be defined for analytic functionals; from this point of view, our last Theorem shows one of the crucial properties of hyperfunctions.

On the other hand flabbiness, i.e. the property which allows to extend a given hyperfunction from any open set to all of $I\!\!R$, is an extremely important property in the theory of differential equations, since it allows to define hyperfunctions on all of $I\!\!R$, regardless of the singularities which they may present at the boundary of their original domain: this is in striking contrast with what happens with differentiable functions, or even with distributions, which may be too rigid to extend beyond a given point. Another important point, as we shall soon see, is the possibility of using the sheaf $\mathcal{B}$ to construct flabby resolutions for sheaves of solutions to systems of differential equations.

To illustrate the power of dealing with hyperfunctions, we prove here a well known result of Sato, [142], concerning the surjectivity of differential operators on spaces of hyperfunctions.

Theorem 1.3.2 *Let*

$$P(x, \frac{d}{dx}) = \sum_{i=0}^{m} a_i(x) \frac{d^i}{dx^i}$$

be a linear finite order differential operator with real analytic coefficients on an open set Ω in $I\!\!R$, and suppose that $a_m \not\equiv 0$. Then, for every f in $\mathcal{B}(\Omega)$, there exists a solution u in $\mathcal{B}(\Omega)$ to the differential equation

$$P(x, \frac{d}{dx})u(x) = f(x).$$

Moreover, any solution u can be extended to a hyperfunction solution defined on any open set containing Ω and on which all the coefficients a_i and f are defined. In particular, if all of the coefficients and f are defined on all of $I\!\!R$, then any solution u defined on Ω can be extended to a hyperfunction solution u^* on $I\!\!R$.

Proof. Let V be a complex neighborhood of Ω to which all coefficients a_i can be analytically continued. By taking V suitably, we can make sure that both V and $V \setminus \Omega$ are simply connected and that all zeroes in V of the leading coefficient a_m are actually contained in Ω. Let now $F \in \mathcal{O}(V \setminus \Omega)$ be a defining function of f. By the well known Cauchy's existence theorem, and since a_m has no zeroes in $V \setminus \Omega$, there is a solution $U \in \mathcal{O}(V \setminus \Omega)$ to the (complex analytic) differential equation

$$P(z, \frac{d}{dz})U(z) = F(z).$$

It is now easily seen that $u = [U]$ gives the required solution. We now proceed to prove the possibility of extending the solution u to a solution of the same equation to all of $I\!\!R$ (we are assuming here that both f and the a_i are defined on $I\!\!R$, and that u is any solution of $P(x, d/dx)u = f$ on some real open set Ω). By what we have just proved (the existence part of the theorem), it will be sufficient to consider the case of homogeneous equations such as $P(x, d/dx)u(x) = 0$. Let

V be the same complex neighborhood of Ω we have used in the first part of the proof, and set
$$W = V \setminus (I\!\!R \setminus \Omega).$$
It is easily seen that W is still a complex neighborhood of Ω, and we denote therefore by $U \in \mathcal{O}(W\backslash\Omega)$ a representative for the solution u of the homogeneous equation on Ω. Note that $P(z, d/dz)U(z)$ is holomorphic on all of W. We will now show that there exists a function $N(z)$ holomorphic on W such that the function
$$P(z, d/dz)U(z) - P(z, \frac{d}{dz})N(z)$$
can be continued analytically on V; this immediately implies that the hyperfunction u^* defined on all of $I\!\!R$ as the equivalence class of $U - N$ gives the required extension, and will therefore conclude the proof of the theorem. To construct such a function N we will use once again the Mittag-Leffler Theorem (Theorem 1.2.1$'$). To begin with, choose two open sets V_1 and V_2 such that $V_1 \cup V_2$ coincides with V, $V_2 \cap (I\!\!R \setminus \Omega) = \emptyset$, and $V_1 \setminus (I\!\!R \setminus \Omega)$ is simply connected and contains no zeroes of $a_m(z)$. Again by Cauchy's existence theorem, we obtain that there exists a function N_0, holomorphic in $V_1 \setminus (I\!\!R \setminus \Omega)$, solution of the equation

$$P(z, \frac{d}{dz})N_0(z) = P(z, \frac{d}{dz})U(z).$$

We can now apply Mittag-Leffler Theorem to find $N_1 \in \mathcal{O}(V_1)$ and $N_2 \in \mathcal{O}(V_2)$ such that, on $V_1 \cap V_2'$,
$$N_0(z) = N_2(z) - N_1(z).$$
If we now define the function
$$N(z) = \begin{cases} N_0(z) + N_1(z), z \in V_1 \setminus (I\!\!R \setminus \Omega) \\ N_2(z), z \in V_2' \end{cases},$$
we immediately see that N is holomorphic on W and that the theorem is therefore completely proved. $\qquad\square$

The reader will notice how this theorem is more powerful than the Cauchy existence theorem which it employs, as it does not require the non-vanishing of the leading coefficient of the differential operator. In the next chapter we will slightly refine this result to discuss the location of the singularities of u with respect to the singularities of f and to the set of zeroes of a_m; in so doing we will be able to prove a preliminary version of the celebrated Sato's Fundamental Principle.

We have seen that the operator $P(x, d/dx)$ is surjective on the space of hyperfunctions; we have therefore looked at P as an operator between vector spaces, but we may actually look at it as a sheaf homomorphism, i.e. we have the following simple fact:

Proposition 1.3.1 *The operator $P(x, d/dx)$ is a sheaf endomorphism of the sheaf $\mathcal{A}$ of real analytic functions and of the sheaf $\mathcal{B}$ of hyperfunctions.*

Proof.   We recall that, given two sheaves of $\mathcal{C}$-vector spaces on a topological space X, say $\mathcal{F}$ and $\mathcal{G}$, a sheaf homomorphism is a collection of $\mathcal{C}$-linear maps

$$h_U : \mathcal{F}(U) \longrightarrow \mathcal{G}(U),$$

indexed on the family $\{U\}$ of all open sets of X, such that the diagram

$$
\begin{array}{ccc}
\mathcal{F}(U) & \longrightarrow & \mathcal{G}(U) \\
\downarrow & & \downarrow \\
\mathcal{F}(V) & \longrightarrow & \mathcal{G}(V)
\end{array}
$$

commutes for any inclusion $V \hookrightarrow U$. In our case, if we consider the sheaf $\mathcal{A}$ of real analytic functions, then the result is an obvious consequence of the definition of partial differential operators. As for the case of the sheaf of hyperfunctions, one immediately obtains the same result if one considers that $P(z, d/dz)$ actually acts as a sheaf endomorphism on the sheaf of germs of holomorphic functions, and therefore the same results applies to hyperfunctions.   □

From general properties of sheaves and their endomorphisms (which can however be immediately derived by their definitions), we obtain the following Corollary, of some independent interest:

Corollary 1.3.1 *Differential operators with real analytic coefficients do not enlarge the support of hyperfunctions.*

Proof.   Immediate from Proposition 1.3.1, and from the general fact that sheaf endomorphisms do not enlarge the supports of the sections on which they act.
                                    □

We have mentioned the importance of flabbiness in extending hyperfunction solutions of differential equations; flabbiness has, however, one other important consequence, which may be worthwhile mentioning. As the reader no doubt knows, one of the features which have made the theory of infinitely differentiable functions (and, as a consequence, the theory of distributions) so successful, is the existence of the so called cut-off functions, i.e. the existence of compactly supported functions. This fact, which sheaf theorists summarize by saying that the sheaf of C^∞ functions is a fine sheaf, is crucial in the study of differential equations, and plays a substantial role in the development of the study of singularities of distributions (we shall come back to this in our next chapter, where microlocalization will be discussed). Despite the attractiveness of hyperfunctions, however, this feature is missing in this case (and this was probably one of the reasons for not using real analytic functions as test functions, before Sato). In particular, given a hyperfunction f, it is not possible to truncate it to a compactly supported one (in the case of distributions, one achieves this result by multiplying the given distribution by a suitable cut-off function), not even if one allows some modifications in a small neighborhoods of the sets in which the support has to be restricted (which it has necessarily to be done in the distribution case). One has, however, the following remarkable result, which we state and prove for hyperfunctions but which, as it is obvious from the proof itself, is actually a result valid for any flabby sheaf.

Proposition 1.3.2 *Let f be a hyperfunction on an open set Ω, and let $supp(f) = A_1 \cup A_2$, with $A_{1,2}$ two closed sets. Then there exist hyperfunctions f_1 and f_2 on Ω, with $supp(f_i) = A_i, i = 1, 2$, such that $f = f_1 + f_2$. The same result holds true if the support of f is given a locally finite decomposition in closed sets.*

Proof. We will give the proof for the case of two open sets A_1 and A_2, for which we set $E_i = \Omega \setminus A_i$. It is indeed sufficient to define the hyperfunction f_1 by the position

$$f_1 = \begin{cases} f & \text{on } E_2 \\ 0 & \text{on } E_1 \end{cases}.$$

Note that this position is well given, since, on the intersection $A_1 \cap A_2$, f is actually zero. We therefore have a hyperfunction on $A_1 \cup A_2$. Since $\mathcal{B}$ is a flabby sheaf, such a hyperfunction extends to all of Ω (and we will still call it f_1). It is $supp(f_1) \subseteq A_1$, and the result follows if we simply define $f_2 = f - f_1$. $\square$

We wish to remark that this result will be essentially used as a substitute for the existence of partitions of unity; as it is well known, partitions of unity are a crucial tool in the theory of fine sheaves (of which the sheaf of infinitely differentiable functions is a good example); the above result constitutes the analogue for flabby sheaves. In a sense, even if no canonical way is provided, some advantage is gained by the fact that we will be able to obtain decompositions in which no modification is necessary near boundaries (which is on the other hand necessary when dealing with the customary partitions of unity).

Our next result has been already proved in the previous section, even though we did not use any sheaf terminology there:

Proposition 1.3.3 *$\mathcal{A}$ is a subsheaf of $\mathcal{B}$.*

Proof. It is sufficient to show that, for every open subset Ω of $\mathbb{R}$, we have an inclusion of $\mathcal{A}(\Omega)$ in $\mathcal{B}(\Omega)$, which is consistent with the restriction maps (i.e. that the restriction, in the sense of hyperfunctions, of a real analytic function is still a real analytic function, namely the restriction

$$\rho^\Omega_{\Omega'} : \mathcal{B}(\Omega) \longrightarrow \mathcal{B}(\Omega'), \quad \Omega' \subseteq \Omega$$

induces the restriction on the subgroups

$$\rho^{\Omega}_{\Omega'|\mathcal{A}(\Omega)} : \mathcal{A}(\Omega) \longrightarrow \mathcal{A}(\Omega')$$

of analytic functions). To prove such an inclusion we resort to the use of the function ϵ which we have defined in the previous section. Recall that the space of real analytic functions is defined as the inductive limit of spaces of holomorphic functions, and let φ be a holomorphic function on some complex neighborhood U of Ω, which defines (through the inductive limit procedure) a real analytic function which we will still call φ. We can now associate to φ the function $\epsilon\varphi$, holomorphic on $U \setminus \Omega$, which defines an object in the quotient space $\mathcal{O}(U \setminus \Omega)/\mathcal{O}(U)$; but if we now take the inductive limit, on all open neighborhoods of Ω, of the equivalence class of $[\epsilon\varphi]$, we obtain the hyperfunction which we associate to φ, in our sheaf homomorphism. The fact that this map is well defined (i.e. independent on all the arbitrary choices we made) and that it is actually a sheaf homomorphism is now immediate, and this concludes the proof. $\Box$

It becomes therefore interesting to study where a hyperfunction fails to be real analytic. Indeed, we say that a hyperfunction f (which, by flabbiness, we may assume to be defined on all of $I\!\!R$) is real analytic on Ω if it belongs to $\mathcal{A}(\Omega)$; if $f = [F]$, its being real analytic really depends on the lack of singularities of the pair of holomorphic functions $F = (F^+, F^-)$ which represents f. More specifically, one has that a hyperfunction $f = [F]$ on Ω is real analytic on Ω if and only if both F^+ and F^- extend analytically across Ω, to holomorphic functions in some complex neighborhood of Ω itself. The reader will note that, by the definition of $\mathcal{B}$, the real analyticity of a hyperfunction f does not depend on the representing function F which has been chosen to test the analyticity.

Real analyticity is obviously a local property (essentially because of Proposition 1.3.3) and therefore, given any f in $\mathcal{B}(\Omega)$, there exists the largest open subset of Ω on which f is real analytic. The complement of such a subset will be called the **singular support** of the hyperfunction f, and will be denoted by the symbol sing supp (f). The choice of the terminology may look somewhat suspicious, but we will show later in this section that the singular support of a hyperfunction is actually the support of a specific section of a different sheaf.

Example 1.3.1 It is obvious that sing supp $(\delta) = \{0\}$; indeed, the delta function is identically zero (and therefore real analytic) cannot be analytically continued.

Example 1.3.2 An analogous argument shows that, for any positive integer n, one has sing supp $(\delta^{(n)}) = \{0\}$.

Example 1.3.3 The Heaviside hyperfunction, too, has singular support equal to the origin, since $H(x)$ is locally constant outside the origin. In different words, the function $\log(-z)$ can be analytically continued across the real axis everywhere except at the origin.

Example 1.3.4 The singular support of the hyperfunctions $[\epsilon]$ and $[\epsilon]$ is the empty set, since they are both real analytic functions over the real line. Indeed, $[\epsilon]$ can be represented by the holomorphic pair $(1,0)$, and both the function 1 and the function 0 can be analytically continued across the real line. A similar argument holds for $[\epsilon]$.

Example 1.3.5 If F is a function holomorphic on, say, the upper half plane, then its boundary value $F(x+i0)$ is a hyperfunction whose singular support coincides with the closed set of all points of $I\!R$ across which F cannot be analytically continued. Indeed, $F(x+i0) = (F,0)$, and 0 can always be analytically continued everywhere.

Example 1.3.6 Let F be a meromorphic function on a complex neighborhood U of a real open set Ω, and assume that all the poles of F lie in Ω. Then we have defined the hyperfunction f.p.(F) as the holomorphic pair $(F/2, -F/2)$. From what precedes, it is clear that the singular support of f.p.(F) coincides with the set of poles of F. In particular, sing supp f.p.$(1/x) = \{0\}$.

Example 1.3.7 The singular support of the hyperfunctions $x_{\pm}^{\mu}$ which we have defined in Example 1.2.11 is, of course, the origin. However, one can check that sing supp $((x+i0)^{\mu})$ is the empty set if μ is a nonnegative integer, and is the origin otherwise. The same is true for $|x|^{\mu}$.

The fact that singular supports are defined in a local way allows us to immediately obtain the following result:

Proposition 1.3.4 *Local operators do not enlarge the singular support of hyperfunctions.*

Proof. This follows immediately from the fact that local operators (both of finite and of infinite order) act as sheaf homomorphisms on both $\mathcal{A}$ and $\mathcal{B}$. Therefore, if a hyperfunction is real analytic at a point x_0, then so is its result after the application of a local operator. □

Because the singular support of a hyperfunction is actually the support of a section of a different sheaf on which, too, local operators with analytic coefficients act, one immediately obtains the previous result as a corollary of a general fact from sheaf theory; this fact will be clear later in this section.

Proposition 1.3.4 shows that if P is a local operator, and f is real analytic, then $P(f)$ is, still, real analytic. The inverse of this property is usually referred to as the "ellipticity" of the operator. We wish to conclude this section with a brief (and somehow preliminary) discussion of the ellipticity phenomenon as it applies to the sheaf of hyperfunctions.

Let us begin with a finite order m constant coefficients differential operator $P(d/dx)$. We begin by stating the following simple fact, whose proof can be found in [142]:

Proposition 1.3.5 *The dimension of the space of hyperfunction solutions to the differential equation*

$$P(\frac{d}{dx})f(x) = 0$$

is m. More precisely, let Ω be any interval of the real line: then

$$dim\{f \in \mathcal{B}(\Omega) : P(\frac{d}{dx})f(x) = 0\} = m.$$

As a consequence of this result, one deduces that, as in the case of distributions, all constant coefficients operators can be considered to be elliptic. To this purpose, we say that an operator T acting on the sheaf $\mathcal{B}$ is an elliptic operator if the following condition is satisfied for any f in $\mathcal{B}(\Omega)$:

if $T(f) = 0$, then f is real analytic on Ω.

This condition is often referred to as the analytic hypoellipticity, to distinguish it from the differentiable hypoellipticity, in which the setting is the sheaf of Schwartz distributions and the fact that f belongs to the kernel of T only implies that f is actually differentiable on Ω. From this point of view, it is clear that, a priori at least, analytic ellipticity is rather stronger than differentiable ellipticity, in which a stronger hypothesis is requested (f is a distribution rather than a hyperfunction) and a weaker conclusion is obtained (differentiability rather than analyticity).

We can indeed prove the following result:

Theorem 1.3.3 *Every linear constant coefficient finite order m differential operator is elliptic.*

Proof.   Let Ω be any open set in $\mathbb{R}$. Then, on each connected component of Ω, there are m linearly independent real analytic solutions to the equation $P(d/dx)f = g$, with g real analytic (this is just the well known Euler Fundamental Principle for solutions of ordinary differential equations). But, in view of Proposition 1.3.5, there are exactly m linearly independent hyperfunction solutions to that same equation. Therefore the space of hyperfunction solutions in Ω coincides with the space of real analytic solutions in Ω, which concludes the proof. $\qquad\square$

The reader will note that, by the linearity of the operators we are working with, the ellipticity condition can be somewhat relaxed, and one can see that it is actually equivalent to the request that every hyperfunction solution to $P(d/dx)f = 0$ is actually real analytic. Such a condition immediately implies (and is therefore equivalent to) the ellipticity of the operator $P(d/dx)$.

In Theorem 1.3.3, we have requested the hypothesis of finite order on the differential operator. In view of the fact that we are restricting our attention to the one variable case, we can remove that hypothesis and obtain the following result:

Theorem 1.3.4 *All local operators acting on the sheaf of hyperfunctions on $\mathbb{R}$ are elliptic.*

Proof. See [103]. $\qquad\square$

We will use this result in the sequel to prove that our representation of hyperfunctions supported at the origin as derivatives of the Dirac delta function can actually be extended to the case of any general hyperfunction.

The situation becomes a little more interesting if one considers variable coefficient operators. To begin with, we shall restrict our attention to linear finite order differential operators with real analytic coefficients.

Let

$$P(z, d/dz) = \sum_{j=0}^{m} a_j(z) \frac{d^j}{dz^j}$$

be a differential operator with holomorphic coefficients and of finite order m: we recall that the index of a differential operator $P(z, \frac{d}{dz})$ acting on the space $\mathcal{O}(V)$ of holomorphic functions on an open set V in $\mathbb{C}$ is defined as the difference

$$\text{Ind}(P) = \dim \text{Ker}(P) - \text{codim Im}(P).$$

We have the following result, originally due to Komatsu [142].

Theorem 1.3.5 *Let V be an open set in $\mathbb{C}$. Then the differential operator $P(z, \frac{d}{dz})$ has index*

$$\text{Ind}(P) = mc(V) - \sum_z ord_z a_m(z),$$

where $c(V)$ denotes the Euler-Poincare' characteristic of V (i.e. $c(V)$ is the difference $b_0(V) - b_1(V)$ of the two first Betti numbers of V) and $ord_z a_m(z)$ denotes the order of zero at z of the leading coefficient $a_m(z)$.

Proof.   To begin with let us remark that, in this particular case, the topological characteristic $c(V)$ can also be interpreted as the number of connected components of V minus the number of compact connected components of its complement in $\mathbb{C}$. This will be the interpretation which we will use for $c(V)$ in the rest of the proof. Now note that if $m = 0$, then $\text{Ker}(P) = 0$, and then one easily finds a basis for the complement of $\text{Im}(P)$ which is constituted by exactly $\sum_z ord_z a_0(z)$ polynomials. If we now consider the case in which $P(z, d/dz) = d/dz$, then from the Mittag-Leffler Theorem one has that

$$\text{Ker}(P) = H^0(V,\mathbb{C}), \text{ and } \text{Coker}(P) = H^1(V,\mathbb{C}).$$

Since $b_0(V) = dim H^0(V,\mathbb{C})$ and $b_1(V) = dim H^1(V,\mathbb{C})$, the result follows immediately also in this case. It is a general fact that the index of the product of operators is the sum of the indices, and so from what we have proved it follows also the result for the operator $P(z, d/dz) = d^m/dz^m$. Finally, [142], it is possible to show that in a Banach space a lower order perturbation does not change the index of an operator; since $\mathcal{O}(V)$ is an inductive limit of Banach spaces in a standard fashion [223], we can conclude the result for the general operator $P(z, d/dz)$ acting on $\mathcal{O}(V)$.   □

From this general result we can immediately obtain an extension of our previous Proposition 1.3.5 to the case of variable coefficients operators.

Theorem 1.3.6 *With the notations used before, and Ω any interval in $\mathbb{R}$, we have that*

$$dim\{f \in \mathcal{B}(\Omega) : P(x, \frac{d}{dx})f = 0\} = m + \sum_x ord_x a_m(x).$$

Proof.   This is a typical diagram chase proof; let V be a connected complex neighborhood of Ω to which all the coefficients a_j can be analytically continued; restrict V in such a way as to take both V and $V \setminus \Omega$ simply connected and such that all the zeroes of $a_m(z)$ are contained in Ω (all of these topological restrictions are necessary to simplify the application of Theorem 1.3.5). Consider now the following commutative exact diagram:

$$
\begin{array}{ccccccc}
& & 0 & & 0 & & 0 \\
& & \downarrow & & \downarrow & & \downarrow \\
0 & \longrightarrow & \mathcal{O}^P(V) & \longrightarrow & \mathcal{O}^P(V\setminus\Omega) & \overset{\pi}{\longrightarrow} & \mathcal{B}^P(\Omega) & \longrightarrow & \frac{\mathcal{B}^P(\Omega)}{\pi\mathcal{O}^P(V\setminus\Omega)} \\
& & \downarrow & & \downarrow & & \downarrow \\
0 & \longrightarrow & \mathcal{O}(V) & \longrightarrow & \mathcal{O}(V\setminus\Omega) & \longrightarrow & \mathcal{B}(\Omega) & \longrightarrow & 0 \\
& & \downarrow{\scriptstyle P_V} & & \downarrow{\scriptstyle P_{V\setminus\Omega}} & & \downarrow{\scriptstyle P_\Omega} \\
0 & \longrightarrow & \mathcal{O}(V) & \longrightarrow & \mathcal{O}(V\setminus\Omega) & \longrightarrow & \mathcal{B}(\Omega) & \longrightarrow & 0 \\
& & \downarrow & & \downarrow & & \downarrow \\
& & \frac{\mathcal{O}(V)}{P\mathcal{O}(V)} & \longrightarrow & 0 & \longrightarrow & 0 \\
& & \downarrow \\
& & 0
\end{array}
$$

where $\mathcal{O}^P$ and $\mathcal{B}^P$ denote, respectively, the sheaves of holomorphic and hyperfunction solutions of the associated homogeneous equation. Since P_V and $P_{V\setminus\Omega}$ have both well defined indices, and since both dim $\mathrm{Ker}(P_V)$ and dim $\mathrm{Ker}(P_V\setminus\Omega)$ are finite, we deduce that also P_Ω has index which is given by

$$
\mathrm{Ind}(P_\Omega) = \mathrm{Ind}(P_V\setminus\Omega) - \mathrm{Ind}(P_V) = m + \sum_x ord_x a_m(x).
$$

This concludes the proof of the Theorem. □

It may be worthwhile noticing that (as it should probably be expected) the diagram shows clearly that two types of solutions are represented in $\mathcal{B}^P(\Omega)$: those coming from $\mathcal{O}^P(V\setminus\Omega)/\mathcal{O}^P(V)$, and those coming from $\mathcal{O}(V)/P\mathcal{O}(V)$.

This hyperfunction version of the index theorem allows us to provide a condition for ellipticity which is much weaker than the one given in our previous Theorem 1.3.3.

Theorem 1.3.7 *With all notations as before, the following statements are equivalent:*

(i) $\mathcal{B}^P(\Omega) \subsetneq \mathcal{A}(\Omega)$;

(ii) $a_m(x) \neq 0$ *for all x in Ω;*

(iii) $P(x, d/dx)f \in \mathcal{A}(\Omega)$ *implies $f \in \mathcal{A}(\Omega)$.*

Proof. (i) implies (ii): suppose that a_m vanishes at a point x_0 in Ω; then, by Theorem 1.3.6, there are more than m linearly independent solutions near x_0.

Take, therefore, a non-trivial solution f in $\mathcal{B}^P(\Omega)$ which is identically zero for all $x < x_0$; such a solution, clearly, is not real analytic and this violates (i).

(ii) implies (iii): let us assume Ω to be connected (the argument would otherwise apply to each connected component). Then on Ω there are exactly m linearly independent real analytic solutions of the homogeneous equation associated to P; on the other hand, Theorem 1.3.6 (and hypothesis (ii)) shows that there are exactly m linearly independent hyperfunction solutions to that same equation; this shows that $\mathcal{A}^P(\Omega)$ coincides with $\mathcal{B}^P(\Omega)$. We now use the linearity of the operator to conclude that, for every real analytic function g on Ω, all solutions to $P(x, \frac{d}{dx})f = g$ must be real analytic as well.

(iii) implies (i): this is immediate. □

Let us now look at a few examples in which we will see the concrete meaning of solving a differential equation in the realm of hyperfunctions.

Example 1.3.8 We shall start with the simplest case in which we deal with a constant coefficient differential equation of order one. Take the equation, on all of $I\!\!R$,

$$\frac{df}{dx} = 0.$$

If we consider $f = [F]$, where $F = (F^+, F^-)$, then $df/dx = 0$ translates in

$$\left[\frac{dF}{dz}\right] = 0,$$

which means, by the definition of hyperfunctions, that

$$\frac{dF^+(x+i0)}{dz} - \frac{dF^-(x-i0)}{dz} = H(x),$$

where H denotes, as usual, the Heaviside function. Since F^+ and F^- are holomorphic functions in, respectively, the upper and lower half-plane, their derivative can coincide on the real axis if and only if $F^+ = F^- + \text{const}$. This shows that the hyperfunction solutions to $df/dx = 0$ are exactly the constants as one expects.

Example 1.3.9 Let us proceed to something slightly more complicated, by considering the equation

(1.3.1) $x f(x) = 1.$

Here the order of the equation is zero, but there is a singular point at the origin (where the leading coefficient x vanishes). Note that whatever solution we will

obtain, it has to coincide with $1/x$ for $x \neq 0$, and have a singularity at $x = 0$.
Again, let $f = [F]$. Then equation (1.3.1) becomes

$$[zF(z)] = [\epsilon] = 1,$$

from which one deduces the existence of a function $G(z)$, holomorphic near the
origin, such that

$$zF(z) - G(z) = \epsilon(z).$$

If we now expand $G(z)$ in Taylor series around the origin, we have
$G(z) = c_0 + zG_1(z)$, with $G_1(z)$ holomorphic near zero. This implies

$$F(z) = \frac{\epsilon(z)}{z} + \frac{c_0}{z} + G_1(z),$$

and therefore (using the notations introduced in the previous section)

$$f(x) = \frac{1}{(x + i0)} + c_1\delta(x).$$

The reader will note that the dimension of the space of solutions is, in this case,
one, which is in agreement with Theorem 1.3.6.

Example 1.3.10 We now combine these two examples, to look at a first order
differential equation with a singularity at the origin. Consider the equation

(1.3.2)
$$x\frac{df(x)}{dx} - \alpha f(x) = 0.$$

In view of Theorem 1.3.6, we know that we should be looking for two linearly
independent solutions to (1.3.2). We need to consider three different cases.

(i) $\alpha = 0, 1, 2, \ldots$ (nonnegative integers); then the two independent solutions
are

$$\begin{cases} x^\alpha \\ x_+^\alpha. \end{cases}$$

(ii) $\alpha = -1, -2, \ldots$ (negative integers); then the two independent solutions
are

$$\begin{cases} \delta^{(-\alpha-1)} \\ \text{p.v.}(x^\alpha). \end{cases}$$

(iii) α not an integer; then the two independent solutions are

$$\begin{cases} x_+^\alpha \\ x_-^\alpha. \end{cases}$$

It may be interesting to note that the two solutions which appear in these three cases fit exactly in the scheme described in our comment after Theorem 1.3.6.

Let us now go back to the inclusion of the sheaf $\mathcal{A}$ in the sheaf $\mathcal{B}$. Generally speaking, we know that given a sheaf inclusion $\mathcal{F} \hookrightarrow \mathcal{G}$ on a topological space X, and given the family $\{\Omega\}$ of all open sets of X, the quotient presheaf given by $\{\mathcal{G}(\Omega)/\mathcal{F}(\Omega)\}$ is not necessarily a sheaf. The quotient sheaf of two sheaves, indicated by $\mathcal{G}/\mathcal{F}$, is on the other hand defined as the sheafification, see [64], of the presheaf described above. In the case of the inclusion of $\mathcal{A}$ into $\mathcal{B}$, and at least as far as the case of one real variable is concerned, the situation is simpler, and has a nice geometric interpretation to which we already made reference previously; we have the following result:

Theorem 1.3.8 *The presheaf $\mathcal{B}/\mathcal{A}$ is actually a sheaf, i.e. $\mathcal{B}/\mathcal{A}(\Omega) = \mathcal{B}(\Omega)/\mathcal{A}(\Omega)$. Moreover, for every hyperfunction f on an open set Ω, its singular support is the support of the equivalence class f mod $\mathcal{A}(\Omega)$, considered as an element of $\mathcal{B}/\mathcal{A}(\Omega)$. Finally $\mathcal{B}/\mathcal{A}$ is a flabby sheaf.*

Proof. See e.g. [103].   □

It is rather obvious to note that the quotient sheaf which we have just represented is a description of the singularities of the hyperfunctions (since the real analytic part is eliminated through the passage to the quotient). Still, this description is not fully adequate, and a complete description of the singularities of hyperfunctions will have to wait until we introduce the process of microlocalization in the next chapter.

We have so far proved that the sheaf of real analytic functions is a subsheaf of the sheaf of hyperfunctions; this is an important result, which legitimizes the use of hyperfunctions as generalized functions; a related, more delicate, issue, is the question of how large the sheaf of hyperfunctions is. As it turns out, we know that every Schwartz distribution can be represented as the boundary value of a holomorphic function (and in 1.2 we have proved that some growth conditions are necessary on the holomorphic functions in order for their boundary values to be distributions); since hyperfunctions arise essentially as the correct concept to formalize the notion of such boundary values, one may expect that the sheaf of distributions is actually a subsheaf of the sheaf of hyperfunctions. This is indeed the case, as our next result shows:

Theorem 1.3.9 *The sheaf $\mathcal{D}'$ of Schwartz distributions on $\mathbb{R}$ is a proper subsheaf of the sheaf $\mathcal{B}$ of hyperfunctions on $\mathbb{R}$.*

One could actually show (see e.g. [103]) that a sequence of sheaf inclusions can be given, namely

$$\mathcal{A} \hookrightarrow \mathcal{L}_{1,loc} \hookrightarrow \mathcal{D}' \hookrightarrow \mathcal{B}.$$

The following theorem can be found directly in [103], to which we refer the reader for any further detail on this matter.

Theorem 1.3.10 *If we indicate with j the inclusion in the sheaf of hyperfunctions:*

(i) *for any function f locally integrable in Ω, and any real analytic function φ in Ω, $\varphi j(f) = j(\varphi f)$;*

(ii) *for any differentiable function f, $(d/dx)j(f) = j(df/dx)$,*

(iii) *for any function f, real analytic in a neighborhood of the interval $[a, b]$;*

$$\int_a^b f(x)dx = \int_a^b j(f)(x)dx.$$

The inclusion of the sheaf of distributions in the sheaf of hyperfunctions also raises the issue of the right notion of boundary value; such an issue is immediately disposed of by the following result in which, again, j represents the inclusion of the sheaf of distributions in the sheaf of hyperfunctions:

Theorem 1.3.11 *Let f be a distribution and let $F = (F^+, F^-)$ be a holomorphic function which represents $j(f)$. Then the function $F^+(x + i\epsilon) - F^-(x - i\epsilon)$ converges to f, in the sense of distributions, as ϵ decreases to zero.*

We will come back in future sections to the various notions of boundary values, as we will tackle the issue of extending these concepts to several real variables. As for now, we continue this section with one of the most important results which, as the reader will see, provide the link between the modern theory of hyperfunctions and its prehistory. We need to recall a preliminary result of Köthe [138], [139] on analytic functionals in the complex plane: we remind the reader that if K is a compact set in $\mathbb{C}$, then the space of germs of holomorphic functions on K is defined as the topological inductive limit

$$\mathcal{O}(K) = \underset{U}{\mathrm{ind}\ \lim}\ \mathcal{O}(U),$$

where U ranges over a cofinal family of open neighborhoods of K.

Theorem 1.3.12 *Let K be a compact set in the complex plane and V an open set containing K. Then*

$$\mathcal{O}(K)' = \frac{\mathcal{O}(V \setminus K)}{\mathcal{O}(V)}.$$

Proof.   See Theorem 1.3.14 below.   □

With the help of this result we can prove the celebrated duality theorem for hyperfunctions:

Theorem 1.3.13 *Let K be a compact set in $\mathbb{R}$, and let $\mathcal{A}(K)$ be the space of germs of real analytic functions defined on a neighborhood of K. Then the dual $\mathcal{A}(K)'$ can be identified with the space $\mathcal{B}_K(\mathbb{R})$ of hyperfunctions supported in K, by means of the inner product*

$$< \varphi, f >= \int_\Omega \varphi(x) f(x) dx,$$

where f is a hyperfunction, φ is a real analytic function and Ω is a real neighborhood of K on which φ is defined.

Proof.   Immediate consequence of Theorem 1.3.12.   ☐

The interest of this theorem is that it establishes that hyperfunctions can be interpreted as non compactly supported real analytic functionals; moreover, at least when they are compactly supported, a fully developed theory of Fourier transform already exists, and one might hope to develop it and extend it to the case of general hyperfunctions. This has actually been done by Kawai, originally in [130], and we will come back to this at a later stage.

Theorem 1.3.13 has been stated, as of now, in the framework of linear spaces, with the duality being understood in that sense; one may, however, question whether the isomorphism which it describes may actually be a topological isomorphism, and, in that case, which topology would we be considering on the space of hyperfunctions. As a matter of fact, unlike what happens for the theory of distributions (which is heavily based on the theory of topological vector spaces), the theory of hyperfunctions has more of a linear space flavor, and indeed, so far, we have not even touched upon the problem of endowing $\mathcal{B}(\Omega)$ or $\mathcal{B}_K$ with a natural topology. Such a problem, however, will be crucial in the next section, and we therefore wish to address it right away. We refer the reader to [223] for all notations and definitions concerning topological vector spaces.

To begin with, we shall show that, for any compact set K in $\mathbb{R}$, the space $\mathcal{B}_K$ of hyperfunctions supported by K can be given a quite natural Frechet space topological structure. This is a consequence of the following result, which is nothing but a strengthening and modification of Proposition 1.2.3.

Theorem 1.3.14 *Let K be a compact set in $\mathbb{R}$ and let $\mathcal{B}_K$ denote the space of hyperfunctions supported by K. Define a map*

(1.3.3) $$T : \mathcal{B}_K \longrightarrow \mathcal{O}(\mathbb{C} \setminus K)$$

by setting, for any hyperfunction f supported by K,

$$T(f)(z) = \int_\mathbb{R} \frac{f(x)}{z - x} dx.$$

Then the map T is injective and its image is a closed subspace of the Frechet space $\mathcal{O}(\mathbb{C} \setminus K)$.

Proof. Since T just provides a defining function for f, the injectivity is clear. Let us now turn to the topological part of the result. Suppose $\{f_j\}$ is a sequence in $\mathcal{B}_K$ such that $\{T(f_j)\}$ converges to some holomorphic function F in the Frechet topology of $\mathcal{O}(\mathbb{C} \setminus K)$. We want to prove that the hyperfunction f identified by the pair F, which is clearly supported by K, is such that $T(f) = F$. This will conclude our proof. To do so, just note that $T(f_j)$ can actually be computed using any boundary value representation for f_j. This was already proved in Proposition 1.2.3. so that we can conclude that if $f_j = [F_j]$, then $T(f_j)$ can be written as the integral

$$T(f_j) = \int_\tau \frac{F_j(\zeta)}{z - \zeta} d\zeta.$$

Now, by hypothesis, $\{F^j\}$ converges uniformly (on each compact set which does not intersect the real axis) to F, and therefore this concludes the proof. $\quad\square$

As a consequence of this result we can identify $\mathcal{B}_K$ with a closed subspace of the Frechet space $\mathcal{O}(\mathbb{C} \setminus K)$, and this make therefore the space $\mathcal{B}_K$ a Frechet space in itself.

It is immediate to notice that the Frechet structure which we have just given to the space of compactly supported hyperfunctions is a natural consequence of the fact that such hyperfunctions can essentially be considered as equivalence classes of functions holomorphic outside a compact set. Also, the reader will note that, in defining the injection of $\mathcal{B}_K$ in the Frechet space of holomorphic functions, we have used a very specific representative for the hyperfunctions involved, essentially based on the standard representative introduced in Proposition 1.2.3. As it is shown in [103] (where the several variables case is treated as well), this particular choice bears no consequences. This fact is particularly obvious in the one variable case, as the following result shows:

Proposition 1.3.6 *Let $\{f_j\}$ be a sequence of hyperfunctions supported in the same compact set K, Ω a real open neighborhood for K and, finally, V a complex neighborhood of $\Omega \setminus K$. Let $\{F_j\}$ be a sequence of functions holomorphic in $\mathcal{O}(V)$ such that $[F_j] = f_j$. Then if $\{F_j\}$ converges to F in the Frechet topology of $\mathcal{O}(V)$, one has that $\{f_j\}$ converges to $f = [F]$ in B_K.*

Proof. The proof is immediate from the direct computation of $T(f_j)$ which one may obtain from using their boundary value representation. $\quad\square$

An immediate corollary of this argument is the following complement to Proposition 1.3.1:

Proposition 1.3.7 *Every linear differential operator with real analytic coefficients acts continuously on the space $\mathcal{B}_K$. Moreover, every local operator acts continuously on the same space.*

Proof.   For the case of finite order differential operators, the proof is an immediate consequence of the definition of the action of differential operators on hyperfunctions and of the definition of the topology on the space of compactly supported hyperfunctions. In the case of infinite order differential operators, we need to recall that such operators act continuously on the space of sections of holomorphic functions, in view of the estimates on the coefficients; indeed, for F a holomorphic function, the norm of $P(D)F$ only depends on the norm of F itself. One then applies the same argument as in the case of finite order to conclude the continuity on the space of hyperfunctions.   ·   □

Let us take this opportunity to go back briefly to one of the important first remarks we made on hyperfunctions; in section 2 we have shown how every hyperfunction supported at the origin could be written up as, formally, a series of derivatives of the Dirac δ function; the previous result immediately implies that the formal series which one constructs is actually convergent in the topology of the space $\mathcal{B}\{0\}$ of hyperfunctions supported at the origin.

There is an important consequence of the way we have defined the topology on the space of compactly supported hyperfunctions, and we wish to point it out in a separate proposition, since it will become necessary later on:

Proposition 1.3.8 *For any compact K in the real line, the space B_K, endowed with the topology described before, is a reflexive space.*

Proof.   Indeed since, for any open complex set U, the space $\mathcal{O}(U)$ is a Montel space, so is the space $\mathcal{B}_K$. Its reflexivity then follows from standard arguments in topological vector spaces.   □

The discussion on reflexivity brings us back to Theorem 1.3.13, of which we can now give a stronger and more interesting formulation, based on the topology we have just described. Before doing so, let us quickly recall the topology of the space of analytic functionals. As the reader may remember, given a real compact set K, the space $\mathcal{A}(K)$ of germs on K of real analytic functions is defined as the inductive limit $\lim \mathcal{O}(U)$ of the spaces of holomorphic functions on a cofinal set $\{U\}$ of open complex neighborhoods of K. As a consequence one sees that $\mathcal{A}(K)$ is a limit of Frechet space, which is naturally endowed with an LF topology; as such, as it is well known, $\mathcal{A}(K)$ is not a Frechet space, even though many of its properties closely resemble those of Frechet spaces. In particular, even though one cannot define a topology on an LF space through the notion of convergent sequences, we have the following result:

Proposition 1.3.9 *A sequence $\{\varphi_j\}$ of germs in $\mathcal{A}(K)$ converges to a germ φ in $\mathcal{A}(K)$ if and only if $\varphi_j(z)$ converges uniformly to $\varphi(z)$ in a complex neighborhood of K.*

Proof. This is an immediate, and general, consequence of the definition of the inductive limit topology of $\mathcal{A}(K)$.   □

We can now state the topological version of Theorem 1.3.13 as follows:

Theorem 1.3.15 *For every compact set K in $\mathbb{R}$, one has the following topological isomorphism:*

$$\mathcal{A}(K)' = \mathcal{B}_K.$$

Proof. This result follows from the preceding propositions in a standard way. The reader is referred to [103].   □

This result is of course of great theoretical interest (besides being a natural completion of Köthe duality theorem), especially as it will allow us to define a notion of Fourier transform for compactly supported hyperfunctions, and thus to develop some of the tools necessary for the approach to differential equations on hyperfunctions along the ideas in Ehrenpreis' Fourier Analysis in Several Complex Variables [52].

Even more interestingly, from this point of view, the converse is true. Indeed, one has the following result:

Theorem 1.3.16 *For any compact set K in $\mathbb{R}$, one has the following topological isomorphism:*

$$\{\mathcal{B}_K\}' = \mathcal{A}(K).$$

As a corollary we get that, since $\mathcal{B}_K$ is an FS-space, then its dual $\mathcal{A}(K)$ is a DFS-space, which fact entails some interesting consequences (see e.g. [102] for some applications and a hyperfunction version of the Ehrenpreis-Palamodov Fundamental Principle). One may also note that, as it was done in [103], Theorem 1.3.16 actually implies Theorem 1.3.15, in view of Proposition 1.3.8 on the reflexivity of the space of compactly supported hyperfunctions. Note, however, that from Theorems 1.3.15 and 1.3.16 we could also directly conclude that both $\mathcal{B}_K$ and $\mathcal{A}(K)$ are reflexive spaces.

We now point out one last result on compactly supported hyperfunctions; this allows us to display a behavior of hyperfunctions which is completely different from the corresponding behavior of Schwartz distributions.

Proposition 1.3.10 *Let $K \subseteq L$ be two connected compact sets on the real line. Then the natural inclusion $\mathcal{B}_K \hookrightarrow \mathcal{B}_L$ has a dense image.*

Proof. In view of the Hahn-Banach theorem, it is sufficient to show that if a functional φ in $\{\mathcal{B}_L\}' = \mathcal{A}(L)$ vanishes on $\mathcal{B}_K$, then it is identically zero. Let therefore $a \epsilon K$, so that, for any multi-index α, $D^\alpha \delta(x - a)$ belongs to $\mathcal{B}_K$. Then, if $<\ ,\ >$ denotes the duality bracket induced by Theorem 1.3.16, one has:

$$0 =< D^\alpha \delta(x - a), \varphi(x) >=< \delta(x - a), (-D)^\alpha \varphi(x) >= (-1)^{|\alpha|} \varphi^{(\alpha)}(a),$$

and thus, by the uniqueness of analytic continuation, $\varphi = 0$ everywhere, which concludes the proof.   □

So far, we have described the topology of compactly supported hyperfunctions, and we have noted how their structure is not too different from what we are already used to; this is essentially the consequence of two facts: the representation of hyperfunctions in terms of a well defined standard holomorphic function, and the duality expressed by Köthe's theorem. Unfortunately, the situation is not at all this simple when we try to consider the case of non-compactly supported hyperfunctions. As a matter of fact, one can say that it is not possible to endow the space $\mathcal{B}(\Omega)$ with any reasonable topology! The intuitive reason for this state of things is essentially the fact that for noncompactly supported hyperfunctions there is no way of choosing a standard defining function. Even the zero hyperfunction has no canonical way to be represented, unlike what happened in the compact case. In addition, as it has been pointed out both in [103] and in [123], we do not expect to be able to ever come out with a reasonable topology for $\mathcal{B}(\Omega)$, since its dual would have to somehow represent the space of "real analytic functions with compact support". This lack of a good topological structure will have rather negative consequences, as it will be immediately apparent in the next few pages.

1.4 Hyperfunctions of Several Variables: Basics

In this section we will try to develop and describe the first fundamental notions in the theory of hyperfunctions on $I\!\!R^n$. The reader will immediately recognize that a totally different level of complexity becomes necessary here, and some of the reasons for this are sketched in the historical appendices to the chapter itself. While section 1.2 only required the most elementary notions from one variable complex analysis, and section 1.3 used the fundamental definitions from sheaf theory, this present section, and the next one, will rely heavily upon the theory of relative cohomology with coefficients in a sheaf, as well as on some delicate properties in the theory of several complex variables.

In order to help the reader's intuition, and in keeping with our original plans, we have restricted our attention to the case of hyperfunctions defined on the Euclidean n-dimensional space; more generally (and the reader is referred to [123] for this) one can extend all of our definitions and arguments to the situation, however useful and important, would have eliminated the clarity and accessibility which, we hope, will make our approach successful.

Let us begin by pointing out some natural attempts to define hyperfunctions in, say, two variables, and to show the difficulties which are linked to these approaches. If nothing else, this will motivate the need for the introduction of a more complex point of view.

A first, ill fated, approach, would consist in a blind mimicking of the definition which we have given in one variable; let K be a compact set in $\mathbb{R}^2$, Ω a real open set which contains K, and V be a complex neighborhood of Ω. Then the quotient $\mathcal{O}(V \setminus K)/\mathcal{O}(V)$, which would be suppose to define the space of hyperfunctions supported in K, would actually vanish for any choice of V and K, in view of the celebrated Hartogs' theorem on the removability of compact singularities of holomorphic functions in several complex variables. This circumstance, crucial in the theory of several complex variables, essentially relies on the fact that (in several complex variables) there are open sets which are not holomorphy domains (in contrast with what happens in the single variable case); thus, essentially, the failure of a direct analog of the single variable treatment is reduced to the fact that the Mittag-Leffler theorem does not hold for every open set in $\mathbb{C}^n$. In cohomological terms, it is no longer true, when several variables are considered, that $H^1(V, \mathcal{O}) = 0$. As it will be clear in the sequel, we actually don't need such a vanishing, which will be replaced by a much subtler result (Malgrange's theorem, as it will turn out).

An alternate and more successful attempt consists in looking back at the way in which distributions of two variables are defined; we are then tempted to define, for a pair of open sets Ω_1 and Ω_2, the space of hyperfunctions on $\Omega_1 \otimes \Omega_2$ as the completion of the tensor product $\mathcal{B}(\Omega_1) \otimes \mathcal{B}(\Omega_2)$; to do so let us consider such a tensor product:

(1.4.1)
$$\mathcal{B}(\Omega_1) \otimes \mathcal{B}(\Omega_2) =$$

$$= \left(\frac{\mathcal{O}(V_1 \setminus \Omega_1)}{(\mathcal{O}(V_1))} \right) \otimes \left(\frac{\mathcal{O}(V_2 \setminus \Omega_2)}{\mathcal{O}(V_2)} \right)$$

$$= \frac{\mathcal{O}(V_1 \setminus \Omega_1) \otimes \mathcal{O}(V_2 \setminus \Omega_2)}{\mathcal{O}(V_1) \otimes \mathcal{O}(V_2 \setminus \Omega_2) + \mathcal{O}(V_1 \setminus \Omega_1) \otimes \mathcal{O}(V_2)}.$$

Unfortunately, as we have thoroughly described in the previous section (but see also [103]) the spaces of hyperfunctions which we need in order to construct this tensor product cannot be endowed with any natural locally convex topology, and so the idea of just completing the previous quotient space seems to fail. Still, if one looks at the quotient which we have just defined, one realizes that a reasonable guess for the space of hyperfunctions on a cartesian product of two open sets is given by what we could call the functional interpretation of the completion of the tensor product (1.4.1); namely, one may suggest the following definition for the space of hyperfunctions on $\Omega_1 \times \Omega_2$:

$$\mathcal{B}(\Omega_1 \times \Omega_2) = \frac{\mathcal{O}(V_1 \setminus \Omega_1) \times \mathcal{O}(V_2 \setminus \Omega_2)}{\mathcal{O}(V_1 \times (V_2 \setminus \Omega_2)) + \mathcal{O}((V_1 \setminus \Omega_1) \times V_2)}.$$

The reader will immediately note that in the previous expression the numerator and the denominator of the quotient are the completions of, respectively, the numerator and the denominator of the quotient in (1.4.1). As we will see in a while, this description actually turns out to be a correct definition for the space of hyperfunctions in two variables, but it immediately shows some fundamental shortcomings; to begin with, its description heavily relies on the use of a specific coordinate system; this is not so much of a problem when dealing with the case of Euclidean spaces, but might become a burden when dealing with more general real analytic manifolds; on the other hand, the definition which we have proposed does not seem amenable to computations, and it might prove quite difficult to show that the object thus constructed is a sheaf on the Euclidean space.

Coming back to our representation, let us restrict our attention, for the moment, to the case in which $\Omega_1 = \Omega_2 = I\!\!R$ and therefore we can take $V_1 = V_2 = \mathcal{C}$. Then we immediately see that our proposed definition of the space $\mathcal{B}(I\!\!R^2)$ is as follows:

$$\mathcal{B}(I\!\!R^2) = \frac{\mathcal{O}((\mathcal{C} \setminus I\!\!R) \times (\mathcal{C} \setminus I\!\!R))}{\mathcal{O}(\mathcal{C} \times (\mathcal{C} \setminus I\!\!R)) + \mathcal{O}((\mathcal{C} \setminus I\!\!R) \times \mathcal{C})}.$$

This is very much in agreement with the following argument (which is just a more direct version of the tensor product approach); let f and g be two hyperfunctions on $I\!\!R$ with defining functions F and G, holomorphic functions in $\mathcal{C}$ outside the real axis. Then one might try to define a hyperfunction $h(x, y) = f(x)g(y)$ by means of the holomorphic function $H(z, w) = F(z)G(w)$. It is then clear that H is holomorphic on $(\mathcal{C} \setminus I\!\!R) \times (\mathcal{C} \setminus I\!\!R)$, and that it is defined modulo functions in $\mathcal{O}(\mathcal{C} \times (\mathcal{C} \setminus I\!\!R)) + \mathcal{O}((\mathcal{C} \setminus I\!\!R) \times \mathcal{C})$, in view of the fact that both F and G are defined modulo entire functions. This brings us back to our original definition for bivariate hyperfunctions. We should also point out that this particular definition is by no means restricted to the case of two variables; as a matter of fact, one may easily envision the following generalization to the case of several real variables. Let $\Omega_1, \Omega_2, \ldots, \Omega_n$ be n open sets in $I\!\!R$, and let $V_1, V_2, \ldots, V_n$ be n complex neighborhoods of them. Then, if we set

$$V \# \Omega = (V_1 \setminus \Omega_1) \times \ldots \times (V_n \setminus \Omega_n)$$

and

$$V_j^* = (V_1 \setminus \Omega_1) \times \ldots \times (V_{j-1} \setminus \Omega_{j-1}) \times V_j \times (V_{j+1} \setminus \Omega_{j+1}) \times \ldots \times (V_n \setminus \Omega_n),$$

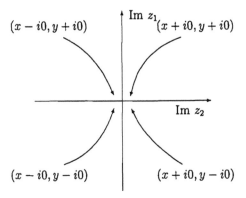

$$(x - i0, y + i0) \qquad \text{Im } z_1 \qquad (x + i0, y + i0)$$

$$\text{Im } z_2$$

$$(x - i0, y - i0) \qquad (x + i0, y - i0)$$

Figure 1.4.1

we can reasonably define the space of hyperfunctions on $\Omega_1 \times \ldots \times \Omega_n$ by the following quotient:

$$B(\Omega_1 \times \ldots \times \Omega_n) = \frac{\mathcal{O}(V \# \Omega)}{\sum_{j=1}^n \mathcal{O}(V_j^*)}.$$

There is, however, at least one other natural way to introduce hyperfunctions in two variables, and it is through the notion of boundary values of holomorphic functions, which, as we have seen, are one of the key concepts that hyperfunctions were designed to deal with.

Let us go back to our example of the hyperfunctions $f = [F]$ and $g = [G]$; by using the notations which we have introduced in the previous sections, we have that $f(x) = F(x + i0) - F(x - i0)$, while $g(x) = G(x + i0) - G(x - i0)$; it is therefore quite natural to consider the product $(fg)(x, y) = f(x)g(y)$ which can be thought of as the sum of four different boundary values as follows:

$$(fg)(x, y) = H(x+i0, y+i0) - H(x-i0, y+i0) + H(x-i0, y-i0) - H(x+i0, y-i0)$$

where the holomorphic function H is defined as the product of F and G, and it is therefore holomorphic on $(\mathbb{C} \setminus \mathbb{R}) \times (\mathbb{C} \setminus \mathbb{R})$. This representation is correctly depicted in the following figure, which illustrates the four directions which, in $\mathbb{C}^2$, allow the approach to $\mathbb{R}^2$.

The reader will immediately note how this same approach would allow us to think of hyperfunctions in n variables as the sum of 2^n boundary values of holomorphic functions. As we will see, this point of view is not too far from the truth, even though it only holds a partial view of truth itself.

In particular, one is immediately struck by the deeper meaning that the notion of boundary value assumes when dealing with several variables. Indeed,

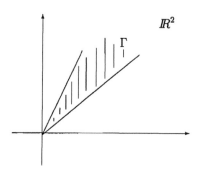

Figure 1.4.2

in one variable the variable $z = x + iy$ only admits two "directions" to approach its real part; y may converge to zero from the positive direction or from the negative direction; when we consider two variables such as (z, w), then one has several possibilities; while our example above (and figure 1.4.1) only refer to the four possibilities (2^n possibilities, in general) which correspond to the four quadrants of $I\!R^2$ (to the 2^n orthants of $I\!R^n$, in general), it is not difficult to realize that other more complex situations may arise, if the orthants are replaced by other more general cones in the appropriate space.

These remarks lead us to define some special open sets which play a crucial role in what follows, and we take this chance to describe the necessary notations for what follows. In so doing, we move to the general case of n variables, since, after discussing possible motivations, we are now ready to provide the correct definitions.

Let g be an open cone in $I\!R^n$, with vertex at the origin. Then for every open set Ω in $I\!R^n$ we can define a **wedge** in $\mathbb{C}^n$ as the open set

$$\Omega \times i\Gamma.$$

as shown in the following set of pictures.

Such an object, as we said, is said to be a wedge, and the cone Γ is called its opening; a wedge is of course a special case of a tubular neighborhood. Given two cones Γ and Γ' with vertex at the origin of $I\!R^n$, we say that Γ' is relatively compact in Γ, and we write $\Gamma' << \Gamma$, if the closed set

$$\Gamma' \cap S^{n-1},$$

where S^{n-1} denotes the unit sphere in $I\!R^n$, is contained in the interior of Γ. We also say, in this case, that Γ' is a proper subcone of Γ (note that this expression, though widely used, is somewhat misleading, since it seems to imply that Γ' is

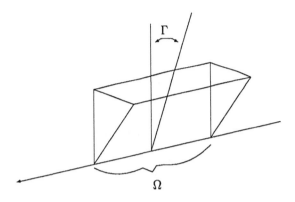

Figure 1.4.3

separated from Γ, while we know that they both have vertex at the origin, and therefore they get infinitesimally close one to another).

Our next concept is that of **infinitesimal wedge** (which generalizes the notation $x \pm i0$); given an open set Ω and an open cone Γ in $I\!R^n$, we say that a complex n-dimensional open set U is an infinitesimal wedge of type $\Omega + i\Gamma 0$ if U is contained in the wedge $\Omega + i\Gamma$ and, for every proper subcone Γ' of Γ and every $\epsilon > 0$ there exists $\delta > 0$ such that

$$U \supset \Omega_\epsilon + i(\Gamma' \cap \{|y| < \delta\}),$$

where Ω_ϵ denotes the set obtained by Ω by shrinking it by ϵ. In this case, see figure 1.4.4, we say that Ω is the edge of the infinitesimal wedge.

We note, see again figure 1.4.4, that an infinitesimal wedge is nothing but an open set contained in the wedge $\Omega + i\Gamma$ which, asymptotically in the vicinity of the edge Ω, approaches a wedge of opening Γ; in the sequel, the symbol $\Omega + i\Gamma 0$ will denote any such a set, since it will never be necessary the specification of the precise set; similarly, the space $\mathcal{O}(\Omega + i\Gamma 0)$ denotes the set of functions which are holomorphic in some infinitesimal wedge of type $\Omega + i\Gamma 0$.

In this respect, we need to quote an important result due to Bochner, and known as the Bochner tube Theorem [103], which will be useful in what follows:

Theorem 1.4.1 *Let U be a connected open set in $I\!R^n$, and let $U^\wedge$ be its convex hull. Then every function F holomorphic in $I\!R^n + iU$ extends analytically to $I\!R^n + iU^\wedge$. As a consequence, if Γ is a cone as above and $\Gamma^\wedge$ is its convex hull, any function F holomorphic in $\Omega + i\Gamma 0$ can be holomorphically extended to the infinitesimal wedge $\Omega + i\Gamma^\wedge 0$.*

Proof. We refer the reader to [103] for detailed proofs of this and related statements.   □

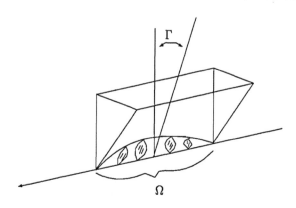

Figure 1.4.4

We can now proceed to provide one further intuitive definition for hyperfunctions in several variables; this will turn out to be the closest to the precise definition, and will provide us with the final intuitive vision of hyperfunctions, before plunging into the formal treatment based on relative cohomology.

Consider N open convex cones in $I\!R^n$ with vertex at the origin, $\Gamma_1, \Gamma_2, \ldots, \Gamma_N$; we are taking the Γ_j to be convex in view of Theorem 1.4.1. Let F_j be holomorphic functions defined, for $j = 1, \ldots, N$, on the infinitesimal wedges $\Omega + i\Gamma_j 0$, for Ω an open set in $I\!R^n$. Then the formal sum

$$(1.4.2) \qquad\qquad f(x) = \sum_{j=1}^{N} F_j(x + i\Gamma_j 0)$$

defines a hyperfunction on Ω.

It is clear that the functions F_j which appear in (1.4.1), and which we can call the **defining functions** for the hyperfunction f are far from being uniquely determined, and actually the cones Γ_j themselves are not uniquely determined at all. It should also be clear that the definition of the space of hyperfunctions $\mathcal{B}(\Omega)$ which is somehow provided by (1.4.1) is still ambiguous and, at best, incomplete.

In what follows, however, we will regard (1.4.1) as an operational way to define hyperfunctions, and we will soon prove that every hyperfunction can be represented, as in (1.4.1), as the sum of a finite number of boundary values of holomorphic functions.

We now proceed to the rigorous introduction of hyperfunctions on $I\!R^n$; this will prove to be a non trivial step, and the reader may wish to skip this approach, on a first reading, and try to rely on our previous intuitive explanations, at least until more familiarity is acquired. What follows heavily relies on the cohomological properties of holomorphic functions in several complex variables, as well as on the theory of derived sheaves, from which we will only recall the strictly necessary facts.

Definition 1.4.1 *Let S be a sheaf on a topological space X, and let $S \subset X$ be a locally closed subset. The n-th derived sheaf of S is the sheaf $\mathcal{H}_S^n(S)$ associated to the presheaf given, on X, by the collection*

$$(1.4.3) \qquad\qquad \{H_{S \cap U}^n(U, S)\}.$$

One easily notes that, by its very definition, the n-th derived sheaf is a sheaf on the topological space X; however, as a sheaf, it is localized on S, in the sense that one can show that $\mathcal{H}_S^n(S)(U) = 0$ whenever U and S are disjoint. As it is usual in the theory of sheaves, it is not easy to describe a sheaf as the sheaf associated to a non-complete presheaf, since it becomes quite difficult to visualize the sections of such a sheaf. In this particular case, the situation is not easier but we get a partial help by the following general result, for which proof we refer the reader to [103]:

Theorem 1.4.2 *The presheaf indicated in (1.4.2) is a sheaf in the following cases:*

(a) *when $n=0$;*

or

(b) *when*

$$(1.4.4) \qquad\qquad H_S^j(S) = 0 \; \text{for} \; j = 0, 1, \ldots, n-1.$$

In this latter case its sections can be obtained as follows:

$$\mathcal{H}_S^n(S)(U) = H_{S \cap U}^n(U, S).$$

It is then clear that the possibility of easily describing the spaces of sections of derived sheaves rests on the possibility of proving the vanishing described in (1.4.4). In sheaf theoretical terms, we say, in such a case, that S is of codimension greater than or equal to n relative to the sheaf S. More specifically, if one could prove the vanishing of $H_S^j(S)$ for all j different from n, then we would say that the set S is purely n-codimensional with respect to S. Thus, pure codimensionality becomes a key issue for the description of sections of derived sheaves. We will be interested in showing, following Sato's original approach in [197], that $\mathbb{R}^n$, as a locally closed subset of $\mathbb{C}^n$, is purely codimensional with respect to the sheaf $\mathcal{O}$ of germs of holomorphic functions. This fact is in some sense the best extension which one can hope to make of the Mittag-Leffler Theorem which, as we saw, does not hold any more in several variables. To prove Sato's result, which will become the basis for the formal definition of hyperfunctions in several variables, we will need some cohomological facts on the sheaf of holomorphic functions in several variables; such facts are by now standards, but they were far from so when originally proved by Martineau [156] (it should be noticed that the original papers by Sato only bear a minimal amount of details concerning the proofs; so, even if he was undoubtedly the discoverer of these theorems, one should credit people such as Martineau, Harvey, Schapira, Komatsu as being the first to actually provide complete proofs).

Theorem 1.4.3   *(a) Let $K = K_1 \times \ldots \times K_n$ be a cylindrical closed set, and V a Stein domain. Then*

$$(1.4.5) \qquad H^j_{K \times V}(\mathbb{C}^n \times V, \mathcal{O}) = 0, \ for \ j \geq n+1.$$

(b) Let Ω be an n-dimensional real open set, and V a Stein domain. Then

$$(1.4.6) \qquad H^j_{\Omega \times V}(\mathbb{C}^n \times V, \mathcal{O}) = 0, \ for \ j \geq n+1.$$

Proof.   See [103].   □

Theorem 1.4.4 *Let K and L be two closed analytic polyhedra, and V a Stein domain. Then*

$$(1.4.7) \qquad H^j_{(L \setminus K) \times V}(\mathbb{C}^n \times V, \mathcal{O}) = 0, \ for \ 0 \leq j \leq n-1.$$

Proof.   See [103].   □

We are now ready to prove the pure codimensionality of $\mathbb{R}^n$ in $\mathbb{C}^n$, with respect to the sheaf $\mathcal{O}$. This fundamental result is due to Sato, but the proof which we are supplying is due to Kaneko [103].

Theorem 1.4.5 *The embedding $\mathbb{R}^n \hookrightarrow \mathbb{C}^n$ is purely n-codimensional with respect to the sheaf $\mathcal{O}$.*

Proof.   We will directly calculate the stalks of the sheaves $\mathcal{S}^j := H^j_F(\mathcal{O})$, for $F = \mathbb{R}^n$, and we will prove that they vanish for j different from n. Since, for any j, the sheaf $\mathcal{S}^j$ is homogeneous with respect to real shifts, we will simply consider the stalk at the origin. Let us begin with the case of $j \geq n+1$. In this case we want to use Theorem 1.4.3 (part b). As a matter of fact, if U is an arbitrary Stein neighborhood of the origin, we immediately deduce that, for $j \geq n+1$, it is $H^j_{\mathbb{R}^n \cap U}(U, \mathcal{O}) = 0$. Therefore, by taking the inductive limit, we deduce that the stalk at the origin of $\mathcal{S}^j$ vanishes. The case of $j \leq n-1$ is slightly more complicated, since we need to use, in this case, Theorem 1.4.4, which requires us to show that for a fundamental system $\{U\}$ of neighborhoods of the origin, one has that the set $\mathbb{R}^n \cap U$ can be thought of as a difference $L \setminus K$ of two closed analytic polyhedra. As shown in [103], this can be easily done by noticing that the real axis can be expressed as the set

$$\mathbb{R} = \{z \in \mathbb{C} : |e^{iz}| \leq 1, |e^{-iz}| \leq 1\};$$

now, one can set
$$f(z) = 1 - (z_1^2 + \ldots + z_n^2),$$

and define the two closed analytic polyhedra

$$L = \mathbb{R}^n \cap \{z \in \mathbb{C}^n; |z_1| \leq 1, \ldots, |z_n| \leq 1\},$$

$$K = L \cap \{z \in \mathbb{C}^n : |e^{f(z)}| \leq 1\}.$$

Finally we consider the fundamental system of open neighborhoods of the origin given by the dilations $\{rU : r > 0\}$ of the open set

$$U = \{z \in \mathbb{C}^n : |z_1| < 1, \ldots, |z_n| < 1, Re(f(z)) > 0\}.$$

It is immediate to verify that U is a complex neighborhood of the origin and that
$$L \setminus K = \mathbb{R}^n \cap U,$$

so that Theorem 1.4.4 allows us to conclude the proof of the pure codimensionality. $\qquad\square$

We are finally ready to provide the formal definition for the sheaf of hyperfunctions on the Euclidean n-dimensional real space;

Definition 1.4.2 *The sheaf $\mathcal{B}$ of hyperfunctions on the topological space $\mathbb{R}$ is defined as the n-th derived sheaf of $\mathcal{O}$, namely*

$$\mathcal{B} = \mathcal{H}_{\mathbb{R}}^n(\mathcal{O}).$$

In view of Theorem 1.4.2 on the description of the global sections of the derived sheaves, and because of the pure n-codimensionality of $\mathbb{R}^n$ which we have just proved, we can actually use Definition 1.4.2 to precisely describe the vector spaces of hyperfunctions on open sets of $\mathbb{R}^n$. Indeed, we have the following immediate consequence of our results and definitions:

Theorem 1.4.6 *Let Ω be an open set in $\mathbb{R}^n$, and let V be any complex neighborhood of Ω. Then*

$$(1.4.8) \quad \mathcal{B}(\Omega) = H^n(V, V \setminus \Omega; \mathcal{O}) = H^n(\mathbb{C}^n, \mathbb{C}^n \setminus \Omega; \mathcal{O}) = H_\Omega^n(\mathbb{C}^n, \mathcal{O}).$$

Proof.　　The first equality is just a consequence of our remarks preceding the statement of the Theorem itself. The second one follows from the excision theorem for relative cohomology (see e.g. [64]). The third is just the definition.

$\square$

Our next goal, in this section, is to employ Definition 1.4.2, together with Theorem 1.4.6, to get a more concrete description of hyperfunctions; as it will turn out, these relative cohomology classes may indeed be seen as boundary values of functions holomorphic on infinitesimal wedges; this will allow us to recover the intuitive definition which we provided earlier in this section. Concurrently, this approach will allow us to visualize hyperfunctions with families of holomorphic functions, thus making easier our task of working with them.

Let us start with a first representation of the relative cohomology group indicated in (1.4.8). Consider the pair $(V, V \setminus \Omega)$, and its relative covering $(\mathcal{V}, \mathcal{V}')$ given by

$$\mathcal{V} = \{V_0, V_1, \ldots, V_n\},$$

$$\mathcal{V}' = \{V_1, \ldots, V_n\},$$

where

$$V_0 = V,$$

and

$$V_j = V \cap \{z \in \mathbb{C}^n : Im(z_j) \neq 0\}.$$

Since the open sets V_j are Stein (and their intersections are Stein as well), and in view of Oka-Cartan's Theorem, one has that the covering $(\mathcal{V}, \mathcal{V}')$ is acyclic and therefore one can apply Leray's Theorem, jointly with Theorem 1.4.6, to conclude that

$$\mathcal{B}(\Omega) = H^n(\mathcal{V}, \mathcal{V}'; \mathcal{O}).$$

The reader interested in the details of the previous argument should review the description of spectral sequences as provided in Chapter III.

The space of hyperfunctions can therefore be expressed as a Čech cohomology group, which we can now try to explicitly calculate. To begin with, let us recall that, by definition, and using the standard notations, it is

$$(1.4.9) \quad H^n(\mathcal{V}, \mathcal{V}'; \mathcal{O}) = \frac{Ker\{\delta^n : C^n(\mathcal{V}, \mathcal{V}'; \mathcal{O}) \longrightarrow C^{n+1}(\mathcal{V}, \mathcal{V}'; \mathcal{O})\}}{Im\{\delta^{n-1} : C^{n-1}(\mathcal{V}, \mathcal{V}'; \mathcal{O}) \longrightarrow C^n(\mathcal{V}, \mathcal{V}'; \mathcal{O})\}}$$

and we therefore have to compute the kernel and the image of the coboundary mappings. It is immediate to see that, because of the fact that the cardinality of $\mathcal{V}$ is $n+1$, then $C^{n+1}(\mathcal{V}, \mathcal{V}'; \mathcal{O})$ vanishes, so that the kernel of δ^n coincides with $C^n(\mathcal{V}, \mathcal{V}'; \mathcal{O})$; this last vector space, in turn, is easily seen to be isomorphic to the space of sections

$$\mathcal{O}(V_0 \cap V_1 \cap \ldots \cap V_n).$$

Let us now try to describe the image of δ^{n-1}; once again, in view of the definition of alternating cochains, we can see that

$$Im(\delta^{n-1}) = \sum_{j=1}^{n} \mathcal{O}(V_0 \cap V_1 \cap \ldots \cap \hat{V}_j \cap \ldots \cap V_n).$$

Finally, if we recall the original definition of the open sets which make up our covering, we see that the following equalities hold:

$$V_0 \cap V_1 \cap \ldots \cap V_n = V \cap (\mathbb{C} \setminus \mathbb{R})^n,$$

and

$$V_0 \cap V_1 \cap \ldots \cap \hat{V}_j \cap \ldots \cap V_n = V \cap \{z \in \mathbb{C}^n : Im(z_k) \neq 0 \, for \, k \neq j\}.$$

As a consequence of these equalities, together with the computations of the kernel and the image of the coboundary mappings, we can finally conclude that, for Ω an open set in $\mathbb{R}^n$, one has the following explicit description of the space of hyperfunctions

$$\mathcal{B}(\Omega) = \frac{\mathcal{O}(U \cap (\mathbb{C} \setminus \mathbb{R})^n)}{\sum_{j=1}^{n} \mathcal{O}(V_1 \cap \ldots \cap \hat{V}_j \cap \ldots V_n)}.$$

It is not useless to remark that, in the case in which Ω is the real plane $\mathbb{R}^2$, and V the complex plane $\mathbb{C}^2$, one obtains the same description which we have intuitively given at the beginning of this section:

$$\mathcal{B}(\mathbb{R}^2) = \frac{\mathcal{O}((\mathbb{C} \setminus \mathbb{R}) \times (\mathbb{C} \setminus \mathbb{R}))}{\mathcal{O}(\mathbb{C} \times (\mathbb{C} \setminus \mathbb{R})) + \mathcal{O}((\mathbb{C} \setminus \mathbb{R}) \times \mathbb{C})}.$$

We now want to push a little further the amount of information which this relative covering of $(V, V \setminus \Omega)$ may provide us with. As a matter of fact, we now will modify this relative covering in order to show how it is always possible to express a hyperfunction on $\mathbb{R}^n$ as a sum of boundary values of 2^n holomorphic functions; this will allow us to work with (classes of) holomorphic functions when dealing with hyperfunctions, even though it will be necessary to precisely understand the holomorphic translation of the equivalence relation which underlies this process.

Let us therefore define, for any $k = 1, \ldots, n$

$$V_{+k} = V \cap \{z \in \mathbb{C}^n : Im(z_k) > 0\},$$

$$V_{-k} = V \cap \{z \in \mathbb{C}^n : Im(z_k) < 0\}.$$

One immediately sees that, with the notations used before,

$$V_k = V_{+k} \cup V_{-k}, k = 1, \ldots, n.$$

We can now consider the relative covering of the pair $(V, V \setminus \Omega)$ defined by

$$\mathcal{V} = \{V_0, V_{\pm 1}, \ldots, V_{\pm n}\},$$

$$\mathcal{V}' = \{V_{\pm 1}, \ldots, V_{\pm n}\}.$$

With the same arguments discussed before, and noticing that

$$V_{+k} \cap V_{-k} = \emptyset,$$

it is not too difficult to prove that the cochain spaces are given as follows:

$$C^{n+1}(\mathcal{V}, \mathcal{V}'; \mathcal{O}) = 0,$$

$$C^n(\mathcal{V}, \mathcal{V}'; \mathcal{O}) = \bigoplus_{\varepsilon_1 = \pm 1, \ldots, \varepsilon_n = \pm 1} \mathcal{O}(V_0 \cap V_{\varepsilon_1 \cdot 1} \cap \ldots \cap V_{\varepsilon_n \cdot n}),$$

and

$$C^{n-1}(\mathcal{V}, \mathcal{V}'; \mathcal{O}) = \bigoplus_{\substack{\varepsilon_1 = \pm 1, \ldots, \varepsilon_n = \pm 1 \\ k = 1, 2, \ldots, n}} \mathcal{O}(V_0 \cap V_{\varepsilon_1 \cdot 1} \cap \hat{V}_{\varepsilon_k \cdot k} \cap \ldots \cap V_{\varepsilon_n \cdot n}).$$

From this characterization of the cochain spaces, it is immediate to derive the isomorphism

$$Ker(\delta^n) \cong \bigoplus_{\varepsilon_1 = \pm 1, \ldots, \varepsilon_n = \pm 1} \mathcal{O}(V_0 \cap V_{\varepsilon \cdot 1} \cap \ldots \cap V_{\varepsilon_n \cdot n}).$$

Similarly, one can see that the coboundary operator $\delta = \delta^{n-1}$ is defined (with, hopefully, obvious choice of symbols) by

$$(\delta\varphi)_{\varepsilon_1 \cdot 1, \ldots, \varepsilon_n \cdot n} = \sum_{k=1}^{n} (-1)^{k-1} \varphi_{\varepsilon_1 \cdot 1, \ldots, \hat{\varepsilon}_k \cdot k, \ldots, \varepsilon_n \cdot n}.$$

where φ is a relative $(n-1)$-cochain. Since, now,

$$V_0 \cap V_{\varepsilon_1 \cdot 1} \cap \ldots \cap V_{\varepsilon_n \cdot n} = V_{\varepsilon_1 \cdot 1} \cap \ldots \cap V_{\varepsilon_n \cdot n} = V \cap \{\mathbb{R}^n + i\Gamma_{\varepsilon_1, \ldots, \varepsilon_n}\},$$

where $\Gamma_{\varepsilon_1, \ldots, \varepsilon_n}$ is the open convex cone (with vertex at the origin) defined in $\mathbb{R}^n$ by

$$\Gamma_{\varepsilon_1, \ldots, \varepsilon_n} = \{y = (y_1, \ldots, y_n) \in \mathbb{R}^n : \varepsilon_j \cdot y_j > 0 \text{ for any } j = 1, \ldots, n\},$$

we deduce the following important isomorphism:

$$B(\Omega) = H^n(\mathcal{V}, \mathcal{V}'; \mathcal{O}) \cong \frac{\bigoplus_{\varepsilon_1 = \pm 1, \ldots, \varepsilon_n = \pm 1} \mathcal{O}(V \cap (\mathbb{R}^n + i\Gamma_{\varepsilon_1, \ldots, \varepsilon_n}))}{Im(\delta)}.$$

Such an isomorphism expresses the fact that every hyperfunction on the open set Ω can be represented by 2^n holomorphic functions in the "numerator" of the previous isomorphism, where two representations give the same hyperfunction if and only if their difference is a coboundary according to the definition we have given of δ^{n-1}.

Let now Γ be one of the open cones defined before; every holomorphic function f on $R^n + i\Gamma$ gives a hyperfunction in a canonical way, via the following sequence of homomorphisms:

$$(1.4.10) \qquad \mathcal{O}(V \cap (R^n + i\Gamma)) \longrightarrow$$

$$\longrightarrow \bigoplus_{\varepsilon_1 = \pm 1, \ldots, \varepsilon_n = \pm 1} \mathcal{O}(V \cap (R^n + i\Gamma_{\varepsilon_1, \ldots, \varepsilon_n}) \longrightarrow \mathcal{B}(\Omega).$$

Definition 1.4.3 *Let f be a holomorphic function on an open set of the form $V \cap (R^n + i\Gamma)$. Then we denote by $b_\Gamma(f)$ the hyperfunction obtained as in (1.4.10), and we say that $b_\Gamma(f)$ is the boundary value of f along the cone Γ. We will sometimes write $b_\Gamma(f) = f(x + i\Gamma 0)$.*

An almost immediate consequence of these arguments and definitions is the following reformulation of the definition of hyperfunction (we will see later on how to make this into an even more general definition):

Proposition 1.4.1 *With the notations given before we have the following isomorphism:*

$$(1.4.11) \qquad \mathcal{B}(\Omega) = \sum_{\varepsilon_1 = \pm 1, \ldots, \varepsilon = \pm 1} b_{\Gamma_{\varepsilon_1, \ldots, \varepsilon_n}} (\mathcal{O}(U \cap (R^n + i\Gamma_{\varepsilon_1, \ldots, \varepsilon_n}))).$$

Proof. Immediate application of the explicit descriptions we have given above.
□

It may be interesting to note that in the case in which $n = 2$, $\Omega = R^2$, and $V = \mathbb{C}^2$, the description of hyperfunctions which is given in Proposition 1.4.1 amounts to say that if a hyperfunction f is represented by a holomorphic function F in $\mathcal{O}((\mathbb{C} \setminus R)^2)$, then $f(x, y) = F(x + i0, y + i0) - F(x + i0, y - i0) - F(x - i0, y + i0) + F(x - i0, y - i0)$, which is in full accordance with what we already intuitively proposed at the beginning of this section.

One different, yet important, way to represent hyperfunctions as sum of boundary values of holomorphic functions relies on a different representation of the cohomology groups described above, which only employs $n + 1$ angular domains. The algebraic basis for this decomposition, which we want to describe, is the following

Proposition 1.4.2 *Let α be a vector in R^n and define the open half-space $E_\alpha = \{ y \in R^n : \alpha \cdot y > 0 \}$, where $\alpha \cdot y$ denotes their real Euclidean inner product.*

Consider now $n + 1$ vectors $\alpha^0, \alpha^1, \ldots, \alpha^n$ in $\mathbb{R}^n$, and set $E_j = E_{\alpha^j}$ for $j = 0, 1, \ldots, n$; suppose now that these vectors satisfy

(1.4.12) $E_0 \cup E_1 \cup \ldots \cup E_n = \mathbb{R}^n \setminus \{0\}.$

Then the following statements are true:

(a) *$E_0 \cap E_1 \cap \ldots \cap E_n = \emptyset$;*

(b) *any n of the vectors above are linearly independent over $\mathbb{R}$; as a consequence the intersection of the corresponding half spaces is a proper open convex cone;*

(c) *define, accordingly to (b), the cone*

$$\Gamma_j = E_0 \cap \ldots \cap \hat{E}_j \cap \ldots \cap E_n;$$

then its dual cone

$$\Gamma_j^o = \left\{ \sum_{k \neq j} \beta_k \alpha^k : \beta_k \geq 0 \right\}$$

is a closed convex set generated by the α^k with $k \neq j$, and α^j belongs to $-Int(\Gamma_j^o)$. As a consequence, the dual cones $\Gamma_0^o, \ldots, \Gamma_n^o$ give a decomposition of the dual space of $\mathbb{R}^n$, a real Euclidean space of dimension n itself, consisting of closed convex pyramids;

(d) *For any j and k in $0, 1, \ldots, n$, the following equality holds:*

$$\Gamma_j + \Gamma_k = E_0 \cap \ldots \cap \hat{E}_j \cap \ldots \cap \hat{E}_k \cap \ldots \cap E_n.$$

Proof.

(a) Assume the intersection is not empty and let α be a nonzero vector in it; then $-\alpha$ would not belong to the union $E_0 \cup \ldots \cup E_n = \mathbb{R}^n \setminus \{0\}$, which would give a contradiction.

(b) Suppose that $\alpha^1, \ldots, \alpha^n$ are linearly dependent. Then there would exist a nonzero vector y such that, for any $j = 1, \ldots, n$, it is $\alpha^j \cdot y = 0$. Then, one immediately sees that both $\alpha^0 \cdot y > 0$ and $\alpha^0 \cdot y \leq 0$ give a contradiction.

(c) We shall prove the statement for the case $j = 0$. Because of part (b), we can find nonzero coefficients $\beta_1, \ldots, \beta_n$ such that

$$\alpha^0 = \sum_{j=1}^{n} \beta_j \alpha^j.$$

We shall prove that all these coefficients are strictly negative. Indeed, if β_1 were positive, one might consider the system of linear equations in y given by

$$\begin{cases} \alpha^1 \cdot y = \frac{1}{\beta_1} > 0, \\ \alpha^j \cdot y = \frac{1}{n}|\beta_j| > 0, j = 2, \ldots, n. \end{cases}$$

Because of (b), such a system as a solution, but then one gets

$$\alpha^0 \cdot y \geq \beta_1 \alpha^1 \cdot y - \sum_{j=2}^{n} |\beta_j| \alpha^j \cdot y = \frac{1}{n} > 0,$$

and therefore $y \in E_0 \cap \ldots \cap E_n$, which would contradict (a).

(d) The inclusion of $\Gamma_j + \Gamma_k$ in the intersection of the half-spaces is obvious. Moreover, since both sides are open convex cones, it really suffices to prove that a halfspace of the form $\alpha \cdot y \geq 0$ containing the left-hand side, always contains the right-hand side. But now, by definition, such an α must be an element of the dual cone $(\Gamma_j + \Gamma_k)^o$; however,

$$(\Gamma_j + \Gamma_k)^o = \Gamma_j^o \cap \Gamma_k^o,$$

and, in view of (c), this is equal to the common sides of the two cones

$$\{\sum_{i \neq j, k} \beta_i \alpha^i \geq 0, \beta_i \geq 0\}.$$

This proves immediately our statement. $\qquad\qquad\qquad\qquad\qquad\qquad$ □

We can now use this result to construct a different relative covering of the pair $(V, V \setminus \Omega)$, which will allow us to see hyperfunctions as sums of $n + 1$ boundary values. Given a nonzero vector α in $I\!R^n$, consider as before the half-space

$$E_\alpha = \{y \in I\!R^n : \alpha \cdot y > 0\}.$$

Take now $n + 1$ vectors $\alpha_1, \alpha_2, \ldots, \alpha_n$ in $I\!R^n$ such that, with the notations used in Proposition 1.4.2, it is

$$E_1 \cup E_2 \cup \ldots \cup E_n = I\!R^n \setminus \{0\},$$

and set

$$V_j = V \cap (I\!R^n + iE_j), j = 0, 1, \ldots, n.$$

Then the covering $(\mathcal{V}, \mathcal{V}')$ given by

$$\mathcal{V} = (V, V_0, V_1, \ldots, V_n), \qquad \mathcal{V}' = (V_0, V_1, \ldots, V_n)$$

is clearly a Stein covering to which Leray's theorem applies, and which can therefore be used to compute the relative cohomology of the pair $(V, V \setminus \Omega)$.

In view of Proposition 1.4.2 (part (a)) we know that no nonzero relative $(n+1)$-cochains exist and therefore any relative n-cochain is necessarily an n-cocycle; such a cochain will always be an alternating collection $F = \{F_j\}$, for $j = 0, 1, \ldots, n$, with

$$F_j \in \mathcal{O}(V_0 \cap \ldots \cap \hat{V}_j \cap \ldots \cap V_n).$$

In order to express the alternating dependence of F_j from its index, we can actually write F as follows:

$$F(z) = \sum_{j=0}^{n} F_j(z) V \bigwedge V_0 \bigwedge \cdots \bigwedge \cdots \bigwedge V_n,$$

where the wedge product only indicates the alternating behavior of the dependence of F_j from its index. A relative $(n-1)$-cochain, on the other hand, can be described, using the same notations, by a collection $G = \{G_{jk}\}$ with

$$G_{jk} \in \mathcal{O}(V_0 \cap \ldots \cap \hat{V}_j \cap \ldots \cap \hat{V}_k \cap \ldots \cap V_n), j, k = 0, \ldots, n.$$

Once again, we will write, for the sake of keeping track of the alternating relations,

$$G(z) = \sum_{j<k} G_{jk}(z) V \bigwedge V_0 \bigwedge \cdots \bigwedge V_j \bigwedge \cdots \bigwedge V_k \bigwedge \cdots \bigwedge V_n.$$

The coboundary of such a cochain is given by the usual alternating formula:

$$\delta(G) = \sum_{j=0}^{n} (\sum_{j<k} (-1)^k G_{jk}(z)) + \sum_{k<j} (-1)^{k+1} G_{kj}(z)) V \bigwedge V_0 \bigwedge \cdots \bigwedge V_j \bigwedge \cdots \bigwedge V_n.$$

If we now set Γ_j to be the intersection of all the E_i with the exception of E_j, and if we recall the conclusions of Proposition 1.4.2, we deduce immediately that the space of hyperfunctions on Ω can be given the following representation:

$$H_{\Omega}^n(V, \mathcal{O}) = \frac{\oplus_{j=0}^{n} \mathcal{O}((I\!\!R^n + i\Gamma_j) \cap V)}{\oplus_{j<k} \mathcal{O}((I\!\!R^n + i(\Gamma_j + \Gamma_k)) \cap V)}.$$

What we have proved so far, shows that hyperfunctions on open sets in $I\!\!R^n$ can actually be seen as formal sums of holomorphic functions, subject to some equivalence relation. Before we conclude this introductory section on hyperfunctions of several variables, we would like to show how we can actually try to define hyperfunctions by starting with such formal sums. We are not going to describe all the necessary details of what we will express, since they can be found, e.g., in [103], but we will try to provide sufficient evidence for our claims.

To begin with, let Ω be an open set in $I\!\!R^n$ and Γ a cone in the dual space $I\!\!R^n$; define (similarly to what we did in one variable)

$$\mathcal{O}(\Omega + i\Gamma 0) = \text{ind}\lim \mathcal{O}(U)$$

where the limit is taken with respect to all infinitesimal wedges of type $\Omega + i\Gamma 0$.

We can now allow Γ to range over the family of all open convex cones (not necessarily strictly convex), and take the direct sum

$$(1.4.13) \qquad \bigoplus_\Gamma \mathcal{O}(\Omega + i\Gamma 0).$$

This is a well defined complex vector space, let us call it X, and we can think of it as the space of all possible boundary values of holomorphic functions, where the boundary values are taken along all possible infinitesimal directions. Of course, if we really want to talk about boundary values, we have to take a quotient which will eliminate the several representations which, in (1.4.13), are given for the same boundary value. In particular, we define the subspace, which we will call Y, generated by all the elements in X of the form

$$F_1(z) + F_2(z) - F_3(z),$$

where each F_j belongs to $\mathcal{O}(\Omega + i\Gamma_j)$ for $\Gamma_1 \cap \Gamma_2 \supseteq \Gamma_3$, and where $F_1(z) + F_2(z) = F_3(z)$ wherever all three functions are defined. It is not difficult to verify that the well defined quotient vector space X/Y is, indeed, isomorphic with the space of hyperfunctions which we have defined previously.

Now that we have given both a formal definition of hyperfunctions in the n-dimensional Euclidean space (via the notion of derived sheaves and relative cohomology), and an intuitive one (via the formal sum of boundary values of holomorphic functions), we can finally give some concrete examples, which will be useful for the reader's understanding of the operations which we will describe in the next section.

Example 1.4.1 The first example is, of course, the Dirac δ function. We know that, in classical distribution theory, the multi-dimensional δ function is defined as the product of n one-dimensional δ functions; we therefore set

$$\delta(x) = \delta(x_1)\delta(x_2)\ldots\delta(x_n) = \prod_{j=1}^n \frac{-1}{2\pi i}\left(\frac{1}{x_j + i0} - \frac{1}{x_j - i0}\right) =$$

$$= \left(-\frac{1}{2\pi i}\right)^n \sum_\sigma \frac{sgn(\sigma)}{(x_1 + i\sigma_1 0)\cdot\ldots\cdot(x_n + i\sigma_n 0)}.$$

In the last representation we have set $\sigma = (\sigma_1, \ldots, \sigma_n), \sigma_j = \pm 1$, and therefore if we denote the σ-th orthant of the space $\mathbb{C}^n$ by Γ_σ, we see that the Dirac's function is the sum of boundary values of functions holomorphic in the wedges $\mathbb{R}^n + i\Gamma_\sigma$. As a matter of fact, this specific property of the Dirac hyperfunction can be used as a motivation for an informal definition of hyperfunctions as sums of boundary values along wedges.

Example 1.4.2 In the case of the Dirac hyperfunction, we have a combination of boundary values from all possible directions. However, one may also consider one-sided boundary values, as in the hyperfunction

$$\frac{1}{x_1 + i0} \cdot \frac{1}{x_2 + i0} \cdot \ldots \cdot \frac{1}{x_n + i0}.$$

More generally, the function $1/z$ which we have used in this case, may be replaced by other holomorphic functions $F_1, \ldots, F_n$ which have singularities on the real axis.

Example 1.4.3 Finally, another simple example which can be given is the following bidimensional hyperfunction (which we will use again in our next section):

$$F(x_1 + i0) \cdot G(x_2 + i0)$$

for

$$F(z) = \sum_{k=0}^{+\infty} e^{2\pi i (2k)! z}, \quad G(z) = \sum_{k=0}^{+\infty} e^{2\pi i (2k+1)! z}.$$

1.5 Hyperfunctions of Several Variables: Main Results

In this last section of the first chapter, we will show how to use hyperfunctions in several variables as if they were actually "generalized functions". In a way, this section is the several variables parallel of section 1.3; still, in its complexity, it will require much more attention on the part of the reader and, unfortunately, the results which we will describe are much less intuitive than the corresponding ones for a single variable.

To begin with, we shall use some basic facts from the theory of sheaf cohomology to obtain the fundamental properties for the sheaf of hyperfunctions. Let us recall that if we have a sheaf homomorphism

$$h : \mathcal{F} \longrightarrow \mathcal{G}$$

between two sheaves $\mathcal{F}$ and $\mathcal{G}$, then such a homomorphism can be uniquely lifted to a homomorphism between the complexes derived from the canonical flabby resolution of the sheaves themselves (see any text on sheaf theory for this fact, e.g. [64] or [123]); since it is known that any flabby resolution (and in particular the canonical one) can be used to determine the cohomology of a sheaf, we deduce the existence of an induced homomorphism

$$h_*^n : H_S^n(X, \mathcal{F}) \longrightarrow H_S^n(X, \mathcal{G})$$

between cohomologies. Such an induced homomorphism, in particular, is an isomorphism if h was, originally, an isomorphism. This simply means that $H^n_S(X, -)$ is a functor.

The first application of this general fact is to define the action of partial differential operators on hyperfunctions; indeed, let $P(z, D_z)$ be a linear partial differential operator with holomorphic coefficients defined on a complex open set U. Then $P(z, D_z)$ induces naturally a sheaf homomorphism, let us call it h, on the sheaf $\mathcal{O}$ of germs of holomorphic functions and therefore, for any real open set Ω contained in U, one defines the action of $P(x, D_x)$ on $\mathcal{B}(\Omega)$ as the induced map h^n_*, acting on the relative cohomology. The reader is invited to go back to section 1.4 and to check how the action is explicitly defined; indeed, this abstract definition we have just given corresponds precisely to have h act directly on the holomorphic functions which define the corresponding hyperfunctions. We should also point out that by taking the inductive limit on a cofinal family of open sets, the action described before actually induces a sheaf homomorphism, so that we can say that linear partial differential operators with real analytic coefficients act as sheaf homomorphisms on the sheaf of hyperfunctions.

Let us incidentally observe how this fact has a remarkable interest in the theory of differential equations; indeed (more details will be given before the end of this section), one can even consider a complex of systems of linear partial differential operators with holomorphic coefficients (i.e. a complex of $\mathcal{D}$-modules as described in Chapter III) defined on the sheaf of holomorphic functions; such a complex induces a complex of differential operators on the sheaf of hyperfunctions, and important consequences can be drawn from this parallel treatment; we recall here the work of Komatsu [140], [144] and of one of the authors [1], [2], [217]; as we have said, we will be back later to this topic.

As a special case of this definition, we see that it is therefore possible to multiply hyperfunctions by real analytic functions, and that we can take derivatives of such hyperfunctions; from a practical point of view, we will seldom regard hyperfunctions as cohomology classes, but we will rather look at them as formal sums of holomorphic functions; from this point of view, the definitions of derivative and of multiplication by a real analytic functions are natural and need no further explanation.

As we already pointed out when dealing with hyperfunctions in one variable, the issue of the product of hyperfunctions is a more delicate one, which we postpone to the next chapter; it will suffice to say, for the time being, that the product of two hyperfunctions is not always defined (as, however, physicists know fairly well, and as it also happens in the case of distributions).

The next important issue we want to tackle is the definition of definite integrals for compactly supported hyperfunctions or, even more, for hyperfunctions which are at least sufficiently smooth at the boundary of the integration set. Since the issue is not totally evident, we will first look at an intuitive definition, based on the naive notion of hyperfunctions as sums of boundary values. The precise cohomological description of this process will be given in a later stage.

Let $K \subseteq \Omega$ be a real compact set with piecewise smooth boundary, and let f be a hyperfunction defined on Ω; we would like to give a meaning to the symbol

$$\int_K f(x)dx;$$

as we already did in the case of one variable, we will assume (even though this is not strictly necessary) that f is, actually, real analytic near the boundary ∂K of K. By adopting the intuitive definition of hyperfunction which we have given before, we can assume that f can be expressed as a finite formal sum of boundary values such as

$$f(x) = \sum_{j=1}^{N} F_j(x + i\Gamma_j 0),$$

at least near K. Of course, because of the analyticity hypothesis on f near the boundary of K, we can assume that each F_j has been chosen so to be analytically continuable to the real axis where f is real analytic. We can now define, for each $j = 1, \ldots, N$, some integration paths $K + i\epsilon_j = \{x + i\epsilon_j(x) : x \in K\}$, where the continuous functions $\epsilon_j(x)$ are defined on K so that:

$$\begin{cases} \epsilon_j(x) = 0, & \text{if } x \in \partial K, \\ x + i\epsilon_j(x) \in K + i\Gamma_j 0, & \text{if } x \in Int(K). \end{cases}$$

We are now in the position to define the integral of f on K by the following setting:

$$\int_K f(x)dx = \sum_{j=1}^{N} \int_{K+i\epsilon_j} F_j(z)dz.$$

One might observe that in this definition we have made a certain number of arbitrary choices, which might alter the result of the integration; as a matter of fact, the independence of the integral from the choice of the integrating paths is a consequence of Cauchy's theorem, but the more delicate issue is the independence on the choice of the boundary value representation for f. That the integral does not depend on the choice of such representation is a consequence of a classical theorem in complex analysis, whose history is strictly intertwined with the history of hyperfunctions: the edge of the wedge theorem. This theorem gives the right condition for a sum of boundary values to represent the zero hyperfunction, and one may deduce from it that if two representations are given for the same hyperfunction, then their difference (representing the zero hyperfunction) has to satisfy certain conditions which imply, in particular, that its integral vanishes. Since the history and the developments of the edge of the wedge theorem are so much related to the developments of the theory of hyperfunctions, we refer the reader to our last historical appendix to this chapter; in there we have given several versions of the edge of the wedge theorem; it will not be difficult, for the reader, to extract the version which would be needed here.

In order to proceed further, we now show how real analytic functions form a subsheaf of the sheaf of hyperfunctions: in so doing, we will also obtain a different characterization of the integration process on hyperfunctions. Our first result is nothing but a reformulation of the n-codimensionality of the sheaf $\mathcal{B}$ together with Theorem 1.4.3 (some work would be needed to make this evident, and the reader is referred to section 1.4).

Theorem 1.5.1 *Let K be a closed set in $\mathbb{R}^n$. The space $\mathcal{B}_K$ of hyperfunctions with support contained in K can be represented by the relative cohomology group*

$$H_K^n(\mathbb{C}^n, \mathcal{O}) = H^n(\mathbb{C}^n, \mathbb{C}^n \setminus K; \mathcal{O}).$$

In particular, if $K = K_1 \times \ldots \times K_n$ is a cylindrical set, then for any cylindrical neighborhood $U = U_1 \times \ldots \times U_n$, the space above can be represented (with the notations already used in section 1.4) by:

$$\frac{\mathcal{O}(U \# K)}{\sum_{j=1}^n \mathcal{O}(U_j^*)}.$$

Let us point out that, in this theorem, K is not necessarily a compact set, but it suffices that it is closed; we will actually always use the case in which K is indeed compact.

As a consequence of Mayer-Vietoris theorem, we also obtain the following result:

Theorem 1.5.2 *Let K be a closed set in $\mathbb{R}^n$ which is the union of finitely many cylindrical closed sets $K_j, j = 1, \ldots, N$. Then every hyperfunction f whose support is contained in K can be decomposed as the sum*

$$f = f_1 + \ldots + f_N,$$

where the hyperfunctions f_j have supports contained in K_j. Moreover, if an open set Ω is the union of finitely many open rectangular solids (i.e. products of open intervals), then every hyperfunction on $\mathcal{B}(\Omega)$ can be extended to a hyperfunction on $\mathbb{R}^n$ whose support is contained in the closure of Ω.

We can now define the notion of definite integral for a hyperfunction as follows: let us start with the case in which f is a hyperfunction supported by a cylindrical real compact set $K_1 \times \ldots \times K_n$. If g is an entire function, then one define the cohomological integral

$$\int_{\mathbb{R}^n} f(x)g(x)dx$$

as follows: take U to be a cylindrical neighborhood of K and choose F to be a function in $\mathcal{O}(U \# K)$, which represents f in the cohomology group $H_K^n(\mathbb{C}^n, \mathcal{O})$; now set

$$\int_{\mathbb{R}^n} f(x)g(x)dx = (-1)^n \oint_{\gamma_1} \ldots \oint_{\gamma_n} F(z)g(z)dz_1 \ldots dz_n,$$

where the curves are paths in $U_j \setminus K_j$ which encircle the compact sets K_j once in the positive direction. If, on the other hand, the support of the hyperfunction f is not contained in a cylindrical compact set, but is covered by the union of a finite number N of such sets, then we can decompose f as in Theorem 1.5.2, i.e. $f = f_1 + \ldots + f_N$, and therefore define

$$\int_{\mathbb{R}^n} f(x)g(x)dx = \sum_{j=1}^N \int_{\mathbb{R}^n} f_j(x)g(x)dx.$$

It is not too difficult (and we leave this to the reader) to verify that this definition of integral coincides exactly with the intuitive one which we have provided earlier; one can also easily show that the cohomological definition does not depend on the many arbitrary choices which have been necessary in the process of the definition.

This definition of integration is useful, in particular, as it allows us to explicitly embed the sheaf of locally integrable functions in the sheaf of hyperfunctions (as a consequence, using the description of distributions as locally finite sums of compactly supported distributions, one obtains the embedding of the sheaf of distributions in the sheaf of hyperfunctions):

Theorem 1.5.3 *The following inclusions are sheaf homomorphisms:*

$$\mathcal{A} \hookrightarrow L_{1,loc} \hookrightarrow \mathcal{D}' \hookrightarrow \mathcal{B}.$$

Proof. We provide here the main ideas of this result which is standard and does not require any special knowledge to be obtained. We begin by describing how to associate a hyperfunction to a compactly supported L_1 function f. If K is a compact set which contains the support of the integrable function f, we define a hyperfunction $j(f)$ in $\mathcal{B}_K$ by setting $j(f) = [G]$, where

$$(1.5.1) \qquad G(z) = (-2\pi i)^{-n} \int_K \frac{f(w)}{(z_1 - w_1) \cdot \ldots \cdot (z_n - w_n)} dw$$

is an element of $\mathcal{O}((\mathbb{CP}^1)^n \# K)$.

It is immediate to verify that j is a $\mathbb{C}$-linear map which respects the supports, i.e. if $f = f_1 + \ldots + f_N$ as in Theorem 1.5.2 then, for each $k = 1, \ldots, N$, it is $supp(j(f_k)) \subseteq K_k$, so that the support of $j(f)$ is actually contained in the support of f. We now know that every function in $L_{1,loc}(\Omega)$ can be decomposed into a locally finite sum

$$f(x) = \sum_t f_t(x)$$

of L_1-functions with compact supports. One can then define

$$j(f) = \sum_t j(f_t),$$

where each $j(f_t)$ is defined as before. Now, the sum on the right hand side of this definition is still locally finite because of the remark we have made concerning the good behavior of j with respect to the supports. We can verify that this map is independent of the way the function f is represented and actually induces a sheaf homomorphism between $L_{1,loc}$ and $\mathcal{B}$, which is an injective homomorphism. Since $\mathcal{A}$ naturally embeds into the sheaf of locally integrable functions we have a chain of inclusions, as stated in the Theorem. The only missing inclusion is the one of $\mathcal{D}'$ into $\mathcal{B}$ (since the inclusion of locally integrable functions into distributions is well known). As for this case, the proof proceeds exactly in the same way with the caveat that (1.5.1) has to be interpreted in the sense of distributions (i.e. it is only formally an integral, but, more properly, it is the action of the distribution f on the Cauchy kernel); also, in this case we recall that any distribution can be expressed as a locally finite sum of compactly supported distributions. This concludes the proof. □

Before we can conclude this section (and this chapter) with some applications of (flabby) sheaf theory to differential equations (for its intrinsic interest, in addition to its motivational role), we need to discuss the flabbiness of the sheaf of hyperfunctions (which we already proved directly in the case of hyperfunctions in one variable) and of some associated sheaves. The proof of the flabbiness of $\mathcal{B}$ is a consequence of some rather deep properties from functional analysis, which we have mentioned when dealing with the topology of the space of hyperfunctions and we will only sketch it here, since we do not wish to duplicate what has already been discussed elsewhere (e.g. in [103]). We devote some time, on the other hand, to a simpler result, namely the flabbiness of the quotient sheaf $\mathcal{B}/\mathcal{A}$. This result, besides its intrinsic interest, will be of great importance in what follows (namely the theory of microfunctions, as described in Chapter II).

To begin with, we will need to state a Lemma, which is essentially due to Grauert [66], even though we only saw it explicitly mentioned and proved in Kaneko's [103]:

Lemma 1.5.1 *Let Ω be an open real set, U a complex neighborhood of Ω and Γ an open convex cone. Let F be a holomorphic function defined on the open set $(\Omega + i\Gamma) \cap U$. Then there exists a function G that is holomorphic on an infinitesimal wedge of type $\mathbb{R}^n + i\Gamma 0$, and is analytically continuable to $\mathbb{R}^n \setminus \bar{\Omega}$, and a function H holomorphic on a neighborhood of Ω such that, wherever they are both defined,*

$$F(z) = G(z) - H(z).$$

We now have the tools to prove the first flabbiness result, as follows:

Theorem 1.5.4 *The quotient sheaf $\mathcal{B}/\mathcal{A}$ is a flabby sheaf.*

Proof.   The proof of this statement goes through the understanding of the way sections of $\mathcal{B}/\mathcal{A}$ are made. As we already mentioned, a quotient presheaf is not, in general, a sheaf, and so the quotient sheaf is only the sheaf associated to the quotient presheaf. In this case however (as we also saw in the case of a single variable), the situation is particularly simple, and we can begin by proving that, for any real open set Ω.

$$(1.5.2) \qquad\qquad \left(\frac{\mathcal{B}}{\mathcal{A}}\right)(\Omega) = \frac{\mathcal{B}(\Omega)}{\mathcal{A}(\Omega)}.$$

This equality is immediately derived from the short exact sequence of sheaves

$$0 \longrightarrow \mathcal{A} \longrightarrow \mathcal{B} \longrightarrow \frac{\mathcal{B}}{\mathcal{A}} \longrightarrow 0.$$

Indeed, by taking the associated long exact cohomology sequence, and recalling that, for any open real set Ω, Malgrange [153] has proved that $H^1(\Omega; \mathcal{A}) = 0$, one obtains (1.5.2). We now use (1.5.2) to prove the flabbiness of the quotient sheaf. Let $[f]$ be an element of $\mathcal{B}/\mathcal{A}(\Omega)$; by (1.5.2) we can represent $[f]$ globally by some f in $\mathcal{B}(\Omega)$. By what we saw in the previous section, we can certainly choose a complex neighborhood U of Ω (which we may take to be Stein without loss of generality) and express f as a sum of boundary values as follows:

$$f(x) = \sum_{\sigma} sgn(\sigma) F_\sigma(x + i\Gamma_\sigma 0), \text{ for } F_\sigma \in \mathcal{O}((\mathbb{R}^n + i\Gamma_\sigma) \cap U).$$

By Lemma 1.5.1 we see that for each F_σ there exist both a function $G_\sigma \in \mathcal{O}(\mathbb{R}^n + i\Gamma_\sigma 0)$ and a function $H_\sigma \in \mathcal{A}(\Omega)$ such that, where this makes sense,

$$F_\sigma(z) = G_\sigma(z) - H_\sigma(z).$$

If we now define g and h to be, respectively, the boundary values of these functions, i.e.

$$g(x) = \sum_{\sigma} sgn(\sigma) G_\sigma(x + i\Gamma_\sigma 0) \in \mathcal{B}(\mathbb{R}^n),$$

and

$$h(x) = \sum \sigma sgn(\sigma) H_\sigma(x) \in \mathcal{A}(\Omega),$$

we get that the element $[g]$ defined in $\mathcal{B}/\mathcal{A}(\mathbb{R}^n)$ is the desired extension of $[f] = [f + h]$. This proves the flabbiness and concludes the proof of the Theorem.   □

Notice also that the flabbiness of $\mathcal{B}$ implies the flabbiness of $\mathcal{B}/\mathcal{A}$ as follows. Consider the commutative diagram

$$
\begin{array}{ccccccccc}
 & & 0 & & 0 & & & & \\
 & & \uparrow & & \uparrow & & & & \\
0 & \longrightarrow & \mathcal{A}(\Omega) & \longrightarrow & \mathcal{B}(\Omega) & \longrightarrow & (\mathcal{B}/\mathcal{A})(\Omega) & \longrightarrow & 0 \\
 & & \uparrow \varrho & & \uparrow \varrho & & \uparrow \varrho & & \\
0 & \longrightarrow & \mathcal{A}(I\!\!R^n) & \longrightarrow & \mathcal{B}(I\!\!R^n) & \longrightarrow & (\mathcal{B}/\mathcal{A})(I\!\!R^n) & \longrightarrow & 0.
\end{array}
$$

As above let $[f]$ be an arbitrary element of $(\mathcal{B}/\mathcal{A})(\Omega)$, where f is a hyperfunction on Ω. Since $\mathcal{B}$ is flabby, one can find an extension f^* of f defined on $I\!\!R^n$. Then the image $[f^*]$ of f^* under the canonical map $\mathcal{B}(I\!\!R^n) \longrightarrow (\mathcal{B}/\mathcal{A})(I\!\!R^n)$ is the required extension of $[f]$.

Let us now recall a result from the theory of sheaves, which we leave unproved (the proof makes use of the Zorn's Lemma, and can be found in [103], Lemma 4.2.2).

Lemma 1.5.2 *Let $\mathcal{F}$ be a sheaf on a topological space X, and let $\{U_j\}$ be an open covering of X. If each sheaf $\mathcal{F}_{|U_j}$ is flabby, so is the sheaf $\mathcal{F}$.*

Theorem 1.5.5 *The sheaf $\mathcal{B}$ of hyperfunctions is a flabby sheaf.*

Proof. In view of Lemma 1.5.2 we will show how its restrictions to the open sets of a suitable covering are indeed flabby. The covering we will use will consist of bounded open sets (e.g. the balls centered at the origin and of radius $j = 1, 2, 3, \ldots$). We will actually show that if Ω is a (connected) bounded open set, then any hyperfunction on Ω can be extended to a hyperfunction on $I\!\!R^n$ whose support is contained in the closure of Ω. Let us then start by considering an exhausting sequence $\{\Omega_k\}$ of open sets for Ω, where each Ω_k is the union of rectangular solids as in the second part of Theorem 1.5.2. From that Theorem we know that every $f_{|\Omega_k}$ has an extension f_k to the whole space, and that the support of such extension is contained in the closure of Ω_k. The (natural) idea is to construct the required extension $\tilde{f}$ as the limit of the sequence $\{f_k\}$. This cannot be achieved in a straightforward manner, since the corrections which the $\{f_k\}$ may need to make the sequence convergent may not provide an extension of the original hyperfunction; this difficulty is essentially related to the fact (which we have already discussed earlier) that the space of hyperfunctions is not at all localizeable; as a matter of fact, if K and L are two compact sets with $K \subseteq L$, then we know that the natural inclusion $\mathcal{B}_K \hookrightarrow \mathcal{B}_L$ has a dense image. The reader is invited to contrast this with the situation which occurs when dealing with Schwartz distributions, where the image is actually a closed subspace; all of the differences in the treatment of convergence and topology can be seen as stemming from this fundamental difference. To overcome this problem, we set

$K_k = \Omega \setminus \Omega_k$, in such a way that K_k is a decreasing sequence of compact sets and

$$supp(f_{k+1} - f_k) \subseteq \bar{\Omega}_{k+1} \setminus \Omega_k \subseteq K_k.$$

As we saw when describing the topology of the space of hyperfunctions, each $\mathcal{B}_K$ has a structure of Frechet space, and the natural inclusions which are induced by the inclusions of the compact sets are continuous as well. Choose (this can be done inductively) a countable family of seminorms $|\cdot|_{k,l}$ which generate the topology of $\mathcal{B}_K$ and such that

$$|f|_{k,l} \leq |f|_{k,l+1}$$

and

$$|f|_{k,l} \leq |f|_{k+1,l}.$$

Choose now g_k in $\mathcal{B}(\Omega)$ such that

$$|f_{k+1} - f_k - g_k| \leq (\frac{1}{2})^k;$$

this can be done because the space of hyperfunctions on the boundary of Ω is dense in each $\mathcal{B}_K$. The proof of the theorem can now be concluded by defining

$$\tilde{f}(x) = f_1(x) + \sum_{k=1}^{+\infty}(f_{k+1} - f_k - g_k).$$

$\square$

Yet another way to see the flabbiness of $\mathcal{B}$ is the following. Let Ω be an open bounded set in $\mathbb{R}^n$. For the triple

$$\mathbb{C}^n \setminus \bar{\Omega} \subset \mathbb{C}^n \setminus \partial\Omega \subset \mathbb{C}^n$$

there is an induced long exact sequence

$$\ldots \to H^n_{\partial\Omega}(\mathbb{C}^n, \mathcal{O}) \to H^n_{\bar{\Omega}}(\mathbb{C}^n, \mathcal{O}) \to H^n_{\bar{\Omega}}(\mathbb{C}^n \setminus \partial\Omega, \mathcal{O}) \to H^{n+1}_{\partial\Omega}(\mathbb{C}^n, \mathcal{O}) \to \ldots$$

Since $\partial\Omega$ is compact, the classical Malgrange Theorem (see e.g. [153]) shows that

$$H^{n+1}_{\partial\Omega}(\mathbb{C}^n, \mathcal{O}) = 0$$

and therefore the flabbiness of $\mathcal{B}$ follows now from Lemma 1.5.2.

We would now like to show some elementary applications of the flabbiness of the sheaf $\mathcal{B}$ to the study of differential equations (both of finite and infinite order). The applications, which are taken mainly from [144] and [217], are also of some interest as they provide a natural introduction to the algebraic treatment of systems of differential equations and therefore to some of the themes which will be discussed in Chapter III.

Let, in the rest of this section, S denote one of the following sheaves on $\mathbb{R}^n$: $\mathcal{A}, \mathcal{B}, \mathcal{D}', \mathcal{E}$ (i.e., respectively, the sheaves of real analytic functions, hyperfunctions, Schwartz distributions and infinitely differentiable functions). Most of what will follow holds true for anyone of these sheaves, and we will point out explicitly whenever a difference arises. Let now $P(D)$ be any $r_1 \times r_0$ matrix of linear differential operators with constant coefficients: by this we simply mean that

$$P(D) = [P_{ij}(D)],$$

and that D indicates the usual differential symbol

$$D = \left(-i\frac{\partial}{\partial x_1}, \dots, -i\frac{\partial}{\partial x_n} \right),$$

where $x = (x_1, \dots, x_n)$ is the variable on the Euclidean space $\mathbb{R}^n$. As we have seen in our previous pages, such a matrix defines a sheaf homomorphism

$$P(D) : S^{r_0} \longrightarrow S^{r_1},$$

whose kernel we denote by S^P, i.e. S^P is the sheaf of solutions in S^{r_0} of the homogeneous system of equations

$$P(D)u = 0;$$

u is of course a vector of solutions.

Denote now by $A = \mathbb{C}[X_1, \dots, X_n]$ the ring of polynomials in n indeterminates, with complex coefficients; if $Q(X)$ is any matrix with entries in A, we will denote by $Q'(X)$ its adjoint matrix ${}^tQ(-X)$. If we go back to our system $P(D)$, and we replace by X_j the element $-i\partial/\partial x_j$, we obtain a matrix $P(X)$ whose adjoint $P'(X)$ defines a $\mathbb{C}$-homomorphism

$$P'(X) : A^{r_1} \longrightarrow A^{r_0},$$

whose cokernel is the finitely generated A-module which will be denoted by

$$M' = \frac{A^{r_0}}{P'(X)A^{r_1}}.$$

As it has been shown by Palamodov [178] (for the more general case, but essentially already through the Hilbert's basis theorem), M' admits a free resolution

$$(1.5.3) \quad 0 \leftarrow M' \leftarrow A^{r_0} \xleftarrow{P_1'(x)} A^{r_1} \xleftarrow{P_2'(x)} A^{r_2} \leftarrow \dots \leftarrow A^{r_{m-1}} \xleftarrow{P_{m-1}'(x)} A^{r_m} \leftarrow 0$$

which terminates for some $m \leq n$. See [3] for a discussion of the complexity of such a procedure.

The first important consequence of (1.5.3) is due to several mathematicians who obtained it independently, and looking at different special cases: we will only mention here some of the mathematicians associated to this result, namely Ehrenpreis, Malgrange, Hörmander, Palamodov, Komatsu, Harvey (see [52], [140], [144] and [217] for the precise references):

Theorem 1.5.6 *If W is a convex open set or a compact convex set in $I\!R^n$, then the sequence of groups of sections*

$$(1.5.4) \quad 0 \to \mathcal{S}^P(W) \to \mathcal{S}(W)^{r_0} \xrightarrow{P(D)} \mathcal{S}(W)^{r_1} \to \dots \xrightarrow{P_{m-1}(D)} \mathcal{S}(W)^{r_m} \to 0$$

is a resolution of $\mathcal{S}^P$.

An immediate consequence of Theorem 1.5.6 (just take a cofinal family of convex neighborhoods of a point and then take the inductive limit) is the fact that we can actually rewrite the sequence at the sheaf level and get

Theorem 1.5.7 *The sequence*

$$(1.5.5) \quad 0 \to \mathcal{S}^P \to \mathcal{S}^{r_0} \xrightarrow{P(D)} \mathcal{S}^{r_1} \xrightarrow{P_1(D)} \mathcal{S}^{r_2} \longrightarrow \dots \mathcal{S}^{r_{m-1}} \xrightarrow{P_{m-1}(D)} \mathcal{S}^{r_m} \to 0$$

is a resolution of the solution sheaf $\mathcal{S}^P$. In the case of hyperfunctions, i.e. when $\mathcal{S} = \mathcal{B}$, resolution (1.5.5) is a flabby resolution.

Remark 1.5.1 Both Theorems 1.5.6 and 1.5.7 remain true if we consider W in $\mathcal{C}^n$ and we take the sheaf $\mathcal{O}$ of germs of holomorphic functions.

Sequences (1.5.4) and (1.5.5) are of course of independent interest, but they become even more full of significance, when one considers the long exact relative cohomology sequence of the sheaf $\mathcal{S}^P$ with respect to the pair $(I\!R^n, K)$, K being a real compact set. In principle, and from general definitions, one has the following sequence:

$$(1.5.6) \quad 0 \longrightarrow H^0_K(I\!R^n, \mathcal{S}^P) \longrightarrow H^0(I\!R^n; \mathcal{S}^P) \longrightarrow H^0(I\!R^n \setminus K; \mathcal{S}^P) \longrightarrow$$
$$\longrightarrow H^1_K(I\!R^n, \mathcal{S}^P) \longrightarrow H^1(I\!R^n; \mathcal{S}^P) \longrightarrow H^1(I\!R^n \setminus K; \mathcal{S}^P) \longrightarrow \dots$$

It is interesting that, essentially as a consequence of Theorem 1.5.7, sequence (1.5.6) can be actually decomposed into short exact sequences. as follows:

Theorem 1.5.8 *With the same notations as above, the following sequences are exact:*

$$0 \to H^0_K(I\!R^n, \mathcal{S}^P) \to H^0(I\!R^n; \mathcal{S}^P) \to H^0(I\!R^n \setminus K; \mathcal{S}^P) \longrightarrow H^1_K(I\!R^n, \mathcal{S}^P) \to 0$$

and, for $p \geq 1$,

$$0 \longrightarrow H^p(I\!R^n; \mathcal{S}^P) \longrightarrow H^p(I\!R^n \setminus K; \mathcal{S}^P) \longrightarrow H^{p+1}_K(I\!R^n, \mathcal{S}^P) \longrightarrow 0.$$

Theorem 1.5.8 has a very interesting series of applications in terms of differential equations: indeed one sees that the vanishing of the relative cohomology can be given important interpretations. For example, Theorem 1.5.8 shows that if the relative cohomology group of order zero vanishes, then we have an injection

$$0 \longrightarrow H^0(\mathbb{R}^n; \mathcal{S}^P) \longrightarrow H^0(\mathbb{R}^n \setminus K; \mathcal{S}^P)$$

which is equivalent to the unique continuation property for solutions of $P(D)u = 0$ from $\mathbb{R}^n \setminus K$ to all of $\mathbb{R}^n$. Similarly, if the first relative cohomology group vanishes we have the surjection

$$H^0(\mathbb{R}^n; \mathcal{S}^P) \longrightarrow H^0(\mathbb{R}^n \setminus K; \mathcal{S}^P) \longrightarrow 0$$

which is equivalent to the existence of such continuation. Of course, if both the 0-th and the first relative cohomology vanish, we have existence and unicity of continuation for the solutions of $P(D)u = 0$, as it is attested by the isomorphism

$$H^0(\mathbb{R}^n; \mathcal{S}^P) \cong H^0(\mathbb{R}^n \setminus K; \mathcal{S}^P).$$

More generally (and the reader is referred to [140], [144] or to [217]), the vanishing of higher order relative cohomologies also has important consequences for the structure of the sheaf of solutions of the system $P(D)u = 0$; such consequences, being less evident, are left to the interested reader.

It becomes therefore interesting and important to determine general conditions which would imply (or be equivalent to) the vanishing of the relative cohomology. To do so, one must consider the dual sequence of (1.5.3), i.e.

$$0 \longrightarrow A^{r_0} \xrightarrow{P(x)} A^{r_1} \longrightarrow A^{r_2} \longrightarrow \ldots \longrightarrow A^{r_m} \longrightarrow 0,$$

which is obviously a semi-exact sequence. Now, by definition, the p-th cohomology group of this complex is what we call the p-th Ext module, i.e. the A-module $Ext_{\mathcal{C}}^p(M, A)$, and the behavior of the relative cohomology can be described in terms of this module. To begin with, we note that, should it be $r_0 = 1$ and $P \neq 0$, then $Ext_{\mathcal{C}}^0(M, A)$ always vanish. More generally, the following algebraic results (which we only formulate for the sheaf of hyperfunctions) can be obtained (their proofs, by now standard, can be found in [144]):

Theorem 1.5.9 $Ext_{\mathcal{C}}^0(M, A) = 0$ if and only if $H_c^0(\mathbb{R}^n; \mathcal{B}^P) = 0$; this last vanishing, in particular, is equivalent to $H_{\{0\}}^0(\mathbb{R}^n, \mathcal{B}^P) = 0$.

Theorem 1.5.10 Take $p \geq 1$. Then $Ext_{\mathcal{C}}^p(M, A) = 0$ if and only if, for any bounded convex subset K of $\mathbb{R}^n$, it is $H_K^p(\mathbb{R}^n, \mathcal{B}^P) = 0$.

In [217] (but see also the subsequent and related papers [11], [12], [137]) it is shown how this same algebraic treatment can be applied to the case of systems of infinite order differential equations. In that case, some extra difficulties arise, which are related to a purely algebraic matter, i.e. the possibility of constructing a finite resolution for the system itself.

In other words, the ring A of polynomials is now replaced by the ring of symbols of infinite order differential equations; such a ring, i.e. the ring of functions which are obtained by replacing

$$D = \left(-i\frac{\partial}{\partial x_1}, \ldots, -i\frac{\partial}{\partial x_n}\right) \text{ with } z = (z_1, \ldots, z_n)$$

turns out to be (see also our section 2 for the one variable case) the ring of holomorphic functions with infraexponential growth, i.e. the ring of entire functions f such that for any $\epsilon > 0$ there exists a positive constant $A_\epsilon = A(\epsilon, f)$ such that

$$|f(z)| \le A_\epsilon exp(\epsilon|z|).$$

As it may be easily imagined (and examples can be easily given, see [217]), there is no general Hilbert sygyzy theorem for such a ring. In [137] Kawai and Struppa (see also the contribution of Meril and Struppa as described in [14]) have given a rather delicate technical condition to ensure that a sygyzy theorem can be restored. This leads to the consideration of the so called "slowly decreasing" systems of infinite order differential equations, for which all the previous theory can be fully restored; even though the problem is algebraic in nature, the only way which has been found so far to deal with it, is strictly analytic. The slowly decreasing condition is rather complicated (even though it was shown to apply to most relevant cases), and its discussion falls outside the scope of this book. We refer the reader to the indicated literature for more details on this question.

1.6 Historical Notes

The birth of the theory of hyperfunctions is particularly fascinating, since (like most theories of deep content and great relevance) it has strong relations with various and diverse preexisting theories, as well as with some quite concrete problems from physics. We have therefore thought it useful to divide this long Appendix into six different sections, each of which deals with a different historical aspect; section 1 will briefly outline the steps which led Sato to the creation of hyperfunctions at the end of the fifties; section 2 will deal with the notion of analytic functional which, from the first steps of the Italian school of Fantappie', to the conclusive results of Grothendieck, Köthe and Martineau, has paved the way for the creation of hyperfunctions; in section 3, we will describe a different forerunner of hyperfunctions; the generalized Fourier transform of Carleman; in section 4 we will go back to analytic functionals, to comment on the giant step which was necessary for Sato in order to extend his ideas to the case of several variables; section 5, on the other hand, will deal with the development of the theory of infinite order differential operators which we have already introduced and, as we shall see, play a crucial role in the theory of hyperfunctions and microfunctions; finally, section 6 will discuss the famous edge of the wedge theorem, its first formulations as well as its relations with the development of the theory of hyperfunctions and microfunctions.

1.6.1 Sato's Discovery

To begin with, Sato's first paper on the theory of hyperfunction was [195], which appeared in the Japanese journal Sugaku; for this reason, the paper was not known outside Japan, and it was only with the appearance of [196] in the Proceedings of the Japan Academy of Sciences that his work became available to western mathematicians. Immediately afterwards, Sato published a much richer account of his new theory in the Journal of the Faculty of Sciences of the University of Tokyo, [197]. It was then that the connections of Sato's work with the previous achievements of mathematics became apparent; but before we get to this topic, let us briefly attempt to recreate the framework of ideas within which Sato was moving at the time of his creation.

It may be said that the origin of Sato's hyperfunction theory lies in the much older uneasiness of mathematicians with the pseudofunctions which engineers and physicists had been using at various stages in history. We only need to recall Fourier uninhibited use of a "series" expression for the delta function in his treatise on the theory of heat [61], or Poisson' s notion of dipole, [188], where the derivative of the delta appears, or finally Heaviside [76], [77], [78] and Dirac [40] with their introduction of various species of operational calculus and with a formal definition of the delta function.

The problem which engineers and physicists alike were concerned with, was the difficulty of dealing, on one hand, with some simple but hard to describe physical objects (impulsive forces or, again, electric dipoles), and on the other hand, with very singular functions. We have mentioned before the British engineer O. Heaviside, who found it useful to introduce the function which, today, carries his name. He needed such a function in order to study some differential equations which arise in the study of signal transmission for telegraphic signals. The problem, of course, was that such a function could not be differentiated according to the usual rules, and yet it became necessary to do so, possibly by changing the meaning of differentiation. To be more precise, the Heaviside function $H(x)$ could be differentiated everywhere except at the origin, but that was exactly the point in which it was interesting to understand the meaning of its derivative H', especially from a physical point of view. It may be worthwhile recalling that the unorthodox use that Heaviside made of differentiations, while giving him correct results, also caused his dismissal from the London Academy of Sciences, [151].

Dirac, on the other hand, carried this approach one step further, by introducing his delta function $\delta(x)$ in such a way as to modify the notion itself of a function; once again, mathematical and physical necessities were pushing for a different notion of function and of differentiation (the development of the notion of function is, after all, one of the lines which one could try to follow in studying the history of modern mathematics).

As it is well known, the first satisfactory solution to the difficulties posed by the objects of Heaviside and of Dirac was due to L. Schwartz who, in 1947 (see his original work [210], but also [151] and [214] for a detailed account of the developments of the theory of distributions) developed a deep and far reaching theory of generalized functions which he called distributions; finally, the equality

$$H'(x) = \delta(x)$$

was fully justified. Within the theory of distributions, as it is well known, the basic viewpoint is essentially based on the vision (inherited by Dirac) of a generalized function as a functional on some space of test functions; in the case of Schwartz's distributions, the space of test functions is the space $\mathcal{D}$ of infinitely differentiable functions with compact support, and therefore the theory of distributions is essentially a theory which can be applied to the study of differential equations on arbitrary differentiable manifolds.

These considerations bring us to some important points in the study of hyperfunctions; as we have seen, even hyperfunctions (at least in the case of compactly supported hyperfunctions) can be seen as functionals on the space of germs of real analytic function; this was not, however, the original approach which Sato followed in his papers and in his work. According to his own recollections [151], Sato was initially unhappy with the fact that the theory of distributions would work in the category of differentiable functions and manifolds; this fact stroke him as unnatural, and he was firmly convinced that the natural space to use as a space of test functions would have to be the space of analytic functions. This point is really worth of attention, since Sato (and his coworkers) have always claimed to be analysts in the sense of classical mathematics (see, in this regard, the introduction to [123]); as we look back to the development of eighteenth century mathematics, we cannot but recall how analyticity was considered the quintessential form of a function (functions, so to speak, had to be expressed as convergent power series, or else there was not even the possibility to work with them; one might even recall Hilbert's address in Paris).

It is however clear that the space of real analytic functions could not be easily taken to be the space of test functions, essentially because we could not consider compactly supported real analytic functions; some other way had to be found, to circumvent this difficulty. It is not easy to see through the first ingenious pages in which Sato develops the theory of hyperfunctions; the results which he proves, as we shall see, are not essentially new, but what is totally new is the spirit and the far reaching approach which he follows; the key point in his first work is the (successful) attempt to give an operational meaning to the notion of boundary value of a holomorphic function. The importance of this notion had already been established in physics (see, for example, the work of Bogoljubov on the edge of the wedge [22]) and it was very well known that all distributions could be expressed as boundary values of holomorphic functions, as we have already mentioned in this Chapter. It was also known, however, that some holomorphic functions did not have boundary values, at least not in the sense of distributions, since the boundary value at the origin of a function such as

$$e^{1/z}$$

is just too much of a singular object to be dealt within the theory of distributions. Still, many problems from the theory of causality and the study of dispersion relations, see e.g. [28], [30], [44], impose the consideration of these more general boundary values, and it was therefore natural for the young Sato, who had originally been a student of the future physics Nobel prize Tomonaga, to look for a precise formalization of such boundary values.

The connections with dispersion relations explain, at least partially, the interest of Sato for a notion which would satisfactorily deal with boundary values of holomorphic functions; still, it is even more interesting to examine the many links between the work of Sato, and what had been done previously in totally different areas of the world.

1.6.2 Analytic Functionals

When the first papers on hyperfunctions were published in English by Sato, it was immediately noted by A. Weil, [196], that one of Sato's first results (his duality theorem) had actually been known already to G. Köthe, who had published a very interesting series of papers on integral representations of analytic functionals [138], [139], which culminated with the duality theorem itself. This comment, which is mentioned directly by Sato in the second part of [196], certainly spurred him towards the production of his totally new approach to the several variables case (in which the duality theorem acquires a much deeper meaning and difficulty). One must also say that this comment only gives part of the story. Indeed, the duality theorem which is usually attributed to Köthe (and rightly so, since he was the first to provide a conclusive proof), has a rather long and by now well established story (see our [218], [219], where more historical details are given on this topic).

It can be safely said that the first to look upon such a problem was the Italian mathematician L. Fantappie', a rather singular student of Volterra and of Severi, who, in 1924, [55], (but see also [57] and, finally, [59], for a complete bibliography on Fantappie's work on the theory of analytic functionals) created a theory of analytic functionals, on the invitation of Severi who had wanted him to understand in a general fashion the action of operators such the differentiation operator.

The original definitions and notations of Fantappie' are rather cumbersome, so we will simplify here his approach, while trying to convey his main ideas. Fantappie' defined the space of ultraregular functions on an open set U of $\mathbb{C}\mathbb{P}^1$ as the space of functions which were holomorphic in U and which would vanish at infinity, when the point at infinity ∞ belongs to U. Fantappie' defined therefore the space $S^{(1)}$ of all ultraregular functions, which (because of its definition) is not even a vector space. A **linear region** R was then introduced as a subset of $S^{(1)}$ closed with respect to the $\mathbb{C}$-linear combinations of its elements (for the sum of two functions to be well defined it was deemed necessary and sufficient that the regions where they were defined would have a non-empty intersection). The first interesting remark of Fantappie' was the fact that there was a bijective correspondence between linear regions R and what he called their **characteristic sets** A, i.e. the intersection of all the regions of $\mathbb{C}\mathbb{P}^1$ where the functions of R are defined. It turns out that A is a proper, non-empty closed subset of $\mathbb{C}\mathbb{P}^1$. Something more can be said; indeed R can be shown to coincide with (A), where, in modern terms, (A) can be defined as

$$\cup(\Gamma^\infty(U(A), \mathcal{O}))$$

with $U(A)$ varying over all possible open neighborhoods of A and Γ^∞ denoting the space of sections of holomorphic functions which vanish at infinity (ultraregular functions).

Finally, Fantappie' defined his **funzionali analitici** (analytic functionals) as the element of the dual of (A). Well, actually this would be to go too far, since, as we have seen, the space (A) was not even a vector space, and even the topology which Fantappie' built for it was not too adequate to a modern treatment of duality. Let us not forget that all of this was taking place in 1924, so that the necessary continuity conditions which had to embedded into the notion of "functional" were suggested to Fantappie' by some famous results of Poincare' on analytic dependence of solutions of analytic Cauchy problems for partial differential equations.

The precise definition of Fantappie' runs as follows. For a given a linear region R, a $\mathbb{C}$-linear map $F : R \longrightarrow \mathbb{C}$ is an analytic functional if:

(a) given y_0, a continuation of y_1, it is $F(y_0) = F(y_1)$;

(b) if $y = y(z, \zeta)$ is holomorphic in z and ζ (actually, some extra conditions of technical nature are necessary), then

$$f(\zeta) = F_z(y(z, \zeta))$$

is holomorphic in ζ, where defined.

The reader will note how these analytic functionals are, mutatis mutandis, essentially our analytic functionals in the sense of $(\mathcal{O}(A))'$. As we have seen in the duality theorem, such a space is isomorphic to the space of hyperfunctions with compact support in A, which, on the other hand, coincide with the space of functions holomorphic in $U \setminus A$, with U some open neighborhood of A in $\mathbb{C}$. This result had already been established, in a rough form, by Fantappie' in [56]. Let us recall the definition of indicatrix of a functional (which we have already used in our treatment of hyperfunctions in the previous pages): given F an analytic functional on A (whether in our modern sense or in Fantappie's sense), we define its indicatrix (sometimes called the Fantappie'-Sato indicatrix) as the holomorphic function

$$u_F(\zeta) = F_z \left(\frac{1}{z - \zeta} \right).$$

It turns out that such an indicatrix is holomorphic except at A and that the following duality theorem holds true, [56]:

Theorem 1.6.1 *Let $A \subseteq \mathbb{C}$ be a compact set and let $\mathcal{O}(A)$ be the space of holomorphic functions defined in some open neighborhood of A. Let F be an analytic functional on $\mathcal{O}(A)$ and let u_F be its indicatrix function. Then, for all y in $\mathcal{O}(A)$, y holomorphic in some open neighborhood M of A, one has:*

$$F(y) = (2\pi i)^{-1} \int_\gamma u_F(t) y(t) dt,$$

where γ denotes a smooth curve which separates the compact A from the set $(\mathbb{C} \cup \{\infty\}) \setminus M$. In other words, the indicatrix u_F represents F via the duality integral.

We believe the reader will notice how Theorem 1.6.1 actually foreruns Köthe's duality theorem; in both cases, indeed, one sees how analytic functionals carried by a compact set A are in a bijective correspondence with the space of holomorphic functions defined on the complement $\mathbb{C} \setminus A$ of A.

In a way, therefore, and with the caution which is always necessary when making historical statements, we can say that Fantappie' had already attributed a dignity to pairs of holomorphic functions having discontinuities along the real line and, therefore, to prehistoric hyperfunctions. He was of course lacking the idea of considering these objects as generalized functions.

The path between the work of Fantappie' and that of Köthe, however, is worthy of a short description, since it provides a case study in the development of mathematical ideas. Indeed, as we have pointed out, Fantappie's approach was mainly a set theoretical approach (even though, at a later stage, he tried to develop a topological description for the spaces he was working with). This was due, partially, to the specific applications which Fantappie' had in mind (interestingly enough, as we shall see, Fantappie' main interest was, for a number of years, concentrated towards a unified treatment of the Cauchy-Kowalewsky phenomenon) and, more fully, to the fact that the spaces which Fantappie' was constructing were essentially limits of Frechet spaces, and dual of such spaces. In the twenties and thirties, when Fantappie' was first developing his approach, no duality theory was known for such spaces, and so no theoretical background was available to help the Italian mathematician develop a more rigorous approach to this study.

With World War II, however, we have one of the major revolutions in twentieth century analysis; the development of distribution theory by L. Schwartz. We wish to briefly describe how this revolution altered the course of events which have (indirectly) led to the birth of hyperfunctions.

As it is well described in Lutzen's historical analysis, [151], a rather unknown "exercise" of Schwartz was instrumental in the development of the theory of distributions. According to Lutzen, Schwartz, immediately after writing his doctoral thesis on series of exponentials, [209], was working in some isolation and decided to try his hand on the problem (which he apparently regarded as devoid of any intrinsic interest) of extending the duality theory for Banach spaces to the case of Frechet spaces. He clearly succeeded in his efforts, and one of the legacies of this period is a groundbreaking paper which he co-authored with J. Dieudonne, [39]. This paper deals with duality theory in Frechet spaces and their limits. This work has proved one of the most influential pieces of works in the theory of topological vector spaces, in view of the large use of Frechet spaces in modern analysis. Fantappie' recognized the importance of this development, and one of his best students, J.S. e Silva, wrote an important doctoral dissertation [211], under the guidance of Fantappie' in 1950. Silva tried to place Fantappie's theory on firm topological grounds. His attempt has been praised, but unfortunately his approach contains some serious flaws which aroused the interest of two mathematicians: Köthe and Grothendieck. They both realized the interest of Silva's program, as well as the shortcomings of his work, and, in the early fifties, wrote a series of papers, [69], [138], [139], in which his ideas were clarified. One can safely say that in these papers lie the foundations for what we now consider the theory of analytic functionals, and certainly contain the first complete proof of the duality theorem. The premature death of Fantappie' in 1954 and his loss of interest for mathematics in the last few years of his life, have unfortunately prevented what could have become a very fruitful collaboration. For the sake of completeness, we may also point out that some extensions of Köthe results were obtained in the early sixties by the German mathematician H. G. Tillmann, [221], [222], who essentially considered the case of functionals carried by subsets of the real line: by that time, as we know, hyperfunctions already existed. Recent historical research has shown that Tillmann was not aware of them and that he was getting closer and closer to a complete formulation of the theory, at least for the one variable case. It has of course to be remarked that before Sato, nobody ever thought of looking at these functionals as generalized functions, and it is probably in this idea which lies the main contribution of Sato (at least for the single variable case, since, as we shall see, the case of several variables is so totally new that no mathematician can claim having been a precursor); on this topic we refer the interested reader to the work of Lutzen [151], or to the original papers by Tillmann.

It might be mentioned that we can probably trace to this series of developments the existence, and the strength, of the French and the Portuguese school of hyperfunction theory and microlocal analysis; it is somehow surprising that so little has remained, on the other hand, in Italy (where only a handful of mathematicians have done serious research in this field) and in Germany.

1.6.3 Generalized Fourier Integrals

We have, so far, only discussed how the theory of analytic functionals was a precursor of the theory of hyperfunctions in one variable. In this section we will discuss Carleman's Fourier theory, as it is exposed in his book [34] (even though some comments are taken from our historical survey [218]).

One of the most intriguing problems of twentieth century analysis has been the attempt to define some kind of Fourier transform for functions which do not decrease at infinity. Some authors (Ehrenpreis among them) even venture to hint [52] that such a theme could be taken as a path through most of the analysis which has been developed in the last sixty years or so. Finding Fourier integral representations for functions of arbitrarily large growth is, indeed, one of the motivations of the celebrated Ehrenpreis-Palamodov Fundamental Principle, [52], [178], which is also one of the themes dear to the Japanese school of Microlocal Analysis (see the references in [123], for example).

Attempts to define Fourier representations for functions which do not belong to L^1 have a long history, and we should at least mention the names of Hahn [72], Wiener [229] and Bochner [20]. From our point of view, however, we would like to fix our attention to the work which, in the thirties, was done by the Swedish mathematician T. Carleman, and which is essentially collected in [34].

Carleman noted a very simple and well known fact: if f is a Lebesgue integrable function on $I\!R$, then its Fourier transform $g(z)$ can be written in the following natural way:

$$g(z) = (2\pi)^{-\frac{1}{2}} \int_{-\infty}^{+\infty} exp(-izy) f(y) dy =$$

$$= (2\pi)^{-\frac{1}{2}} \int_{-\infty}^{0} exp(-izy) f(y) dy + (2\pi)^{-\frac{1}{2}} \int_{0}^{+\infty} exp(-izy) f(y) dy =$$

$$= g_1(z) - g_2(z).$$

If now, in this expression, one allows the real variable z to take on complex values, it is immediate to notice that g_1 is holomorphic in the upper half complex plane and that g_2 is holomorphic in the lower half complex plane, so that one sees that the Fourier transform of an L^1 function can be expressed as the difference of two functions holomorphic on opposite sides of the real line in $\mathbb{C}$.

Carleman then stated two problems which appeared very naturally from the consideration of the identity established above. Is it always possible to decompose a function defined on $\mathbb{R}$ as the difference of two functions holomorphic on the opposite sides of the real axis? And, in the affirmative case, is this decomposition uniquely determined?

It was not too difficult, for Carleman, to prove that both questions can be answered in the affirmative, and that, therefore, there exists some kind of equivalence between functions defined on the real line (we refer to Carleman's [34] for the details) and pairs of holomorphic functions as above or, equivalently, elements of $\mathcal{O}(\mathbb{C} \setminus \mathbb{R})$. Now that the intrinsic interest of these pairs of holomorphic functions was established, Carleman went on to notice that a Fourier representation such as the one given above did actually make sense even for functions which had polynomial growth at infinity, rather than L^1 decay. If, indeed, a function f satisfies a polynomial growth condition of the form

$$\int |f(x)|dx = O(|x|^k), \text{ for some } k \in \mathbb{N},$$

then its classical Fourier transform may not exist, but the functions g_1 and g_2 introduced above are well defined, and holomorphic on complementary halfplanes. We can therefore conclude that not only functions could be represented as pairs of holomorphic functions, but that such pairs could also be considered as some sort of generalized Fourier transform. Up to this point, however, the theory was unsatisfactory (and Carleman was quick to point this out) since it was somehow lacking symmetry; Carleman thus proceeded in trying to show how to treat pairs of holomorphic functions as some sort of generalized functions, thus coming closer than ever to the point of view which would have eventually produced hyperfunctions. It would be inappropriate here to enter the details of the work of Carleman, for which we refer the reader to the original work [34], but we content ourselves by pointing out that Carleman actually succeeded in developing a full-fledged theory of Fourier transform for pairs of holomorphic functions, in such a way that one could even obtain an inversion formula. It has not been explored, as far as we know, the existence of possible connections between Carleman's theory and the theory of Fourier transform for hyperfunctions developed by Kawai in [130].

1.6.4   Hyperfunctions in Several Variables

Up to now, we have mentioned some of the physical motivations which led Sato to the creation of hyperfunctions (essentially the dispersion relations and the study of causality), together with the work of some precursors, which, however,

had confined their ideas to the case of a single variable. This restriction is not
too surprising, in view of the extreme complexity of the technical tools which
become necessary in trying to extend Sato's original ideas to several variables.

As we have seen in this chapter, the first difficulty lies in the fact that
the notion itself of boundary value, which had been the guiding notion in the
development of Fantappie's theory, as well as in Carleman's one, is suddenly
unclear, since there are just too many directions to take into account (and not
just "above" and "below"). But other formidable hurdles seem to forbid the
generalization of the notion of hyperfunction from one to several variables; to
begin with, even the theory of complex analysis in several complex variables was
not, at the beginning of the century, too well developed; also, some of the key
results in one variable, are not true anymore (just think of Hartogs' removable
singularities theorem, or of the vanishing of the first cohomology group with
coefficients in the sheaf of germs of holomorphic functions). In one word, one
may probably say that the main difficulty lies in the fact that it is not true
anymore that any open set in $\mathbb{C}^n$ is a domain of holomorphy.

From what it is known so far, it appears that Sato was led to the discovery
of the right way to define hyperfunctions in several variables by the appearance,
when studying dispersion relations, of families of holomorphic functions which
satisfied "strange" relations, which later on turned out to be just the cocycle
relations needed for the definition of relative cohomology. Sato was, however,
unaware of relative cohomology theories, and so it had to develop completely
from scratches such a theory for the sheaf of germs of holomorphic functions.
Interestingly enough, at almost the same time, Grothendieck was developing a
theory which is almost exactly of the same scope, even though for totally different purposes (Grothendieck was actually setting up his revision of the methods
of Algebraic Geometry, which were totally algebraic in nature, so that one may
say that Algebraic Geometry and Algebraic Analysis share at least this part of
their past).

With respect to these developments, it is interesting to note, [204], that Sato
first realized a hyperfunction as a relative cohomology class back in the Spring-
Summer of 1958. Later on, however, he left for the Institute for Advanced
Studies in Princeton in 1960 where, in the company of L. Schwartz, he explained
to A. Weil (in his office) his hyperfunction theory using relative cohomology.
The reaction of A. Weil did not encourage Sato too much (he was not aware
that Weil was not, at the time, too fond of cohomological methods) and this fact
apparently prevented him from writing the third paper on hyperfunctions, which
would have contained the theory of derived categories, including his treatment
of spectral sequences and hypercohomology. We now know that the equivalent
of this material is contained in Hartshorne's [73]. Incidentally, Sato gave a series
of talks on $\mathcal{D}$-modules in the Kawada Seminar at the University of Tokyo just
before his departure for Princeton.

By 1960, however, the development of hyperfunctions was essentially completed and, in fact, just before leaving Tokyo for Princeton, Sato had already switched to the study of the algebraic treatment of systems of differential equations. We will come back to this topic in a later chapter.

Even though, before Sato and Grothendieck, nobody had developed any notion of relative cohomology, it might be interesting to take a step back to the work of Fantappie'. As we already mentioned (but the interested reader is referred here to [58] and [215]), Fantappie' was essentially interested in applying his theory of analytic functionals to the study of the Cauchy-Kowalewsky theorem, which he wanted to approach from a functional point of view; he actually managed [58], [60], to express the correspondence which associates to the initial conditions the unique solution by means of an analytic functional (whose arguments were, in fact, the initial conditions); in order to compute the solution to a given initial value problem, it was necessary to know the value of the functional for very simple functions (i.e. to know the Fantappie' indicatrix of the functional). The nature itself of this problem led Fantappie' (and some of his later students) to deal with analytic functionals acting on holomorphic functions of several complex variables.

Fantappie' had to struggle very hard trying to understand the correct generalization of his work to the case of several variables; we now know why his task was so hard, but we doubt that his difficulties were fully appreciated at the time, even though a large part of Italian mathematics was concerned with similar problems, from the more geometric point of view of Algebraic Geometry. In the multivariable case, Fantappie' managed to mimic his own approach by considering [58], [215] ultraregular functions defined as analytic functions on $\mathbb{CP}^n$ which would vanish at infinity (the notion of infinity in the case of several complex variables was, itself, a debated one at the time). Even though the initial steps were simple, Fantappie' met his first challenge when trying to prove an analogue of the duality theorem. The first difficulty was in the impossibility of easily defining an indicatrix; the difficulty is of geometric nature, since the obvious kernel which one would like to consider,

$$y(z, \zeta) = \frac{1}{(z_1 - \zeta_1) \cdot (z_2 - \zeta_2) \cdot \ldots \cdot (z_n - \zeta_n)},$$

does not necessarily intersect all functional regions (A). However, even when such an indicatrix is used (Fantappie' called it the antisymmetric indicatrix), no duality theorem can be proved, except under some very stringent hypotheses on A, in which case the result is just a trivial restatement of the one-variable case.

Even more interesting is the phenomenon of the multivaluedness of analytic functionals. Indeed, when integral representations are used to describe analytic functionals, they turn out to be multivalued (even when their indicatrices are not) because of the different integration contours which can be used in the representation formulas. This phenomenon is interesting as well as it must have been unpleasant to Fantappie'; its explanation, however, was too deep and required too much new mathematics for Fantappie'. Indeed, it turns out that it was only with the brilliant work of Martineau, [157], [158], that Fantappie's intuition was proved correct, even though one had to renounce the hope to describe a functional via a holomorphic function (its indicatrix). As Martineau correctly pointed out, the way out is the use of relative cohomology classes of the sheaf of holomorphic functions (and the ultraregularity phenomenon is easily dealt with, by means of a correct choice of representative cocycles). We therefore see how, even though in a primitive way, Fantappie's ideas were on the right track, and how it was really only the cohomological approach which was missing.

We should take this chance to describe the tremendous role which Martineau had in the diffusion of the theory of hyperfunctions (he may be credited as being the single strongest force behind the great development of the hyperfunction/microlocal analysis school in France). This role has been "recognized" by the Japanese school itself, who dedicated to his memory the volume of the Proceedings of the October 1971 Katata Conference [141], which Martineau could not attend because of his premature death. Besides his influential Seminaire Bourbaki of February 1960, in which hyperfunctions are introduced to mathematicians outside Japan, Martineau had already been working for a while on the study of analytic functionals, and their applications to the theory of infinite order differential equations. Such problems (in which he was exploiting sheaf cohomology to provide conclusive answers to questions left open by Fantappie') made him naturally the readiest recipient for the theory of Sato.

1.6.5 Infinite Order Differential Equations

We have seen in this chapter how infinite order differential operators play a rather important role in the development of hyperfunctions; in the next few chapters we will see how essential the study of these operators is, and why they are so strongly intertwined with the theory of hyperfunctions and microfunctions. In this short section of the appendix, we wish to attract the reader's attention to some aspects of mathematics of the early years of this century which are not so widely known, but in which some aspects of these modern theories showed their first appearances. We refer here to S. Pincherle, one of the major Italian characters in turn of the century mathematics, who developed (see [186], [187]) a rather refined theory of infinite order differential operators, which he called "operazioni distributive" (distributive operations), to signify their linear character.

Interestingly enough (and the reader may want to read our [220], in which some of these aspects are treated with some detail), Pincherle's work is actually mentioned by Fantappie' as one of the motivation for his creation of analytic functionals, and, in particular, Pincherle's distributive operations can be viewed as special examples of analytic functionals.

Pincherle's approach to the study of infinite differential operators was, in many ways, similar to the one followed by the workers in microlocal analysis. Indeed, after his trip to Berlin in 1877-78 (where he attended a complex analysis course taught by Weierstrass) Pincherle was led to the study of the inversion of definite integrals, and, in particular, to the consideration of the integral equation of the first kind

$$\int_\gamma k(x,y)\varphi(y)dy = f(x),$$

$(k, f$ given functions, φ an unknown function, γ a curve in the complex plane and x, y in $\mathbb{C}$) which he regarded as an operator

$$A : \varphi \longrightarrow A(\varphi) = f.$$

This functional point of view was not totally new, since it had been successfully employed in the case of the Fourier and the Mellin transform (to quote just two relevant examples), but Pincherle had the intuition to extend the geometrical theory of homographies from the finite dimensional case to the case of infinite dimension, by letting such operators act on the spaces of formal power series and of convergent power series.

As we follow Pincherle's development of his theory (completely described in his [187]), we can see that his operators can be interpreted as the convolution operators associated to analytic functionals with compact carriers. Let us spend a couple of words on this aspect, which may become of interest later on. Let φ be an entire function in one variable (note that all of Pincherle's theory was strictly confined to the single variable case), and let μ be an analytic functional carried by the compact set K (this means that μ belongs to $\mathcal{O}'(K)$, where the compact K is not uniquely defined, since analytic functionals do not constitute a sheaf, and therefore a notion of support is not fully defined, but is replaced by the weaker notion of carrier, see [83]). One can then define a convolution between μ and φ, as follows:

$$\mu * \varphi(z) =< \mu, t \longrightarrow \varphi(z+t) >;$$

such a definition makes $\mu*$ a continuous linear operator from $\mathcal{O}(\mathbb{C})$ into itself. Therefore we see that Pincherle's operations were isomorphic to analytic functionals with compact carriers, i.e., ultimately, to compactly supported hyperfunctions. Even more interesting is the fact that Pincherle proceeded to develop a calculus for these operators, which would allow him to deal with both local and non-local operators (e.g. with the translation operator), in order to eventually solve the inversion problem.

We have not discussed in detail the inversion problem in this chapter, but it is interesting the fact that Pincherle managed to introduce a rather refined calculus for negative powers of the basic differentiation operator $\frac{d}{dz}$, which in some sense is akin to the one developed independently by Heaviside, [77], [78].

1.6.6 The Edge of the Wedge Theorem

As we have pointed out in the last section of this Chapter, a key instrument in the understanding of when a sum of boundary values of holomorphic functions gives the zero hyperfunction is the so called **edge of the wedge theorem**, whose purpose is exactly to determine when such a formal sum gives rise to a unique holomorphic function. As such, the edge of the wedge theorem is a typical holomorphic continuation theorem (see e.g. Vladimirov's [226]) but its interest, at least from our point of view, lies in its many connections with the theory of hyperfunctions, and with the fact that it arose (as hyperfunctions did) from considerations from theoretical physics.

Thus, a complete paper on the edge of the wedge theorem would have to have at least three components: its physical background and significance, its mathematical role, and finally its connections with the theory of hyperfunctions. In this section of the historical appendix, we will only briefly touch upon the history of such a theorem (its original formulation has been greatly expanded and modified in the years immediately following its first statement), and we will naturally discuss its connections with Sato's theory. As far as its physical meaning, we refer the reader to our references [29], [99], [146] while its mathematical importance is discussed within the framework of more general analytic continuation results in [226]. A very interesting and accessible survey on this theorem and related results is also provided by Rudin's series of lectures [191].

It is fair to say that the very first version of the edge of the wedge theorem was formulated, for the very simple case of one complex variable, by Painlevé, back in 1888. This case was already discussed in this chapter as Theorem 1.2.2 and, as we already noticed, it is nothing but a special case of Schwarz's reflection principle. When only one complex variable appears, however, not only the result is fairly immediate, but also no wedge appears to justify its name. In order to find a first extension of Painlevè's theorem to the case of several complex variables, we probably have to wait until 1956, when Bogolyubov first gave a statement (we drew this information from the later work of Vladimirov [227]), justifying the interest of the results with the (at that time) intense researches into the Wightman function and the dispersion relations (interestingly enough, the same material which spurred Sato's ideas, and the same time period as well). That this theorem was of great relevance for theoretical physicists was attested by the many papers on the topic (see our previous references) and to those we refer the reader interested in exploring this relevance.

But let us now prepare the notations which are necessary for the statement of the edge of the wedge theorem in its first version (as given by Bogolyubov); for A any subset of $I\!\!R^n$, we will denote by $T(A)$ the tube with base A, i.e.

$$T(A) = I\!\!R^n \times iA;$$

for x in $I\!\!R^n \setminus \{0\}$, we will write (as in the previous part of this section) $x0$ to denote the direction of the half line through the origin and passing through x; finally, for A again a subset of $I\!\!R^n$, we will denote by $A0$ the set

$$A0 = \{x0 : x\epsilon A \setminus \{0\}\}.$$

We can now state Bogolyubov's theorem:

Theorem 1.6.2 *Let Γ_1 be an open convex cone in $I\!\!R^n$, and set $\Gamma_2 = -\Gamma_1$. For any open set Ω in $I\!\!R^n$ and any complex neighborhood U of it, there exists a complex neighborhood $U' \subseteq U$ such that the following is true: let $f_j \in \mathcal{O}(U \cap T(\Gamma_j)), j = 1, 2$, have the following boundary values*

$$f_j(x + i\Gamma_j 0) = \lim_{\substack{y \to 0 \\ y \in \Gamma_j}} f_j(x + iy),$$

where the limit is intended to be uniform on compact sets containing x and compact subcones containing y. If the limits coincide in the sense that

$$f_1(x + i\Gamma_1 0) = f_2(x + i\Gamma_2 0),$$

then there exists a function f in $\mathcal{O}(U')$ such that

$$f = f_j \text{ on } U' \cap T(\Gamma_j), \ j = 1, 2.$$

Without going into the details of the proof, it might suffice to recall that the original proof (see e.g. [169]) relies, not surprisingly, on Cauchy's integral formula. First one reduces the theorem to the case in which the cone Γ is of special shape (for example the first quadrant). One then applies a suitable analytic transformation and finally uses Cauchy's integral formula to obtain the explicit analytic continuation. This approach is quite natural, especially in view of Painleve's original result, and can be found in Rudin's [191]. It may be worthwhile recalling that this theorem has stimulated many other mathematicians (see our reference list), and so several alternate proofs have been given; all of them, deal with the issue of separate analyticity, and the key ingredient is always Cauchy's integral formula. The most general result in this framework is the so called Malgrange-Zerner theorem (despite the fact that it only appears in Martineau's [159], and that neither Malgrange nor Zerner seem to have published it): in it we need the notion of relatively open cone which is given as follows: a cone Γ in $I\!\!R^n$ is said to be relatively open if it is open in its linear hull (i.e. in the linear subspace of $I\!\!R^n$ spanned by Γ). With this definition we can state Malgrange-Zerner's theorem as follows:

Theorem 1.6.3 *Let Γ_1 and Γ_2 be two relatively open convex cones in $I\!\!R^n$ and let Γ_{12} be the convex hull of their union. For any open set Ω of $I\!\!R^n$ and a complex neighborhood U of its, there exists a complex neighborhood $U' \subseteq U$ of Ω*

such that the following analytic continuation property holds. If f is a continuous function on the set $U \cap (T(\Gamma_1) \cup T(\Gamma_2) \cup \Omega)$ and if its restrictions to $U \cap (T(\Gamma_j))$ are holomorphic in $(\Gamma_j), j = 1, 2$, then there exists a continuous function F on $U' \cap (T(\Gamma_{12}) \cup \Omega)$, such that its restriction to $U' \cap T(\Gamma_{12})$ is holomorphic in $T(\Gamma_{12})$ and that

$$f = F \text{ on } U' \cap (T(\Gamma_1) \cup T(\Gamma_2) \cup \Omega).$$

We easily see, from Theorem 1.6.2 to 1.6.3, an improvement due to the generality of the relative positions of the cones, and of course the weaker assumptions of continuity.

That the continuity hypotheses (i.e. the coincidence of the two boundary values, which in Bogolyubov's statement is to be taken in the space of continuous functions) could be considerably weakened, was immediately apparent; as a matter of fact, it was almost immediately shown that Theorem 1.6.2 could be formulated in the framework of the theory of distributions. Since most distributions of interest in physics arise as boundary values of holomorphic functions, this extension had an impact which was not just mathematically significant. Let us recall the notion of boundary value in the sense of distributions, so to be able to state the distribution version of Bogolyubov's theorem.

Let Ω be as usual an open set in $\mathbb{R}$, and U a complex neighborhood of its; let $\mathcal{D}(\Omega)$ and $\mathcal{D}'(\Omega)$ be, respectively, the spaces of compactly supported infinitely differentiable functions on Ω, and the space of distributions on Ω (the first space endowed with its usual locally convex topology); finally let Γ be an open convex cone; for $f \in \mathcal{O}(U \cap T(\Gamma))$ and $g \in \mathcal{D}(\Omega)$, the integral

$$\int_\Omega f(x + iy)g(x)dx$$

can be defined for any (sufficiently small) y in Γ. If the limit

$$\lim_{\substack{y \to 0 \\ y \in \Gamma}} \int_\Omega f(x + iy)g(x)dx$$

exists, for every g in $\mathcal{D}(\Omega)$, then such a limit defines a distribution which will be denoted by

$$f(x + i\Gamma 0),$$

and this notation is consistent with the same notation used in the case of the limit in the space of continuous functions. We can therefore state the following result (which we provide here directly in the more general version due to Epstein [53], in which the position of the cones is general):

Theorem 1.6.4 *Let Γ_1 and Γ_2 be two open convex cones in $\mathbb{R}^n$. For any open set Ω (and a complex neighborhood U) there exists a complex neighborhood $U' \subseteq U$ such that the following is true: if*

$$f_j \in \mathcal{O}(U \cap T(\Gamma_j)), j = 1, 2,$$

have both distribution boundary values $f_j(x + i\Gamma_j 0)$ such that

$$f_1(x + i\Gamma_1 0) = f_2(x + i\Gamma_2 0),$$

then there exists a function f in $\mathcal{O}(U' \cap T(\Gamma_{12})))$ which coincides with f_j on $U' \cap T(\Gamma_j)$ for $j = 1, 2$.

Note that when $\Gamma_1 = -\Gamma_2$ then the convex hull of their union is all of $I\!\!R^n$, which shows how Theorem 1.6.4 actually contains Theorem 1.6.3.

The next great advance in the theory of the edge of the wedge theorem came with the realization (due to Martineau) that in order to deal with the case in which several cones were appearing, a more abstract approach to the problem was necessary. In particular, the introduction of the notion of boundary value in the sense of hyperfunctions became imperative.

In order to state the results we are interested in, we need first to recall, from section 1.5, the basics on boundary values in the sense of hyperfunctions; let Ω and U be as above and let Γ be an open convex cone in $I\!\!R^n$. As we know, the vector space $\mathcal{B}(\Omega)$ is defined by $H_\Omega^n(U, \mathcal{O})$; however, one may recall the long exact sequence for relative cohomology (analogous to the one which is well known for singular cohomology), which gives

$$\ldots \longrightarrow H^{n-1}(U \setminus \Omega; \mathcal{O}) \longrightarrow H_\Omega^n(U, \mathcal{O}) \longrightarrow H^n(U; \mathcal{O}) \longrightarrow$$

$$\longrightarrow H^n(U \setminus \Omega; \mathcal{O}) \longrightarrow H_\Omega^{n+1}(U, \mathcal{O}).$$

Because of Grauert's Theorem [66], we can assume that the neighborhood U has been chosen to be Stein. We can now use the vanishing of the relative cohomology in dimension different from n, together with Cartan's Theorem B in the above long exact sequence, which shows the vanishing of the n-th cohomology of Stein open sets, to obtain an important isomorphism:

$$H^{n-1}(U \setminus \Omega; \mathcal{O}) \cong B(\Omega).$$

We can therefore consider a holomorphic function $f \in \mathcal{O}(U \cap T(\Gamma))$, and associate to it its hyperfunction boundary value $b_\Gamma(f)$, which is the element of $\mathcal{B}(\Omega)$, defined by the $(n-1)$-cocycle which can be naturally associated to f (see the construction in section 1.5).

At this point, one may define a space of holomorphic functions defined near $I\!\!R^n$, for which two different notions of boundary values exist; a distribution notion and a hyperfunction one. Following Martineau's [156], we define the subspace $\mathcal{O}_\mathcal{D}(U \cap T(\Gamma))$ of $\mathcal{O}(U \cap T(\Gamma))$ to be the subspace of those holomorphic functions whose elements can be continued as distributions to a full complex neighborhood of Ω. Martineau succeeded to show the following two fundamental facts:

Theorem 1.6.5 *A function* $f \in \mathcal{O}(U \cap T(\Gamma))$ *has a distribution boundary value (as defined before) if and only if it actually belongs to the subspace* $\mathcal{O}_D(U \cap T(\Gamma))$. *In this latter case, its distribution boundary values coincide with its hyperfunction boundary value, i.e.*

$$b_\Gamma(f)(x) = f(x + i\Gamma 0).$$

It was using this last result, that Martineau was able to prove its full version of the edge of the wedge theorem (he actually used the hyperfunction version to deduce, as a corollary of Theorem 1.6.4, the distribution version). We conclude this section of the appendix by giving Martineau's hyperfunction edge of the wedge theorem.

Theorem 1.6.6 *Let* $\Gamma_1, \Gamma_2, \ldots, \Gamma_m$ *be open convex cones in* $\mathbb{R}^n$. *Then for any real open set* Ω *and its complex neighborhood* U, *there exists a complex neighborhood* $U' \subseteq U$ *such that the following is true: given functions* f_j *in* $\mathcal{O}(U \cap T(\Gamma_j))$ *such that, in* $\mathcal{B}(\Omega)$,

$$\sum_{j=1}^{m} b_\Gamma(f_j) = 0,$$

then there exist holomorphic functions g_{jk} *in* $\mathcal{O}(U' \cap T(\Gamma_{j,k}))$ *for* $j, k = 1, \ldots, m$, *and* $j = k$, *such that*

$$f_j = \sum_{k \neq j} g_{jk} \quad on \ U' \cap T(\Gamma_j), j = 1, \ldots, m.$$

and

$$g_{jk} + g_{kj} = 0.$$

We point out at this point that the story of the edge of the wedge theorem does not quite ends here; as a matter of fact, the theorem itself is a key element (and motivation) for the introduction of the sheaf of microfunctions; this aspect, however, we will leave to the next chapter, where microfunctions will be treated.

Chapter 2

Microfunctions

2.1 Introduction

The last chapter should have provided the reader with a sufficiently ample background on the general theory of hyperfunctions (even though we restricted our attention to the case of hyperfunctions defined on Euclidean spaces, purposely avoiding the more delicate aspects which arise from the necessity of dealing with general real analytic varieties).

This chapter, on the other hand, will introduce the other great player in the field of Algebraic Microlocal Analysis, namely the sheaf of microfunctions. Even though some of the algebraic machinery which is needed has already been introduced in the previous chapter, we still wish to warn the inexperienced reader since the theory of the sheaf of microfunctions is definitely a delicate issue. The main reason is that microfunctions do not arise naturally as generalized functions on the Euclidean space but, rather, describe the singularities of such generalized functions. In dealing with hyperfunctions, the reader could always rely on its experience with the space of distributions, whose behavior (in many instances) was mimicked by the space of hyperfunctions. Of course the interest in studying hyperfunctions arose exactly from those situations in which the behavior was so markedly different, and we hope we succeeded in conveying this aspect in our previous chapter.

When dealing with microfunctions, however, the reliance on the infinitely differentiable case seems to falter (at least in most of the classical introductions to the subject) and this makes its understanding more difficult.

In this chapter we will try to rely on the infinitely differentiable analogies as much as possible, at least as a way of introducing the topic, and we will then show the power of the sheaf of microfunctions as constructed by Sato. Also in this case, of course, the creation and the advancement of the theory is essentially due to Sato and his coworkers, but we will see that similar ideas were brewing in many different areas of mathematics, so that the interplay between different readings is even more interesting here than it was for hyperfunctions.

The chapter is structured as follows: in section 2.2 we describe different notions of singularities for hyperfunctions, and, in particular, the crucial notion of singular spectrum and its relations to differential operators. This introduction lays the ground for the construction of the sheaf of microfunctions which is done (along the lines of Chapter I) in two steps; the single variable case is treated in section 2.3 while the several variables case is dealt with in section 2.4. Finally, section 2.5 goes back to the study of microlocal operators, which had only been cursorily discussed in the first chapter. Now that microfunctions are available, it is the right moment for a more complete discussion of this topic and of its interest within the theory of differential equations.

This chapter also ends with a historical appendix (which is also an excuse for some alternative treatment of the topics discussed in the main text); this time the appendix deals with the introduction and the role of microfunctions in physics as well as with the treatment which Hörmander has given of a notion very similar to the singular spectrum, namely his notion of analytic wave front set.

With the equipment provided by these two chapters, the reader should now be capable to dwell into the main part of this book, i.e., the algebraic treatment of systems of differential equations.

2.2   Singular Support, Essential Support and Spectrum

In the previous chapter, we have examined hyperfunctions as generalized functions which form a sheaf properly containing the subsheaf of real analytic functions. Thus, in particular, one may want to know the location of the singularities of hyperfunctions, i.e., the set in which a given hyperfunction fails to be real analytic; such a set, which can be easily defined, is the so called **singular support** of a hyperfunction. Its definition (see also Chapter I) reads as follows:

Definition 2.2.1 *Let f be a hyperfunction on an open set U on $\mathbb{R}^n$. The singular support of f, indicated by sing supp (f), is the complement in U of the largest open subset of U on which f is real analytic.*

The reader will note that such a definition is well given, because $\mathcal{A}$ is a subsheaf of $\mathcal{B}$, and therefore there exists the largest open subset of U in which f is real analytic (in other words, the property of being real analytic is a local property).

This definition, however, does not seem particularly useful, since it only pinpoints the singular locus of a hyperfunction, without explaining the differences between various singularities. As an example (which really originates from physics) one may ask why multiplication of distributions (hyperfunctions) is possible in some cases but not in others. It is well known, e.g., that the square of the delta function is not a well defined distribution, while it is possible to consider the square of $1/(x+i0)$. Since real analytic functions do indeed multiply among themselves (and of course we can multiply a real analytic function by a hyperfunction), it is somehow clear that the problem of multiplication resides with the singularity locus of hyperfunctions; on the other hand, the singular support of both the delta and $1/(x+i0)$ consists of the origin, and so a more refined description seems necessary.

It turns out that the key instrument for the understanding of this and other related phenomena, is the notion of microlocalization of a sheaf; by this term one usually refers to the fact that the study of singularities is now removed from the original manifold, and is transferred to its cotangent bundle (or spheric cotangent bundle); the necessity of such a behavior is common knowledge in the community of differential equations scholars, where the symbol of an operator is exactly defined on such bundle. The procedure for such microlocalization is not at all immediate, and may be advantageous to begin by exploring a similar situation, which can be used as a motivational tool, namely the differentiable case. Let us therefore begin by taking a Schwartz distribution f in $I\!R^n$, and let us consider the issue of its differentiability at a point x_0. If f were compactly supported, we could use the Paley-Wiener Theorem for the space $\mathcal{E}'$ of compactly supported distributions to study the differentiability of f in terms of the growth of its Fourier transform. On the other hand, if f is not compactly supported, no such theorem exists (in other words, we do not have a good intrinsic description of the differentiability of f, for the precise reason that it does not in general admit a Fourier transform). The way to approach this situation, therefore, is to exploit the fineness of the sheaf of differentiable functions and to multiply our distribution f by a compactly supported function, say $\chi_{x_0} = \chi$, which is identically one in a neighborhood of x_0, and quickly vanishes outside it (we can take for χ any cutoff function). Then, the product $f \cdot \chi$ is a compactly supported distribution which coincides with f in a neighborhood of x_0. Note that one can actually represent f as the limit, in a suitable topology, of a sequence $\{f \cdot \chi_n\}$ of compactly supported distributions obtained by considering the cutoff functions χ_n associated to the spheres $\mathcal{B}(x_0, \varepsilon_n)$ centered in x_0 and of radius ε_n convergent to zero. If f is infinitely differentiable, then each element of the sequence is infinitely differentiable as well. Also note that while the differentiability of $f \cdot \chi$ depends only on the differentiability of f in a neighborhood of x_0, the differentiability of the sequence of distributions associated to a sequence of spheres of a decreasing radiuses, only depends on the differentiability of f at x_0. One can now exploit the compact support of the sequence to determine its differentiability in terms of the growth of its Fourier transform. Let us recall here, for the sake of completeness, the versions of the Paley-Wiener theorem which we need to employ:

Theorem 2.2.1 (*Paley-Wiener for $\mathcal{E}'$*). *The vector space of compactly sup-*

ported distributions $\mathcal{E}'$ is algebraically isomorphic, via the Fourier transform, to the space of entire functions of exponential type which, on the real axis, have polynomial growth. In other words, the space of compactly supported distributions is isomorphic to the space

$$PW(\mathbb{C}^n) := \{F \in \mathcal{H}(\mathbb{C}^n) : |F(z)| \leq A(1+|z|)^B \exp(B|\mathrm{Im}z|)\}$$

where A and B are positive constants depending on F.

Theorem 2.2.2 *(Paley-Wiener for $\mathcal{D}$). The vector space of compactly supported infinitely differentiable functions $\mathcal{D}$ is algebraically isomorphic, via the Fourier transform, to the space of entire functions F which satisfy the following estimate: there is a positive constant B such that for any integer $M > 0$, there exists a positive constant C_M such that*

$$F(z) \leq C_M(1+|z|)^{-M} \exp(B|\mathrm{Im}z|).$$

It is a highly non-trivial fact, essentially due to Ehrenpreis [52], that the isomorphisms above can be made into topological isomorphisms. This fact has some very important consequences which we will not explore in this setting. We refer the interested reader to Ehrenpreis' fundamental work [52].

Based on the theorems stated above, we realize that we certainly have that the products $f \cdot \chi_n$ satisfies the weaker condition

(2.2.1) $|\widehat{f \cdot \chi_n}(z)| \leq A(1+|z|)^B \exp(B|\mathrm{Im}z|);$

however, if the distribution f is infinitely differentiable at the point x_0, then the products $f \cdot \chi_n$ also are infinitely differentiable, and therefore they satisfy the stronger condition

(2.2.2) $|\widehat{f \cdot \chi_n}(\xi)| \leq C_M(1+|\xi|)^{-M}, \quad \xi \in \mathbb{R}^n.$

This fact gives us a useful insight; if a distribution fails to be differentiable, we can "measure" the degree of this failure by looking at where, in the frequency plane, its Fourier transform fails to satisfy condition (2.2.2). This consideration allows to distinguish, for example, between the delta function and the boundary values of $1/z$, i.e. $1/(x+i0)$ and $1/(x-i0)$. As a matter of fact (we are considering here only the one-dimensional case), one can easily see that the Fourier transform of the delta function (considered as a compactly supported distribution) is given by

$$e^{ix\xi}$$

which, clearly, does not satisfy condition (2.2.2) in either direction (in one dimension, of course, there are only two directions to be considered: $\xi > 0$ and $\xi < 0$). On the other hand the Fourier transform of $1/(x+i0)$ turns out to be (for $H(\xi)$ the Heaviside function)

$$e^{ix\xi} \cdot H(\xi)$$

which satisfies condition (2.2.2) for $\xi > 0$. This shows that the singularity of δ is somehow more serious than the singularity of $1/(x + i0)$. We will see how to use this information to justify the possibility of taking the square of the second and not of the first.

For the case of compactly supported distributions, we can use these natural remarks to introduce the process of microlocalization as follows. As we mentioned before, we are going to consider distributions defined on an open set U of the n-dimensional Euclidean real space, while the microlocalization will take place on the product $U \times \mathbb{R}^n$, to be regarded as the cotangent bundle of U. This approach lends itself to a quick modification when dealing with a more general manifold M (to be assumed differentiable when dealing with distributions, and real analytic when dealing with hyperfunctions). For the next definition, we recall that a cone Γ in $\mathbb{R}^n$ is a subset of $\mathbb{R}^n$ stable under the dilations $\xi \to \rho\xi$ for $\rho > 0$; more generally a subset of $U \times \mathbb{R}^n$ is said to be conic if it is stable under the transformations

$$(x, \xi) \to (x, \rho\xi), \quad \rho > 0.$$

Definition 2.2.2 *Let f be a distribution on U. We say that f is infinitely differentiable in (x_0, ξ_0), $x_0 \in U$, $\xi_0 \in \mathbb{R}^n \setminus \{0\}$, if there is a compactly supported infinitely differentiable function χ which is identically one in a neighborhood of x_0, and there is an open cone Γ_0 in $\mathbb{R}^n$ containing ξ_0 such that:*

for every $M > 0$ there is a non-negative number C_M such that, for any ξ in Γ_0,

$$|(\widehat{\chi f})(\xi)| \leq C_M(1 + |\xi|)^{-M}.$$

The reader will note that the notion of conic set is essential in this setting, and so the cotangent bundle is often replaced by the cosphere bundle. In our flat case this correspond to replacing $T^*U = U \times \mathbb{R}^n$ by the following quotient:

$$S^*U = U \times (\mathbb{R}^n \setminus \{0\})/ \sim$$

where the equivalence relation is given by $(x, \xi) \sim (y, \eta)$ if and only if $x = y$ and $\xi = \rho\eta$ for some positive ρ.

There is a canonical projection π of $T^*U \setminus \{0\}$ onto its quotient S^*U, and a subset Γ of $T^*U \setminus \{0\}$ is said to be conic if and only if it is $\Gamma = \pi^{-1}(\pi(\Gamma))$. We will also say that Γ is conically compact if $\pi(\Gamma)$ is compact (this, of course, does not mean that Γ itself is compact: as a matter of fact, this is never the case).

Definition 2.2.3 *A distribution f on an open set U is said to be infinitely differentiable in a conic subset Γ of $T^*U \setminus \{0\}$ if it is infinitely differentiable in a neighborhood of every point of Γ.*

The set which describes the singularities of the distribution f is therefore the complement of the largest set in which f is differentiable, as given by the following fundamental definition:

Definition 2.2.4 *Let f be a distribution in an open set U. The complement in $T^*U \setminus \{0\}$ of the union of all conic open sets in which f is infinitely differentiable is the* **wave front set** *of f and will be denoted by $WF(f)$.*

Before we go back to the case of hyperfunctions, we would like to give some properties of this notion. In order to do so, we need to take a brief detour through the concept of pseudo-differential operator. This will be, in any case, a useful introduction for the notion of microlocal operators, which we will deal with in section 5 of this chapter.

We begin by constructing what will later be the symbols of the pseudo-differential operators.

The notion of **pseudo-differential operator** is a very natural one, which stems from the time-honored attempt to algebrize analysis. Pseudo-differential operators, which so clearly opened the road to microlocalization, have been called, [224], "... the most important step forward in our understanding of linear partial differential equations since distributions".

To understand the development of what might seem a difficult concept, we will begin with the notion (introduced by F. John in the fifties [100], [101]) of "parametrix" of an elliptic linear partial differential equation.

Let then $P = P(\xi_1, \ldots, \xi_n) = P(\xi)$ be a polynomial with complex coefficients in n real variables, and let

$$P(D) = P(D_1, \ldots, D_n)$$

be the corresponding linear partial differential operator obtained by replacing ξ_j by

$$D_j = -i \frac{\partial}{\partial x_j};$$

we say that $P(\xi)$ is the symbol of $P(D)$. Consider now the equation

$$P(D)u = f,$$

with $f \in \mathcal{D}(\mathbb{R}^n)$ given, and let us look for a solution $u \in \mathcal{D}$. The first, naive, attempt would be to formally apply the Fourier transform, so that

$$P(\xi)\hat{u}(\xi) = \hat{f}(\xi),$$

$$\hat{u}(\xi) = \frac{\hat{f}(\xi)}{P(\xi)}$$

and finally

(2.2.3)
$$u(x) = \left(\frac{1}{2\pi}\right)^n \int \frac{\hat{f}(\xi)}{P(\xi)} e^{ix\cdot\xi} d\xi.$$

The integral on the right hand side of (2.2.3), however, makes usually no sense, because of the zeroes of P.

Assume, however, that $P(D)$ is elliptic as in the following definition.

Definition 2.2.5 *Let* $m = \deg P(\xi)$ *and write* $P(\xi) = P_m(\xi) + Q(\xi)$, *with* $\deg Q(\xi) \leq m-1$ *and* $P_m(\xi)$ *homogeneous of degree* m. *We will call* $P_m(\xi)$ *the principal symbol of* $P(D)$ *and we say that* $P(D)$ *(or* $P(\xi)$*) is elliptic if* $P_m(\xi) \neq 0$ *for all* $\xi \neq 0$. *Since* P_m *is homogeneous its zero-set* $V_{P_m} = \{\xi \in \mathbb{R}^n : P_m(\xi) = 0\}$ *is a cone in* $\mathbb{R}^n$ *known as the characteristic cone of* $P(D)$ *(or* $P(\xi)$*).*

Remark 2.2.1 It is immediate to verify that for $n = 1$ all polynomials are elliptic, while for $n \geq 2$ the class of elliptic polynomials becomes an important, but proper subclass of the class of polynomials. The symbols of both the Laplace and the Cauchy-Riemann operators are elliptic, while the symbol of the heat operator is not.

Remark 2.2.2 It is an easy exercise to verify that if $P(\xi)$ is elliptic then its characteristic variety $V_P = \{\xi \in \mathbb{R}^n : P(\xi) = 0\}$ is compact in $\mathbb{R}^n$.

This simple remark has, in fact, relevant consequences, since we may now consider the integral in (2.2.3) only outside of a compact containing V_P. More precisely, if V_P is contained in the ball centered in the origin and of radius r and χ is a regularizing C^∞ function such that $\chi(\xi) = 0$ for $|\xi| < r$, $\chi(\xi) = 1$ for $|\xi| > r' > r$, then we can define a sort of approximate solution by

(2.2.4)
$$v(x) := \left(\frac{1}{2\pi}\right)^n \int \frac{\hat{f}(\xi)}{P(\xi)} \chi(\xi) e^{ix\cdot\xi} d\xi.$$

How good of an approximation is v to u? It is actually quite good; in fact one immediately sees that

$$P(D)v(x) = f(x) - Rf(x),$$

where R is the operator defined by

$$Rf(x) := \left(\frac{1}{2\pi}\right)^n \int \hat{f}(\xi)(1 - \chi(\xi)) e^{ix\cdot\xi} d\xi,$$

i.e.

$$Rf = h * f$$

with h the Fourier transform of $1 - \chi$.

By its construction $1-\chi$ is compactly supported, and therefore h is (extendible to be) entire of exponential type (just apply the Paley-Wiener theorem mentioned above). Another important point is the fact that

$$|P(\xi)| \geq |P_m(\xi)| - |Q(\xi)| \geq 1$$

for $|\xi|$ large enough, so that χ/P actually defines a tempered distribution on $I\!\!R^n$ which is (by the isomorphism theorem) the Fourier transform of a tempered distribution K. Then (2.2.4) actually reads as

$$v = K * f,$$

and we have therefore obtained that

(2.2.5) $P(D)K = \delta - h$

or equivalently (looking at the corresponding convolution operators)

(2.2.6) $P(D)K = I - R$

where $R : \mathcal{E}' \to \mathcal{E}$ is a continuous linear operator. We will use (2.2.5) or equivalently (2.2.6) as the definition for the notion of a parametrix K for the operator $P(D)$. A parametrix is therefore a distribution K satisfying (2.2.5) or a convolution operator satisfying (2.2.6) with R as above.

Remark 2.2.3 The theory of parametrices is rich (see e.g. [100], [224]). It will suffice here to note that one can use K to construct a fundamental solution for $P(D)$. Indeed, if w is an entire function such that

$$P(D)w = h$$

(and this is easily seen to exist by purely functional analytic arguments) then

$$E = K + w$$

is a fundamental solution since

$$P(D)E = P(D)K + P(D)w = \delta - h + h = \delta.$$

Remark 2.2.4 To summarize, we have seen that a parametrix K is essentially a clever modification of a right inverse of $P(D)$. The next step towards pseudo-differential operators is the generalization of these ideas to the case of differential operators with variable coefficients.

If Ω is an open subset of $I\!\!R^n$, a variable coefficients differential operator on Ω is defined by

$$P(x, D) = \sum_{|\alpha| \leq m} c_\alpha(x) D^\alpha,$$

with $\alpha = (\alpha_1, \ldots, \alpha_n)$, $|\alpha| = \alpha_1 + \cdots + \alpha_n$, $D = (D_1, \ldots, D_n)$ and where the functions $c_\alpha(x)$ are suitably regular in Ω (e.g. c_α could be infinitely differentiable or real analytic in Ω). Then the **symbol** of $P(x, D)$ is the polynomial in $2n$ variables

$$P(x, \xi) = \sum_{|\alpha| \leq m} c_\alpha(x) \xi^\alpha$$

and its principal symbol is defined as

$$P_m(x, \xi) = \sum_{|\alpha| = m} c_\alpha(x) \xi^\alpha.$$

As before one defines

$$V_{P_m} := \{(x, \xi) \in \Omega \times I\!\!R^n : P_m(x, \xi) = 0\},$$

which is now an algebraic variety in $I\!\!R \times I\!\!R^n$ or (more generally) in $T^*\Omega$.

In the case of variable coefficients operators we say that $P(x, D)$ is elliptic if, for every $x_0 \in \Omega$,

$$V_{P_m}(x_0) := \{\xi \in I\!\!R^n : P_m(x_0, \xi) = 0\} = \{0\}.$$

One can then try to replicate the process which was employed in the case of constant coefficients to construct operators K and R such that

$$P(x, D)Kf = f - Rf,$$

with $R : \mathcal{E}'(\Omega) \to \mathcal{E}(\Omega)$ continuous and linear. It is therefore natural to try to construct a kernel $k(x, \xi)$ such that K can be defined by

$$Kf(x) = \left(\frac{1}{2\pi}\right)^n \int \hat{f}(\xi) k(x, \xi) e^{ix \cdot \xi} d\xi.$$

If we set, as before, $v(x) = Kf(x)$, we have for any function $\omega(x)$,

$$P(x, D)(\omega(x) e^{ix \cdot \xi}) = e^{ix \cdot \xi} P(x, D + \xi)(\omega(x)),$$

and, therefore,

$$P(x, D)v(x) = \left(\frac{1}{2\pi}\right)^n \int \hat{f}(\xi) e^{ix \cdot \xi} P(x, D + \xi) k(x, \xi) d\xi,$$

so that we would have

(2.2.7) $$P(x, D)K = I$$

if we could solve

(2.2.8) $$P(x, D + \xi) k(x, \xi) = 1.$$

In order to solve (2.2.8) note that

$$P(x, D + \xi) = P_m(x, \xi) + \sum_{j=1}^{m} P_j(x, \xi, D)$$

where the $P_j(x, \xi, D)$ are differential operators of order j in x, whose coefficients are homogeneous polynomials of degree $m - j$ in ξ. It follows that in order for $k(x, \xi)$ to be a solution of (2.2.8) it is necessary to take it as a sum of functions homogeneous in ξ. But k must also be tempered in the ξ variables (in order for K to appropriately act on the space of distributions) and so our best bet is an infinite series of terms with negative degrees of homogeneity:

$$k(x, \xi) = \sum_{j=0}^{\infty} k_j(x, \xi),$$

and k_j homogeneous of degree $-j$.

We will not push here this attempt which we already know cannot possibly work since (2.2.7) cannot be exactly solved (in view of the zeroes of the symbol of P). Once again we need to eliminate these singularities with some sort of cut-off process (we are dealing here with an elliptic operator). However, since we now have an infinite series, just one cut-off function will not suffice and we will need to define our kernel by

(2.2.9) $$k(x, \xi) = \sum_{j=0}^{\infty} \chi_j(\xi) k_j(x, \xi).$$

We will not give all the details of the construction of the parametrix, our goal being here to motivate our subsequent treatment. It will suffice to say that suitable cut-off functions χ_j can be constructed so that if $k(x, \xi)$ is given by (2.2.9), then

$$P(x, D + \xi) k(x, \xi) = 1 - r(x, \xi)$$

with a function $r(x, \xi)$ which can be explicitly computed and which is such that the operator R defined by

$$Rf(x) = \left(\frac{1}{2\pi}\right)^n \int \widehat{f}(\xi) r(x, \xi) e^{ix \cdot \xi} d\xi$$

is a continuous linear map from $\mathcal{E}'(\Omega)$ to $\mathcal{E}(\Omega)$. This procedure therefore provides us with a parametrix for the case of variable coefficients elliptic operators.

Our discussion so far can be summarized by saying that the attempt to invert a differential operator has lead us to the introduction of a general class of integral operators (of which both parametrices and differential operators are a special case), namely those operators which, for suitable kernels t, can be written as

$$Tf(x) = \left(\frac{1}{2\pi}\right)^n \int \widehat{f}(\xi) t(x, \xi) e^{ix \cdot \xi} d\xi.$$

We will now give a formal definition for such operators, after we deal with some preliminary notions.

Definition 2.2.6 *Let* $K \in \mathcal{D}(\Omega \times \Omega)$ *be a distribution on* $\Omega \times \Omega$. *We say that* K *is* **separately regular** *if both*

$$K : \quad \varphi(x) \to \langle K(x,y), \varphi(y) \rangle$$

and

$$K^t : \quad \psi(y) \to \langle K(y,x), \psi(x) \rangle$$

map $\mathcal{D}(\Omega)$ *into* $\mathcal{E}(\Omega)$.

In addition, such a distribution will be said to be **very regular** if, besides being separately regular, it is C^∞ outside the diagonal of $\Omega \times \Omega$. The study of these kernels is crucial in view of the celebrated Schwartz's kernel theorem (maybe his most important early result on distributions) which shows that for any continuous linear map

$$K : \mathcal{D}(\Omega) \to \mathcal{D}'(\Omega)$$

there is a unique distribution $K(x,y)$ in $\Omega \times \Omega$ such that for all $f \in \mathcal{D}(\Omega)$

$$K(f)(x) = \langle K(x,y), f(y) \rangle.$$

Note that this condition is equivalent to requiring that K maps $\mathcal{D}(\Omega)$ into $\mathcal{E}(\Omega)$ and extends as a continuous linear map of $\mathcal{E}'(\Omega)$ into $\mathcal{D}'(\Omega)$.

Definition 2.2.7 *A very regular operator* K *is said to be* **regularizing** *if it extends as a continuous linear map of* $\mathcal{E}'(\Omega)$ *into* $\mathcal{E}(\Omega)$ *(rather than just* $\mathcal{D}'(\Omega)$*).*

Remark 2.2.5 In order for an operator to be regularizing, it is necessary and sufficient that K is C^∞ in $\Omega \times \Omega$.

Remark 2.2.6 A typical example of regularizing operator is the operator R which we constructed while developing a parametrix for elliptic operators.

The content of Remark 2.2.6 actually originates the following

Definition 2.2.8 *Let* $P(x, D)$ *be a differential operator in* Ω. *A* **parametrix** *for* P *is any very regular operator*

$$K : \mathcal{E}'(\Omega) \to \mathcal{D}'(\Omega)$$

such that

$$KP - I$$

is regularizing.

We have already discussed the importance of the existence of parametrices for the surjectivity of the operator $P(x, D)$. One can also easily prove that if P admits a parametrix, then it is hypoelliptic in the sense that if $P(x, D)f \in C^\infty$, then $f \in C^\infty$.

We are now ready to discuss pseudo-differential operators (or, rather, standard pseudo-differential operators):

Definition 2.2.9 *For any real number m, we denote by*

$$S^m(\Omega, \Omega)$$

the subspace of $C^\infty(\Omega \times \Omega \times I\!\!R^n)$ whose elements $a(x, y, \xi)$ satisfy the following estimate:

*for any compact subset T of $\Omega \times \Omega$, and any α, β, $\gamma > 0$,
there exists $C = C(\alpha, \beta, \gamma; T) > 0$ such that*

$$|D_\xi^\alpha D_x^\beta D_y^\gamma a(x, y, \xi)| \leq C(1 + |\xi|)^{m-|\alpha|} \quad \text{on } T \times I\!\!R^n.$$

The elements of $S^m(\Omega, \Omega)$ are called **amplitudes** of degree less than or equal to m. These amplitudes will be the symbols of our pseudo-differential operators (this is not quite accurate, as we will see later, but we can assume it at least for the moment).

Definition 2.2.10 *Let $a \in S^m(\Omega, \Omega)$. We will denote by*

$$\text{Op}(a) : \mathcal{D}(\Omega) \to \mathcal{D}'(\Omega)$$

the operator defined by

$$\text{Op}(a)(f)(x) := \left(\frac{1}{2\pi}\right)^n \int \left(\int a(x, y, \xi)f(y)e^{i(x-y)\cdot\xi}dy\right) d\xi.$$

Moreover, a linear continuous operator

$$A : \mathcal{E}'(\Omega) \to \mathcal{D}'(\Omega)$$

is said to be a pseudo-differential operator of order m if there exists $a \in S^m(\Omega, \Omega)$ such that

$$A = \text{Op}(a).$$

We will then write $A \in \Psi^m(\Omega)$.

Remark 2.2.7 Any differential operator $P(x, D)$ is a pseudo-differential operator with $a(x, y, \xi) = P(x, \xi)$. Also, any regularizing operator A with kernel $A(x, y)$ is a pseudo-differential operator with

$$a(x, y, \xi) = A(x, y)\chi(\xi)e^{-i(x-y)\cdot\xi}$$

for $\chi \in \mathcal{D}(I\!R^n)$, $\int \chi(\xi)d\xi = 1$.

Conversely, if we set

$$\Psi^{-\infty}(\Omega) = \bigcap_{m \in I\!R} \Psi^m(\Omega),$$

we can show that if $A \in \Psi^{-\infty}(\Omega)$, then A is regularizing.

Before we proceed to relating pseudo-differential operators to the notion of wave-front set, we want to clarify an earlier comment we made regarding amplitudes and symbols. The problem which one obviously needs to deal with is two-fold: on one hand, the role of the variable y seems to be more "technical" than really necessary; on the other hand, a given pseudo-differential operator can be obtained through infinitely many different amplitudes, while we would like to have some sort of one-to-one correspondence between symbols and operators. This can be achieved through a quotient process:

Definition 2.2.11 *For any m, $S^m(\Omega)$ denotes the subspace of $S^m(\Omega, \Omega)$ consisting of amplitudes which are independent of y. Also, we set*

$$S^{-\infty}(\Omega) = \bigcap_{m \in I\!R} S^m(\Omega).$$

Definition 2.2.12 *The space of* **symbols** *of degree less than or equal to m is defined by*

$$\dot{S}^m(\Omega) := \frac{S^m(\Omega)}{S^{-\infty}(\Omega)}.$$

Remark 2.2.8 All of the spaces introduced above can be endowed with suitable Fréchet topologies. Since we will not need them, we refer the interested reader to [224] for further details.

The fact that Definition 2.2.12 is the correct one follows because the map

$$a(x, \xi) \to \operatorname{Op}(a)$$

from $S^m(\Omega)$ to $\Psi^m(\Omega)$, defined by

$$\operatorname{Op}(a)(f)(x) := (2\pi)^{-n} \int \hat{f}(\xi)a(x,\xi)e^{ix\cdot\xi}d\xi$$

actually yields an isomorphism between $\dot{S}^m(\Omega)$ and $\Psi^m(\Omega)/\Psi^{-\infty}(\Omega)$. This last space will be denoted by $\dot{\Psi}^m(\Omega)$.

We can now relate these concepts to the notion of wave-front set of a distribution; to do so we will microlocalize the notion of regularization and regularizing operators (see also Definition 2.2.2).

Definition 2.2.13 *A pseudo-differential operator A in U is said to be regularizing in (x_0, ξ_0), $x_0 \in U$, $\xi_0 \in \mathbb{R}^n \setminus \{0\}$, if there is a compactly supported infinitely differentiable function χ supported in U which is identically one near x_0, and an open cone Γ_0 in $\mathbb{R}^n$ containing ξ_0 such that the symbol $a(x, \xi)$ of A satisfies the following property:*

$$\text{for any } N \geq 0 \text{ and every pair of multi-indices}$$
$$\alpha = (\alpha_1, ..., \alpha_n), \ \beta = (\beta_1, ..., \beta_n), \text{ there exists } C_{\alpha,\beta,N} \geq 0 \text{ such that}$$

$$\sup_x |\partial_x^\alpha \partial_\xi^\beta [\chi(x) a(x, \xi)]| \leq C_{\alpha,\beta,N}(1 + |\xi|)^{-N} \quad \text{for any} \quad \xi \in \Gamma_0.$$

Accordingly, we say that a pseudo-differential operator A on U is regularizing in a conic subset Γ of $U \times (\mathbb{R}^n \setminus \{0\})$ if it is regularizing in the neighborhood of every point of Γ.

Definition 2.2.14 *Let A be a pseudo-differential operator A on U. We call micro-support of A the set $\mu\text{-supp}(A)$ defined as the complement in $U \times (\mathbb{R}^n \setminus \{0\})$ of the union of all conic open sets in which A is regularizing.*

Note that, just like in the case of the wave-front set of distribution, the micro-support of a pseudo-differential operator is a closed conic subset of the set $U \times (\mathbb{R}^n \setminus \{0\})$. The link between these two notions is contained in the following result whose proof appears, for example, in [224]:

Proposition 2.2.1 *Let f be a distribution in U, and let (x_0, ξ_0) be a point of $U \times (\mathbb{R}^n \setminus \{0\})$.*

Then the following properties are equivalent:

(i) $(x_0, \xi_0) \notin WF(f)$;

(ii) there exists a differentiable function $a(x, \xi)$ in $U \times (\mathbb{R}^n \setminus \{0\})$, positive-homogeneous of degree zero, equal to one in some neighborhood of (x_0, ξ_0) such that if A is any properly supported pseudo-differential operator in Ω with symbol $a(x, \xi)$, then $A(f) \in \mathcal{C}^\infty(U)$;

(iii) there exists a conic open neighborhood Γ of (x_0, ξ_0) in $U \times (\mathbb{R}^n \setminus \{0\})$ such that

$$B(f) \in \mathcal{C}^\infty(U)$$

for any properly supported pseudo-differential operator B in U whose microsupport is contained in Γ.

This proposition has several important consequences which we summarize in a single Proposition:

Proposition 2.2.2 *Let U be an open set in $\mathbb{R}^n$. Then:*

(i) *if $f \in \mathcal{D}'(U)$, then the base projection of $WF(f)$ coincides with the singular support of f;*

(ii) *if A is a properly supported pseudo-differential operator on U then A is regularizing in a conic open subset Γ of $U \times (\mathbb{R}^n \setminus \{0\})$ if and only if Af is infinitely differentiable in U for all distributions $f \in \mathcal{D}'(U)$.*

(iii) *let $\varphi : U \to U'$ be a surjective diffeomorphism between two open sets in $\mathbb{R}^n$, and let $\varphi_+ : T^*U \to T^*U'$ and $\varphi_* : \mathcal{D}'(U) \to \mathcal{D}'(U')$ be the induced diffeomorphism; then*

$$WF(\varphi_* f) = \varphi_+(WF(f))$$

for any $f \in \mathcal{D}'(U)$;

(iv) *if A and f are as in (ii), then*

$$WF(Af) \subseteq WF(f) \cap (\mu - supp(A));$$

in particular,

$$WF(Af) \subseteq WF(f).$$

Before proceeding to microlocalize the notions of distribution and wave-front set, we show which are the ideas behind the definition of the trace of a distribution on a submanifold and of the product of two distributions. To this purpose, let us split the coordinate x in $\mathbb{R}^n$ as

$$x = (x', x''), \qquad x' \in \mathbb{R}^p, \ x'' \in \mathbb{R}^{n-p},$$

with x', x'' non-empty. Assume furthermore, the following condition on a distribution $f \in \mathcal{D}'(\mathbb{R}^n)$:

(2.2.10) there exists $\varepsilon > 0$ such that for every $M > 0$ there is $C_M > 0$
such that
$$|\hat{f}(\xi)| \le C_M (1 + |\xi|)^{-M} \quad \text{if } |\xi'| \le \varepsilon |\xi''|,$$
where ξ' and ξ'' are, respectively, the dual variables of x' and x''.

Since, naturally, there exist $C, m > 0$ such that

$$|\hat{f}(\xi)| \le C(1 + |\xi|)^m \quad \text{on } \mathbb{R}^n,$$

then if $|\xi''| \le |\xi'|/\varepsilon$ one has, for some new constant $C > 0$,

(2.2.11) $$|\hat{f}(\xi)| \le C(1 + |\xi'|)^{m+n+1}(1 + |\xi''|)^{-n-1}$$

and (2.2.11) also holds true when $|\xi''| \geq \frac{|\xi''|}{\varepsilon}$, as it follows immediately from (2.2.10). This estimate shows that

$$\hat{f}_0(\xi') := \left(\frac{1}{2\pi}\right)^{n-p} \int \hat{f}(\xi', \xi'') d\xi''$$

is tempered and its inverse Fourier transform is what can be regarded as the **trace** of f on $x'' = 0$. But now, it is not difficult to see that (2.2.10) actually means that

$$WF(f) \cap \{\xi \in I\!\!R^n : \xi' = 0\} = \emptyset.$$

By (iii) of Proposition 2.2.2 we therefore obtain that if $Y \subseteq I\!\!R^n$ is a differentiable manifold, and $f \in \mathcal{D}'(I\!\!R^n)$, then the trace of f along Y (or, if one prefers, the restriction $f_{|Y}$) is well defined whenever the wave-front set $WF(f)$ of f does not intersect the conormal bundle of Y in $I\!\!R^n$.

In particular, if $WF'(f)$ denotes the image of $WF(f)$ under the symmetry $(x, \xi) \to (x, -\xi)$, and since, for f, g distributions on $I\!\!R^n$,

$$f(x) \cdot g(x)$$

is nothing but the trace of $f(x) \otimes g(y)$ along the diagonal of $I\!\!R^n \times I\!\!R^n$, one obtain the following important result:

Proposition 2.2.3 *If $f, g \in \mathcal{D}'(I\!\!R^n)$ satisfy*

$$WF(f) \cap WF'(g) = \emptyset,$$

then the product $f \cdot g$ is a well defined distribution on $I\!\!R^n$.

As our last remark on the differentiable case, we want to microlocalize the sheaf of distributions. We therefore construct the following presheaf on $T^* I\!\!R^n \setminus \{0\}$: for any open set $U \subseteq T^* I\!\!R^n \setminus \{0\}$, we denote by c.sp.$(U)$ its conic span, i.e. the smallest conic set containing U, and we define

$$\mathcal{F}(U) := \frac{\mathcal{D}'(I\!\!R^n)}{\{f \in \mathcal{D}'(I\!\!R^n) : f \in \mathcal{C}^\infty(\text{c.sp.}(U))\}}.$$

It is immediate to verify that the functor

$$U \to \mathcal{F}(U) \qquad .$$

gives a presheaf on $T^* I\!\!R^n \setminus \{0\}$, and the associated sheaf is what will be called **sheaf of microdistributions**. Clearly every distribution $f \in \mathcal{D}'(I\!\!R^n)$ defines a section of this sheaf whose support (in the sense of sheaf theory) is actually the wave-front set of f (we will see later on how the situation for hyperfunctions is quite analogous).

Remark 2.2.9 The reader will have certainly noticed how this analysis is all local in nature and therefore could be repeated for distributions defined on any differentiable manifold.

Remark 2.2.10 An analogous microlocalization process could have been applied to the space of pseudo-differential operators to build, on $T^*\mathbb{R}^n \setminus \{0\}$, the sheaf of pseudo-differential operators.

Now that the reader is acquainted with the infinitely differential case, we will try to see what can be done to modify this approach to fit the case of hyperfunctions. Historically, as we will discuss in the appendix to this chapter, there is a very interesting intertwining between differentiable analysis, analytic analysis, hyperfunction analysis and theoretical physics, and it is not always appropriate to discuss about priorities for concepts which have been arising almost simultaneously in different contexts and with different techniques.

The first remark one can make concerns the first step we took in defining the notion of singularity for a distribution; the key tool is obviously the fineness of the sheaf of distributions, actually of C^∞ functions, which allows us to localize the problem via the use of compactly supported infinitely differentiable functions. This same procedure is clearly unacceptable for the analytic case, since no compactly supported analytic functions exist. There are some ways around this which we will describe in our historical appendix, when dealing with Hörmander' s analytic wave-front set, but we now want to show how the Japanese school of Sato dealt with the problem.

Let us begin with the one variable case, and let therefore $f = [F]$ be a hyperfunction defined on the real line (or on an open subset of it); we know that we can really think of F as a pair of holomorphic functions, say (F^+, F^-), with $F^\pm$ holomorphic in some open sets of, respectively, the upper and the lower half plane. One may then observe that, in order for f to be real analytic, one needs that the "difference of the boundary values" of F^+ and F^- along the real axis is real analytic. One can easily show that for f to be real analytic it is necessary that f admits a representative F for which both F^+ and F^- extend analytically beyond the real axis. Here we see, therefore, that the obstruction to the analyticity of f is given by the obstruction to the holomorphic continuation of the two functions which represent f. Analyticity can therefore be described in terms of points of the real axis (where f may or may not be analytic) and the directions (along which the representatives can be analytically continued); in the one dimensional case there are only two such directions, a positive one (representing the upper half plane) and a negative one (representing the lower half plane). Thus (as before for the differentiability of distributions) the natural locus for the study of the analyticity of hyperfunctions is not the real line (or an open set U of the real line), but its cotangent bundle deprived of the zero section or, even better, the cosphere bundle S^*U. If we confine our attention to the single variable case, we see that S^*U is a trivial bundle whose fiber is S^0 or, better, $iS^0_\infty = \{i\infty, -i\infty\}$, and the following definitions are therefore quite natural:

Definition 2.2.15 *A hyperfunction f is* **microanalytic** *at $x_0 + i\infty$ if f has a representation $f = (F^+, F^-)$ in which F^+ extends analytically in a complex neighborhood of x_0.*

Definition 2.2.16 *A hyperfunction f is* **microanalytic** *at $x_0 - i\infty$ if f has a representation $f = (F^+, F^-)$ in which F^- extends analytically in a complex neighborhood of x_0.*

It is then immediate to reformulate our previous argument by saying that f is real analytic in x_0 if and only if it is microanalytic both at $x_0 + i\infty$ and $x_0 - i\infty$. It is also obvious that these definitions do not depend on the choice of the representative for f. Let us point out that being microanalytic also implies, for a given hyperfunction, the existence of a rather convenient representation, since we immediately have

Proposition 2.2.4 *If the hyperfunction f is microanalytic at the point $x_0 + i\infty$, then we can express it as the boundary value from below of a holomorphic function, i.e. there exists an open set V in the complex plane containing x_0 and a function F, holomorphic in U^- such that*

$$f(x) = F(x - i0),$$

at least in a neighborhood of x_0. The same statement (with obvious modifications) holds if $+i\infty$ is replaced by $-i\infty$.

Definition 2.2.17 *Let f be a hyperfunction on an open set U of the real line. The set of points of f where f is not microanalytic is called the "singular spectrum" of f and is denoted by $S.S.(f)$.*

By our definitions, $S.S.(f)$ is a subset of $S^*\mathbb{R} = \mathbb{R} \times iS^0_\infty$, while the singular support, sing supp(f) defined in Chapter I, is a subset of $\mathbb{R}$.

The following two results are an immediate consequence of the definition given above:

Proposition 2.2.5 *Let $\pi : S^*\mathbb{R} \to \mathbb{R}$ be the natural projection, and let $f \in \mathcal{B}(\mathbb{R})$. Then*

$$\pi(S.S.(f)) = \text{sing supp}(f).$$

Proposition 2.2.6 *Linear differential operators with real analytic coefficients do not enlarge the singular spectrum of a hyperfunction.*

Proof.   Immediate.   □

The notion of singular spectrum is the analog, for hyperfunctions, of the notion of wave-front set (this link will be completely explained in the appendix); one therefore expects some relevance with respect to the problem of the definition of the product of two hyperfunctions. This expectation is fulfilled by

Theorem 2.2.3 *Let f, g be hyperfunctions on $\mathbb{R}$, and assume that $S.S.(f)$ and $S.S.(g)$ have no antipodal points (i.e. there is no $(x,\xi) \in S^*\mathbb{R}$ such that $(x,\xi) \in S.S.(f)$ and $(x,-\xi) \in S.S.(g)$). Then the product $f \cdot g$ is well defined and*

(2.2.12) $$S.S.(f \cdot g) \subseteq S.S.(f) \cap S.S.(g).$$

Proof.   The problem of defining the product $f \cdot g$ is clearly local, and we therefore show how to define

$$f \cdot g(x)$$

under the condition that there is no $\xi \in iS^0_\infty$ for which

$$(x,\xi) \in S.S.(f), \quad (x,-\xi) \in S.S.(g).$$

Assume

$$f(x) = F^+(x + i0) - F^-(x - i0)$$

and

$$g(x) = G^+(x + i0) - G^-(x - i0)$$

with obvious meaning for the symbols. Since S^0_∞ only consists of two points we can consider the following cases:

(i) $x \notin \pi(S.S.(f))$, in which case we have no restrictions on g;

(ii) f is microanalytic in $x_0 + i\infty$, and g is, as well, microanalytic in $x_0 + i\infty$.

(There are of course two other cases which are, however, totally symmetric). In the first case, f is real analytic at x and therefore we have nothing to prove. In case (ii), both f and g can be represented (Proposition 2.2.4) as boundary values from below:

$$f(x) = F^-(x - i0),$$
$$g(x) = G^-(x - i0),$$

for suitable F^- , G^-. Then one defines

$$f \cdot g(x) := (F^- \cdot G^-)(x - i0).$$

It is immediate to verify that such product is well defined, and that (2.2.12) holds.   □

Remark 2.2.11 It is immediate to verify that such product satisfies all the usual algebraic properties of functional products.

Example 2.2.1

(i) $S.S.(\delta) = \{0\} \times iS^0_\infty$; indeed, $1/z$ does not have an analytic extension either way through the real axis;

(ii) $S.S.(\delta^{(n)}) = \{0\} \times iS^0_\infty$, for the same reason as in (ii);

(iii) $S.S.(H) = \{0\} \times iS^0_\infty$, since H is defined as the difference of the boundary values of $\log z$ (in the complex sense) for which no analytic continuation across the origin can be given;

(iv) $S.S.\left(\dfrac{1}{x \pm i0}\right) = \{0 \pm i\infty\}$, since $\dfrac{1}{x \pm i0}$ is actually the boundary value from one side.

Example 2.2.2 As a consequence of Theorem 2.2.3, and of Example 2.2.1 we obtain the well known fact that the Dirac delta cannot be multiplied by itself or by $1/(x \pm i0)$, while it makes sense to define

$$\left(\frac{1}{(x+i0)}\right)^m := \frac{1}{(x+i0)^m}.$$

When dealing with distributions, we have discussed (Proposition 2.2.2) the invariance of the notion of wave-front set with respect to diffeomorphism. We will now prove a similar result for the real analytic category. We still confine our attention to the case of one variable:

Definition 2.2.18 *Let U, U' be two open sets in $\mathbb{R}$; a real analytic function $\phi : U \to U'$ is said to be a **real analytic isomorphism** if it is a bijection and $\phi'(x) \neq 0$ at any point $x \in U$.*

Definition 2.2.19 *Let $\phi : U \to U'$ be a real analytic isomorphism, and let $f = [F^+, F^-] \in \mathcal{B}(U')$. Then the pull-back $\phi^* f$ is the hyperfunction on U defined by*

$$\phi^* f = (F^+(\phi), F^-(\phi)) \quad \text{if} \quad \phi'(x) > 0,$$

and by

$$\phi^* f = (-F^-(\phi), -F^+(\phi)) \quad \text{if} \quad \phi'(x) < 0.$$

It is an immediate consequence of the definition the fact that the following formulas hold:

$$\text{supp}(\phi^* f) = \phi^{-1}(\text{supp}(f))$$

$$\text{sing supp}(\phi^* f) = \phi^{-1}(\text{sing supp}(f))$$

$$S.S.(\phi^* f) = (\phi^{-1} x^t(\phi'))(S.S.(f)),$$

where

$$\phi^{-1} x^t(\phi') : U' \times iS_\infty^0 \to U \times iS_\infty^0$$

maps $(\phi(x), \xi)$ to $(x, \phi' \cdot \xi)$.

We are now on the brink of microlocalizing the sheaf of hyperfunctions, to obtain the sheaf of microfunctions. This will be done in sections 2.3 and 2.4; before doing so, however, we need to study the singularities of hyperfunctions of several variables.

Since, as we have seen, $\mathcal{B}$ is a sheaf, we recall the definition of the support of a hyperfunction f as the complement of the largest open set on which f vanishes, while its singular support is the complement of the largest open set on which f is real analytic.

We need to observe, however, that unlike what happens in the case of a single variable, it is not necessarily obvious to check whether or not a hyperfunction vanishes or is analytic in a neighborhood of a point. A good characterization can however be obtained as a (highly non-trivial) consequence of the flabbiness of the sheaf $\mathcal{B}$. We refer the reader to [103], but we quote here the result which is of interest to us:

Proposition 2.2.7 *Let U be an open set in $\mathbb{R}^n$, Γ a cone in $\mathbb{R}^n$ and let $f(x) \in \mathcal{B}(U)$ be represented as a unique boundary value*

$$f(x) = F(x + i\Gamma 0).$$

Then f is zero (or real analytic) in U if and only if F is zero (or analytically continuable) in U.

Proof. This Proposition is nothing but a very special case of the Edge of the Wedge Theorem.   □

We can now proceed to attempt to analyze in a microlocal fashion the notion of analyticity for a hyperfunction. Clearly, when only one variable is involved, analyticity (or lack of) can be interpreted in terms of continuation across the real axis, either from above or from below, so that only **two** directions are involved, and this is reflected in the fiber S_∞^0 of $S^* \mathbb{R}$. When the number of variables grows, however, S_∞^0 is replaced by S_∞^{n-1} and the spheric cotangent bundle becomes

$$S^* \mathbb{R}^n = \mathbb{R}^n \times iS_\infty^{n-1}.$$

Definition 2.2.20 *Let* $f \in \mathcal{B}(\mathbb{R}^n)$. *We say that* f *is* **microanalytic** *at* $(x, i\xi\infty) \in S^*\mathbb{R}^n$ *or, equivalently, at* $(x, \xi) \in \mathbb{R}^n \times S^{n-1}$ *if* f *admits a representation as sum of boundary values (see Chapter 1)*

$$f(x) = \sum_{j=1}^{N} F_j(x + i\Gamma_j 0)$$

with F_j *holomorphic functions defined on infinitesimal wedges*

$$\mathbb{R}^n + i\Gamma_j 0$$

such that, for any $j = 1, \dots, N$,

$$\Gamma_j \cap \{y \in \mathbb{R}^n : \langle \xi, y \rangle < 0\} \neq \emptyset.$$

The set of all points in $\mathbb{R}^n \times iS_\infty^{n-1}$ *(i.e. in* $S^*\mathbb{R}^n$*) where* f *is not microanalytic is called the* **singular spectrum** *of* f *and is denoted by* $S.S.(f)$.

Note that, by its very definition, $S.S.(f)$ is a closed subset of $S^*\mathbb{R}^n$ whose projection onto $\mathbb{R}^n$ should indeed be exactly sing supp(f). That this is the case is indeed once again a non-trivial consequence of the Edge of the Wedge Theorem and we quote it for future reference.

Proposition 2.2.8 *Let* $\pi : S^*\mathbb{R}^n \to \mathbb{R}^n$ *be the canonical projection. Then for every hyperfunction* f *on* $\mathbb{R}^n$ *one has*

$$\pi(S.S.(f)) = sing\ supp(f).$$

Note that, in general, $\pi^{-1}(\text{sing supp}(f))$ is much larger than $S.S.(f)$.

Example 2.2.3

(i) $S.S.(\delta) = \{0\} \times iS_\infty^{n-1}$;

(ii) $S.S.\left(\dfrac{1}{x_j + i0}\right) = \{x_j = 0\} \times \{\xi^{(j)}\}$, where $\xi^{(j)} = (0, 0, \dots, i\infty, \dots, 0)$, the $i\infty$ being at the j-th position;

(iii) $S.S.\left(\dfrac{1}{x_j + i0} + \dfrac{1}{x_k + i0}\right) = \{x_j = 0\} \times \{\xi^{(j)}\} \cup \{x_k = 0\} \times \{\xi^{(k)}\}$.

We would like to show that a product among hyperfunctions can be defined if the singular spectra are suitably placed with respect to each other.

To do so, we state the following result, whose proof is postponed to section 5 (see also in [103], [123]). We need a preliminary definition:

Definition 2.2.21 *If* Γ *is a subset of* $S^*\mathbb{R}^n$, *then the symbol* Γ^0 *denotes the set of its images under the map*

$$(x, \xi) \to (x, -\xi),$$

where $(x, \xi) \in \Gamma$.

Theorem 2.2.4 *Let* f, g *be hyperfunctions on* $\mathbb{R}^n$, *and suppose that*

$$S.S.(f) \cap (S.S.(g))^0 = \emptyset.$$

Then the product $f \cdot g$ *is well defined and it is*

$$supp(f \cdot g) \subseteq supp(f) \cap supp(g).$$

2.3 Microfunctions of One Variable

In this short section we will finally microlocalize the sheaf of hyperfunctions, much in the same way in which we have microlocalized the sheaf of distributions. This section is devoted to the case of one variable, in which all results and proofs are quite transparent.

We recall that in Chapter I we studied the singularities of hyperfunctions by looking at the (flabby) sheaf $\mathcal{B}/\mathcal{A}$ which is however, still a sheaf on $\mathbb{R}$ (or $\mathbb{R}^n$) and therefore cannot contain any information on the spectra of hyperfunctions.

The crucial idea, here, consists in pushing up $\mathcal{B}/\mathcal{A}$ from $\mathbb{R}$ to $S^*\mathbb{R}$. The standard way in which this is obtained goes through the introduction of preliminary sheaf of, we could say, microanalytic functions. To do so, consider the basis of open sets for $S^*\mathbb{R}$ given as follows: for U_1, U_2 open sets in $\mathbb{R}$, we consider their disjoint union

$$U_1 \amalg U_2 := (U_1 \times \{i\infty\}) \cup (U_2 \times \{-i\infty\}) \subseteq \mathbb{R} \times iS^0_\infty$$

which is clearly an open set in $S^*\mathbb{R}$ and is such that (for π again the canonical projection $\pi : S^*\mathbb{R} \to \mathbb{R}$)

$$(\pi^{-1}\mathcal{B})(U_1 \amalg U_2) = \mathcal{B}(U_1) \oplus \mathcal{B}(U_2).$$

Now we define on $S^*\mathbb{R}$ the sheaf of microanalytic functions by setting

$$\mathcal{A}^*(U_1 \amalg U_2) := \{f \in \mathcal{B}(U_1) : S.S.(f) \cap (U_1 \times \{i\infty\}) = \emptyset\} \oplus$$

$$\oplus \{f \in \mathcal{B}(U_2) : S.S.(f) \cap (U_2 \times \{-i\infty\}) = \emptyset\},$$

for any pair of open sets U_1, U_2 in $\mathbb{R}$. The reader will immediately note that an element of

$$\mathcal{A}^*(U_1 \amalg U_2)$$

is essentially a pair of boundary values of holomorphic functions, one from below, and the other from above. And so we give our fundamental definition:

Definition 2.3.1 *The sheaf $\mathcal{C}$ of* microfunctions *is defined by*

(2.3.1)
$$\mathcal{C} := \frac{\pi^{-1}\mathcal{B}}{\mathcal{A}^*}.$$

Note that since both $\pi^{-1}\mathcal{B}$ and $\mathcal{A}^*$ are sheaves on $S^*\mathbb{R}$, so is $\mathcal{C}$. Note also that by (2.3.1) we mean that $\mathcal{C}$ is the sheaf associated to the presheaf

$$\Gamma \to \frac{(\pi^{-1}\mathcal{B})(\Gamma)}{\mathcal{A}^*(\Gamma)}, \quad \Gamma \subset S^*\mathbb{R} \text{ open}.$$

Indeed, at least in general, the sheaf built as quotient of two sheaves is <u>not</u> a sheaf per se, but needs to sheafified. In our case, as it will become soon apparent, the situation is simpler (at least as long as we are dealing with just one variable).

Remark 2.3.1 It is an immediate consequence of the previous definition that the following short sequence of sheaves on $S^*\mathbb{R}$ is exact:

(2.3.2)
$$0 \to \mathcal{A}^* \to \pi^{-1}\mathcal{B} \to \mathcal{C} \to 0.$$

This sequence is known as Sato's fundamental sequence.

Theorem 2.3.1 $\mathcal{C}$ *is a flabby sheaf and, for any open set* $\Gamma \subset S^*\mathbb{R}$,

(2.3.3)
$$\mathcal{C}(\Gamma) = \frac{(\pi^{-1}\mathcal{B})(\Gamma)}{\mathcal{A}^*(\Gamma)}.$$

Proof.   There is, of course, a map

$$T : (\pi^{-1}\mathcal{B})(\Gamma) \to \mathcal{C}(\Gamma),$$

whose kernel is $\mathcal{A}^*(\Gamma)$. To prove (2.3.3) it will be enough to show that T is actually surjective. Note that the flabbiness of $\mathcal{C}$ follows immediately from (2.3.3) and the flabbiness of $\mathcal{B}$. Assume, to this purpose, that $\Gamma = U \times \{i\infty\}$ (this causes no loss of generality). The proof which follows is essentially a Mittag-Leffler type argument. Let then $\{U_j\}$ be a locally finite covering of U and consider $f \in \mathcal{C}(U)$. Suppose that f is defined locally as $f_j \mod \mathcal{A}^*(U_j \times \{i\infty\})$ for some $f_j \in \mathcal{B}(U_j)$. By the way we have defined $\mathcal{A}^*$, there are holomorphic functions F_j such that

$$f_j(x) = F_j(x + i0) \mod \mathcal{A}^*((U_j \times \{i\infty\})$$

and from the compatibility condition

$$f_{jk} = F_k(x + i0) - F_j(x + i0) \in \mathcal{A}^*((U_j \cap U_k) \times \{i\infty\})$$

we deduce that f_{jk} is in fact a real analytic function, and, therefore,

$$\{f_{jk}\} \in Z^1(U, \mathcal{A}).$$

Essentially by Mittag-Leffler Theorem we deduce the existence of functions g_j in $\mathcal{A}(U_j)$ such that

$$f_{jk}(x) = g_k(x) - g_j(x).$$

The global hyperfunction f on U can now be constructed by patching up the collection

$$F_j(x + i0) - g_j(x),$$

and this concludes the proof of the theorem. □

Remark 2.3.2 For any open set $U \subseteq \mathbb{R}$, one can define the spectral decomposition mapping

$$\mathrm{sp} : \mathcal{B}(U) \to \mathcal{C}(\pi^{-1}(U))$$

as follows

$$\mathcal{B}(U) \to (\pi^{-1}\mathcal{B})(\pi^{-1}(U)) = \mathcal{B}(U) \oplus \mathcal{B}(U) \to \mathcal{C}(\pi^{-1}(U))$$

$$f \to f \oplus f \to f \mathrm{mod}\ \mathcal{A}^*(U \times \{i\infty\}) \oplus f \mathrm{mod}\ \mathcal{A}^*(U \times \{-i\infty\}).$$

It is immediate to verify that this spectral decomposition mapping induces (globally) the following short exact sequences:

$$0 \to \mathcal{A}(U) \to \mathcal{B}(U) \xrightarrow{\mathrm{sp}} \mathcal{C}(\pi^{-1}(U)) \to 0$$

and (locally)

$$0 \to \mathcal{A} \to \mathcal{B} \to \pi_*\mathcal{C} \to 0,$$

where the last sequence of sheaves is on $\mathbb{R}$, and π_* denotes the push down (also called direct image) of a sheaf under π.

2.4 Microfunctions of Several Variables

In this section we show how to extend to the case of several variables the construction of the sheaf of microfunctions. We will not give all the necessary proofs, for which we refer the reader to [103], [123], [206].

Our first approach to the study of singularities of hyperfunctions of several variables was given in Theorem 1.5.4, where we proved that the sheaf $\mathcal{B}/\mathcal{A}$ is a flabby sheaf and that, for any open set $U \subseteq \mathbb{R}^n$,

$$\left(\frac{\mathcal{B}}{\mathcal{A}}\right)(U) = \frac{\mathcal{B}(U)}{\mathcal{A}(U)}.$$

In order to lift this sheaf to $S^*\mathbb{R}^n = \mathbb{R}^n \times iS^{n-1}_\infty$ we introduce, as we have already done in section 2.3, the sheaf $\mathcal{A}^*$ of microanalytic functions on $S^*\mathbb{R}^n$.

Definition 2.4.1 *The sheaf $\mathcal{A}^*$ is the subsheaf of $\pi^{-1}\mathcal{B}$ associated to the presheaf defined by*

$$S^*\mathbb{R}^n \supset U \times \Gamma \to \{f \in \mathcal{B}(U) : S.S.(f) \cap (U \times \Gamma) = \emptyset\}.$$

It is clear that the presheaf introduced in Definition 2.4.1 clearly represents hyperfunctions which are microanalytic in given directions; moreover, by the definition of inverse image of a sheaf, it also follows that if U and Γ are connected, then

$$\mathcal{A}^*(U \times \Gamma) = \{f \in \mathcal{B}(U) : S.S.(f) \cap (U \times \Gamma) = \emptyset\}.$$

We can define the sheaf of microfunctions as the quotient sheaf

$$\mathcal{C} := \frac{\pi^{-1}\mathcal{B}}{\mathcal{A}^*}$$

so that, by definition, the following is an exact sequence of sheaves on $S^*\mathbb{R}^n$:

(2.4.1) $$0 \to \mathcal{A}^* \to \pi^{-1}\mathcal{B} \to \mathcal{C} \to 0.$$

Now the germ of a microfunction at a point (x, ξ) is simply the residue class of a germ of a hyperfunction at the point x, modulo the germs of hyperfunctions which are microanalytic at (x, ξ). Note that $(\pi^{-1}\mathcal{B})_{x+i\xi\infty} = \mathcal{B}_x$.

Before being able to push down sequence (2.4.1) to $\mathbb{R}^n$ we will need to take a detour on the classical notion of "plane wave decomposition", essentially introduced by F. John in [101], and first used in connection with the theory of hyperfunctions by Kataoka in [95] (see also [103]). The classical starting point is the well known decomposition of the delta function as

(2.4.2) $$\delta(x) = \frac{(n-1)!}{(-2\pi i)^n} \int_{S^{n-1}} \frac{d\xi}{(x\xi + i0)^n}$$

which, of course, only makes sense in the framework of the theory of hyperfunctions. The classical, formal, proof for (2.4.2) is obtained as follows: by inverse Fourier transform one has

$$\delta(x) = \frac{1}{2\pi} \int_{\mathbb{R}^n} e^{ix\omega} d\omega.$$

Now write $\omega = r\xi$ (polar coordinates with $r > 0$, $\xi \in S^{n-1}$) and integrate by parts with respect to r to obtain

$$\frac{1}{2\pi} \int_{\mathbb{R}^n} e^{ix\omega} d\omega = \frac{1}{2\pi} \int_{S^{n-1}} d\xi \int_0^{+\infty} e^{ix\xi r} r^{n-1} dr =$$

$$= \frac{(n-1)!}{(-2\pi i)^n} \int_{S^{n-1}} \frac{d\xi}{(x\xi + i0)^n},$$

where in the last step we have used the fact that, for $\text{Im}(y) > 0$, it is

$$\int_0^{+\infty} e^{iyr} r^{n-1} dr = \frac{(n-1)!(-1)^n}{i^n y^n}.$$

It is clear that (2.4.2) provides an integral decomposition of the hyperfunction $\delta(x)$ in terms of the boundary values which represent it. In order to extend this idea to all hyperfunctions, one needs to introduce the kernel

$$W_0(z,\xi) := \frac{(n-1)!}{(-2\pi i)^n} \cdot \frac{1}{(z\xi)^n}$$

for $(z,\xi) \in \mathbb{C}^n \times \mathbb{C}^n$,

$$z\xi := z_1\xi_1 + \ldots + z_n\xi_n,$$

and its "twisted" version

$$W(z,\xi) := \frac{(n-1)!}{(-2\pi i)^n} \frac{(1-iz\xi)^{n-1} - (1-iz\xi)^{n-2}(z^2 - (z\xi)^2)}{(z\xi + i(z^2 - (z\xi)^2))^n},$$

where now ξ is a real unit vector.

Remark 2.4.1 The reason for introducing the twisted kernel (due to Kashiwara [107]) is that W_0 has singularities along the real hyperplane $x\xi = 0$, while the only real singularities of W appear at the origin.

Let now Γ be an open convex cone, and denote by Γ^0 its dual (hence a closed convex cone); set

$$W(z,\Gamma^0) := \int_{\Gamma^0 \cap S^{n-1}} W(z,\xi) d\xi.$$

It is easy to verify that $W(z,\Gamma^0)$ is a holomorphic function on an infinitesimal wedge of type $\mathbb{R}^n + i\Gamma 0$ which induces a hyperfunction $W(x,\Gamma^0)$ on $\mathbb{R}^n$. This hyperfunction allows us to give a precise (and generalizable) proof of (2.4.2) as follows:

Proposition 2.4.1 *Let Γ_j^0, $j = 1, \ldots, N$, be a covering of S^{n-1} such that the Γ_j^0's have no common interior. Then*

(2.4.3)
$$\delta(x) = \sum_{j=1}^N W(x, \Gamma_j^0).$$

Proof. First we note that the right hand side of (2.4.3) provides a uniquely defined hyperfunction on $\mathbb{R}^n$ regardless of the choice of the cones Γ_j^0. We can therefore choose the representation which makes our computations simplest. For

$$\sigma = (\sigma_1, \ldots, \sigma_n), \qquad \sigma_j = \pm 1,$$

define
$$\mathrm{sgn}(\sigma) := \sigma_1 \sigma_2 \dots \sigma_n$$
and the σ-th orthant
$$\Gamma_\sigma := \{\eta \in I\!\!R^n : \sigma_j \eta_j > 0, j = 1, \dots, n\}.$$
First consider the case of $W_0(z, \Gamma^0)$. Since
$$\frac{(n-1)!}{(-2\pi i)^n} \int_{\bar{\Gamma}_\sigma \cap S^{n-1}} \frac{d\xi}{(z\xi)^n} = (\frac{1}{2\pi})^n \int_{\bar{\Gamma}_\sigma \cap S^{n-1}} d\xi \int_0^{+\infty} e^{izr\xi} r^{n-1} dr =$$
$$= (\frac{1}{2\pi})^n \int_{\bar{\Gamma}_\sigma} e^{iz\omega} d\omega = \frac{\mathrm{sgn}(\sigma)}{(2\pi)^n} \int_0^{\sigma_1 \infty} e^{iz_1 \omega_1} d\omega_1 \dots \int_0^{\sigma_n \infty} e^{iz_n \omega_n} d\omega_n =$$
$$= \left(\frac{1}{2\pi}\right)^n \cdot \frac{\mathrm{sgn}(\sigma)}{z_1 \dots z_n},$$
we then obtain, by taking the boundary values, that
$$\sum_\sigma W_0(x, \bar{\Gamma}_\sigma) = \sum_\sigma \left(\frac{-1}{2\pi i}\right)^n \frac{\mathrm{sgn}(\sigma)}{z_1 \dots z_n}_{|z=x+i\Gamma_\sigma},$$
which, by definition, is exactly $\delta(x)$.

If we now look at the case of $W(z, \bar{\Gamma}_\sigma)$, one can see that, for suitable boundary integrals $B_\sigma(z)$, it is
$$W(z, \bar{\Gamma}_\sigma) = W_0(z, \bar{\Gamma}_\sigma) + B_\sigma(z),$$
where the $B_\sigma(z)$ can be explicitly computed (see [95], [103], [107]). When the sum over σ is taken, however, and the boundary values are considered, the terms $B_\sigma(x)$ add to zero and we therefore obtain
$$\sum_\sigma W(x, \bar{\Gamma}_\sigma) = \sum_\sigma W_0(x, \bar{\Gamma}_\sigma) = \delta(x).$$

$\square$

We are now ready for the proof of the following fundamental result:

Theorem 2.4.1 *Let U, Γ be open sets in $I\!\!R^n$ and iS_∞^{n-1} respectively. Then*

(2.4.4)
$$\mathcal{C}(U \times \Gamma) = \frac{\mathcal{B}(U)}{\mathcal{A}^\bullet(U \times \Gamma)}.$$

In particular, the sequence

(2.4.5)
$$0 \to \mathcal{A}(U) \to \mathcal{B}(U) \overset{\mathrm{sp}}{\to} \mathcal{C}(\pi^{-1}(U)) \to 0$$

is exact, and therefore
(2.4.6)
$$0 \to \mathcal{A} \to \mathcal{B} \to \pi_* \mathcal{C} \to 0$$

is an exact sequence of sheaves. Moreover, if f is a hyperfunction, then

(2.4.7)
$$S.S.(f) = \mathrm{supp}(\mathrm{sp}(f)).$$

Proof. We first note that all the assertions follow from (2.4.4). Indeed (2.4.5) is nothing but a restatement of (2.4.4) for $\Gamma = iS_\infty^{n-1}$. Similarly, (2.4.6) follows by (2.4.5) by taking inductive limits with respect to U. Finally, (2.4.7) follows from the definition of singular spectra of hyperfunctions.

We will therefore limit ourselves to the proof of (2.4.4) for the special case $\Gamma = S^{n-1}$, since the general case presents some additional technical difficulties. To do so, it will be sufficient to prove the exactness of

$$0 \to \mathcal{A}^*(U \times iS_\infty^{n-1}) \to \mathcal{B}(U) \to \mathcal{C}(U \times iS_\infty^{n-1}) \to 0$$

and in particular we only need to prove the surjectivity of

$$\mathcal{B}(U) \to \mathcal{C}(U \times iS_\infty^{n-1}).$$

Let then $f \in \mathcal{C}(U \times iS_\infty^{n-1})$. By definition of $\mathcal{C}$, one can consider the map

$$T : f \to \int_{\mathbb{R}^n} f(y) W(x - y, \xi) dy$$

which is easily seen to be a (non-surjective) sheaf homomorphism

$$T : \mathcal{C} \to \frac{\mathcal{B}}{\mathcal{A}}.$$

The image $T(f)$ belongs to $\mathcal{B}/\mathcal{A}(U \times iS_\infty^{n-1})$; let $g \in \mathcal{B}(U \times iS_\infty^{n-1})$ be a global representative of $T(f)$. Such a g exists in view of Theorem 1.5.4, and we will now show that the hyperfunction

$$g(x) := \int_{S^{n-1}} g(x, \xi) d\xi$$

is a representative for f in the quotient (2.4.4).
Take then $(x, \xi) \in U \times S^{n-1}$ and suppose f can be represented, in a neighborhood $U_0 \times S_0^{n-1}$ of (x, ξ) by a compactly supported hyperfunction $\tilde{f}$. Then the function

$$h(x, \xi) := g(x, \xi) - T(\tilde{f})$$

is analytic in $U_0 \times S_0^{n-1}$; by taking the boundary value we obtain that

$$g(x) - \tilde{f}(x)$$

is analytic and g is the required representative. $\qquad\square$

We conclude this section by stating, without a proof, the fundamental result on microfunctions due to Kashiwara (see e.g. [103] for a sketch of the proof):

Theorem 2.4.2 *The sheaf $\mathcal{C}$ of microfunctions is flabby.*

Remark 2.4.2   Before we begin a new section on microlocal operators, we will now give a more cohomological construction of the sheaf $\mathcal{C}$ of microfunctions: the reader will need to refer to our following chapters for some of the concepts we use here. Let us begin with the following diagram:

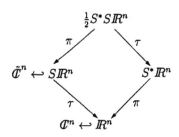

where $\frac{1}{2}S^*S\mathbb{R}^n$ is a half of the fiber product of $S\mathbb{R}^n$ and $S^*\mathbb{R}^n$ over $\mathbb{R}^n$ defined by

$$\frac{1}{2}S^*S\mathbb{R}^n = \{(x,\bar{\xi},\bar{\eta}) : \langle\xi,\eta\rangle \geq 0\},$$

$\bar{\xi} = x+i\xi\infty$, $\bar{\eta} = x+i\eta 0$, and where $\tilde{\mathbb{C}}^n$ is the disjoint union $\tilde{\mathbb{C}}^n = (\mathbb{C}^n\backslash\mathbb{R}^n)\sqcup S\mathbb{R}^n$, a blow-up in $\mathbb{C}^n$ along $\mathbb{R}^n$. Then $\tau^{-1}\mathcal{O}$ is a sheaf over $S\mathbb{R}^n$, and $S\mathbb{R}^n$ is purely 1-codimensional for $\tau^{-1}\mathcal{O}$, i.e. $\mathcal{H}^j_{S\mathbb{R}^n}(\tau^{-1}\mathcal{O}) = 0$ for $j \neq 1$, (see [123], Proposition 2.1.1). Moving clockwise in the above diagram, $\pi^{-1}\mathcal{H}^1_{S\mathbb{R}^n}(\tau^{-1}\mathcal{O})$ is a sheaf over $\frac{1}{2}S^*S\mathbb{R}^n$. Then the sheaf $\pi^{-1}\mathcal{H}^1_{S\mathbb{R}^n}(\tau^{-1}\mathcal{O})$ is purely $(n-1)$-codimensional for the map $\tau : \frac{1}{2}S^*S\mathbb{R}^n \to S^*\mathbb{R}^n$, that is $R^j\tau_*(\pi^{-1}\mathcal{H}^1_{S\mathbb{R}^n}(\tau^{-1}\mathcal{O})) = 0$ for $j \neq n-1$, (see [123], Proposition 2.1.2'). The abstract definition of $\mathcal{C}$ is therefore

$$\mathcal{C} = R^{n-1}\tau_*(\pi^{-1}\mathcal{H}^1_{S\mathbb{R}^n}(\tau^{-1}\mathcal{O}))$$

where the symbol $R^j\tau_*$ denotes the higher direct image functor.

From this observation, one may expect the sheaf $\mathcal{C}$ to be somehow the sheaf of "cotangential hyperfunctions". Then the definition should be $\mathcal{H}^n_{S^*\mathbb{R}^n}(\pi^{-1}\mathcal{O})$. As we will see, the above definition of $\mathcal{C}$ coincides with $\mathcal{H}^n_{S^*\mathbb{R}^n}(\pi^{-1}\mathcal{O})$. To verify this, we apply Leray spectral sequence to the projection map $S^*\mathbb{R}^n \times \tilde{\mathbb{C}}^n \to \tilde{\mathbb{C}}^n$. That is, for $\tilde{U}$ and V^* open sets in $\tilde{\mathbb{C}}^n$ and $S^*\mathbb{R}^n$, respectively, we consider the induced Leray spectral sequence of relative cohomology

$$E_2^{p,q} = H^p_{S\mathbb{R}^n}(\tilde{U}, R^q\tau_*(\pi^{-1}(\tau^{-1}\mathcal{O})))$$

abutting to $H^n_{S\mathbb{R}^n}(\tilde{U}\times V^*, \pi^{-1}(\tau^{-1}\mathcal{O}))$. Since V^* can be taken to be contractible,

$$H^p(V^*, \tau^{-1}\mathcal{O}) = 0 \quad \text{for } p \neq 0.$$

Hence

$$R^q \pi_*(\pi^{-1}(\tau^{-1}\mathcal{O})) = 0 \quad \text{for} \quad q \neq 0$$

and

$$R^0 \pi_*(\pi^{-1}(\tau^{-1}\mathcal{O})) \approx \tau^{-1}\mathcal{O}.$$

Therefore the above $E_2^{p,q}$-term vanishes for $q \neq 0$. Namely,

$$E_2^{p,0} = H_{S\mathbb{R}^n}^p(\tilde{U}, \tau^{-1}\mathcal{O}).$$

Furthermore, the purely 1-codimensionality of $S\mathbb{R}^n$ for $\tau^{-1}\mathcal{O}$ implies the following isomorphism:

$$\pi^{-1}\mathcal{H}_{S\mathbb{R}^n}^1(\tau^{-1}\mathcal{O}) \overset{\approx}{\to} \mathcal{H}_{\pi^{-1}(S\mathbb{R}^n)}^1(\pi^{-1}(\tau^{-1}\mathcal{O})) \approx \mathcal{H}_{\frac{1}{2}S^*S\mathbb{R}^n}^1(\pi^{-1}(\tau^{-1}\mathcal{O})).$$

In order to obtain the desired isomorphism, we need to compute the higher direct image

$$R^p \tau_*(\mathcal{H}_{\frac{1}{2}S^*S\mathbb{R}^n}^1(\pi^{-1}(\tau^{-1}\mathcal{O}))).$$

Since, from Remark 3.2.1, we have

$$0 = E_2^{p-2,2} \to E_2^{p,1} \to E_2^{p+2,0} = 0,$$

we deduce that the abutment $R^{p+1}(\tau_* \mathcal{H}_{\frac{1}{2}S^*S\mathbb{R}^n}^0)(\pi^{-1}(\tau^{-1}\mathcal{O}))$ is isomorphic to $E_\infty^{p,1} \approx E_2^{p,1}$. As we noted, $\pi^{-1}\mathcal{H}_{S\mathbb{R}^n}^1(\tau^{-1}\mathcal{O})$ is purely $(n-1)$-codimensional for τ. Hence we need to consider only $p = n - 1$ in the above $E_2^{p,1}$-term. Then the abutment becomes

$$R^n(\tau_* \mathcal{H}_{\frac{1}{2}S^*S\mathbb{R}^n}^0)(\pi^{-1}(\tau^{-1}\mathcal{O})).$$

Since $\frac{1}{2}S^*S\mathbb{R}^n$ is $\tau^{-1}(S^*\mathbb{R}^n)$, we have

$$R^n(\tau_* \mathcal{H}_{\frac{1}{2}S^*S\mathbb{R}^n}^0)(\pi^{-1}(\tau^{-1}\mathcal{O})) \approx R^n(\mathcal{H}_{S^*\mathbb{R}^n}^0 \tau_*)(\tau^{-1}(\pi^{-1}\mathcal{O})),$$

where $(\pi^{-1}(\tau^{-1}\mathcal{O})) \approx (\tau^{-1}(\pi^{-1}\mathcal{O}))$. Consequently, we have

$$R^n \mathcal{H}_{S^*\mathbb{R}^n}^0(\pi^{-1}\mathcal{O}) = \mathcal{H}_{S^*\mathbb{R}^n}^n(\pi^{-1}\mathcal{O}),$$

obtaining the isomorphism

$$\mathcal{C} = R^{n-1}\tau_*(\pi^{-1}\mathcal{H}_{S\mathbb{R}^n}^1(\tau^{-1}\mathcal{O})) \approx \mathcal{H}_{S^*\mathbb{R}^n}^n(\pi^{-1}\mathcal{O}).$$

2.5   Microlocal Operators

In Chapter I we have briefly discussed infinite order differential operators, and their action on the sheaf of hyperfunctions. As we have mentioned in our historical notes, the study of infinite order differential operators (and their inversion) is a classical attempt to reduce analysis to algebra and to create some sort of operational calculus. As we shall see in our subsequent chapters the reason for the introduction of infinite order differential operators is even stronger when one deals with the classification of systems of differential equations. In this section we will go beyond infinite order differential operators and we will discuss a larger class of operators, namely those operators which act as endomorphisms on the sheaf $\mathcal{C}$ of microfunction; such operators will be called microlocal operators.

The reader who has followed us up to now will have no difficulty in recognizing (under this new name) the pseudo-differential operators discussed earlier on in this chapter. As usual, we begin by treating the one-variable case, where things are more explicit. We will then move on to case of several variables.

In Chapter I we have seen that if we denote by D the differentiation operator

$$D = \frac{d}{dx},$$

then it is possible to consider infinite series of the form

$$\sum_{k=0}^{+\infty} a_k(x)D^k, \quad a_k \in \mathcal{A}$$

as long as

$$\lim_{k \to \infty} \sqrt[k]{\sup_{z \in K} |a_k(z)| k!} = 0$$

on any compact K, where $a_k(z)$ are holomorphic extensions of $a_k(x)$. The consideration of these operators has already showed how much richer is the theory of hyperfunctions in comparison with the theory of distributions. Our next step consists in defining and studying negative powers of D.

Let us remind the reader, here, that we have shown, essentially in Chapter I, that infinite order linear differential operators with real analytic coefficients preserve microanalyticity, and they therefore define not only an endomorphism of the sheaf $\mathcal{B}$ but also an endomorphism of the sheaf $\mathcal{C}$. Moreover, these operators are defined on an object $[f]$ in $\mathcal{C}$ by considering their action on a hyperfunction f which represents $[f]$, and, finally, their action on f is defined by having the operator act on a holomorphic representative F of f. We will work in a similar way to define negative powers of D.

Let U be an open set in the complex plane, and let $F \in \mathcal{O}(U)$; then we can define

$$D^{-1}F(z) := \int_a^z F(z)dz$$

where $a \in U$, and the definition of $D^{-1}F(z)$ does not depend on the choice of the path γ connecting a to z, as long as U is simply connected and $\gamma \subseteq U$. This definition readily extends to hyperfunctions: let $f(x) = F_+(x+i0) - F_-(x-i0)$ be a hyperfunction which is real analytic in a neighborhood of a point a. Then one can define

$$(2.5.1) \quad D^{-1}f := \int_a^x f(x)dx := \left[\int_{\gamma_+(z)} F^+(z)dz \right] - \left[\int_{\gamma_-(z)} F_-(z)dz \right],$$

where $\gamma_\pm(z)$ are paths connecting a and z and contained, respectively, in the upper and lower half-plane (with the exception of a).

Note that $D^{-1}f$ is therefore the hyperfunction defined as the difference of the two equivalence classes on the right hand side of (2.5.1). Note that the necessity of requesting analyticity at a point a stems from the way in which we have defined the integrals of hyperfunctions. More generally, if a is singular for f we can always use the flabbiness of the sheaf of hyperfunctions (see Proposition 1.3.2) and decompose

$$f = f_1 + f_2$$

with $\operatorname{supp}(f_1) \subseteq \{x : x \le a\}$ and $\operatorname{supp}(f_2) \subseteq \{x : x \ge a\}$. This method allows us to define an operator D^{-1} which is defined (at the level of germs) on $\mathcal{B}$ modulo $\mathcal{A}$, so that we have

$$D^{-1} : \frac{\mathcal{B}}{\mathcal{A}} \to \frac{\mathcal{B}}{\mathcal{A}}.$$

Such an operator, on the other hand, is perfectly well defined on the sheaf of microfunctions, since the sheaf $\mathcal{C}$ already incorporates the quotient by $\mathcal{A}$, and so

$$D^{-1} : \mathcal{C} \to \mathcal{C}$$

act as a sheaf homomorphism. One is therefore naturally lead to the consideration of formal operators such as

$$\sum_{k=1}^{+\infty} b_k(z)D^{-k}$$

and to the conditions which one must impose on the sequence $\{b_k(z)\}$ of holomorphic functions for such operators to be well defined. In complete analogy with what we proved in Chapter I for the case of (positive) infinite order differential operators, we can show the following result:

Proposition 2.5.1 *Consider the formal operator*

$$Q(z, D) := \sum_{k=1}^{+\infty} b_k(z)D^{-k}$$

where the coefficients $b_k(z)$ are holomorphic in an open set U of the complex plane. Then the following two conditions are equivalent:

(i) for any $z_0 \in U$, there exists $\delta > 0$ such that for any $F \in \mathcal{O}(V)$, with V a convex set such that $z_0 \in V \subseteq \{|z - z_0| \leq \delta\}$, the series

$$\sum_{k=1}^{+\infty} b_k(z) D^{-k} F(z)$$

converges locally uniformly to an element in $\mathcal{O}(V)$;

(ii) for every compact set $K \subseteq U$,

$$(2.5.2) \qquad \limsup_{k \to \infty} \sqrt[k]{\sup_{z \in K} |b_k(z)| k!} < +\infty.$$

Proof. Just an application of Cauchy's formula as in Proposition 1.2.2. □

Remark 2.5.1 What greatly distinguishes Proposition 2.5.1 from Proposition 1.2.2 (and operators such as $Q(z, D)$ from the usual infinite orders differential operators) is the fact that

$$P(z, D) := \sum_{k=0}^{+\infty} a_k(z) D^k$$

acts as a sheaf homomorphism on the sheaf $\mathcal{O}$ and therefore, as a consequence, on the various sheaves $\mathcal{A}$, $\mathcal{B}$, $\mathcal{C}$. This is absolutely not the case with an operator such as $Q(z, D)$. In particular, even if F were holomorphic on all of U, the series

$$\sum b_k(z) D^{-k} F(z)$$

would still not be necessarily convergent for $z \neq z_0$. As we will see in a moment, however, $Q(z, D)$ still defines a sheaf homomorphism on $\mathcal{B}/\mathcal{A}$ and on $\mathcal{C}$ and the interest of inverting differential operators is therefore one more reason for introducing the notion of microfunctions.

Definition 2.5.1 *A* **pseudo-differential operator,** *or* **microdifferential operator** *is a formal series*

$$Q(z, D) := \sum_{-\infty}^{+\infty} b_k(z) D^k$$

where $\{b_k(z)\}_{-\infty}^{-1}$ satisfies (2.5.2) and $\{b_k(z)\}_{k=0}^{+\infty}$ satisfies (1.2.6). Its **symbol** *is the formal power series*

$$Q(z, \zeta) := \sum_{-\infty}^{+\infty} b_k(z) \zeta^k.$$

Finally, if $b_k(z) \equiv 0$ for $k > m$ and $b_m(z) \not\equiv 0$, we will say that $Q(z, D)$ is of **order** *m.*

Theorem 2.5.1 *Let $Q(z, D)$ be a microdifferential operator of order less than or equal to zero, defined in a complex neighborhood of a real open set U. Then Q defines a sheaf endomorphism on $\mathcal{B}/\mathcal{A}$ and on $\mathcal{C}$.*

Proof. Let $x \in U$ and let f be a germ of a hyperfunction which represents an element $[f]$ in $\mathcal{C}_{(x,\xi)}$ (a similar argument applies to the case of $(\mathcal{B}/\mathcal{A})_x$, and set

$$f(x) = F^+(x + i0) - F^-(x - i0)$$

for some pair of holomorphic functions (F^+, F^-) defined near x. To define

$$Q(x, D)[f],$$

we apply $Q(z, D)$ to both F^+ and F^- (note that the symbol D in $Q(x, D)$ represents D_x while the one in $Q(z, D)$ represents D_z) by choosing initial points of integration $z_0^{\pm}$ sufficiently close to x and within the domain of holomorphy of $F^{\pm}$. We therefore set

$$Q(x, D)[f] := [Q(z, D)F^+(x + i0) - Q(z, D)F^-(x - i0)],$$

where in particular

$$Q(z, D)F^{\pm} = \sum_{k=1}^{+\infty} b_k(z) \underbrace{\int_{z_0^{\pm}}^{z} \cdots \int_{z_0^{\pm}}^{z}}_{k\,\text{times}} F^{\pm}(z)dz.$$

It is obvious that $Q(z, D)F^{\pm}$ are well defined holomorphic functions, but may lose meaning as z moves away from the point x. In this case, for $Q(z, D)F^{\pm}$ to be still defined, one needs to replace $z_0^{\pm}$ with two new points $z_1^{\pm}$. The difference between the two representations can be computed by integration by parts and is given by

$$\sum_{k=1}^{+\infty} b_k(z) \underbrace{\int_{z_0^{\pm}}^{z_1^{\pm}} dz \cdots \int_{z_0^{\pm}}^{z_1^{\pm}}}_{k\,\text{times}} F^{\pm}(z)dz =$$

$$= \sum_{k=1}^{\infty} b_k(z) \int_{z_0^{\pm}}^{z_1^{\pm}} \frac{(z - \zeta)^{k-1}}{(k - 1)!} F^{\pm}(\zeta)d\zeta$$

which, on $\mathbb{R} \cap \{|z - z_0^{\pm}| \leq \delta\} \cap \{|z - z_1^{\pm}| \leq \delta\}$, gives a real analytic function. This concludes the proof. $\square$

We now proceed to deal with the case of several variables and, in so doing, we will consider a larger class of operators, namely the so called microlocal operators (which are more general than microdifferential operators). We need first to recall a few preliminary results on the product and integration of hyperfunctions (see Theorem 2.2.4) and microfunctions.

Proposition 2.5.2 *Let f, g be two hyperfunctions on $\mathbb{R}^n$, and assume*

$$S.S.(f) \cap (S.S.(g))^0 = \emptyset.$$

Then the product $f \cdot g$ is well defined and

(2.5.3) $S.S.(f \cdot g) \subseteq \{(x, \lambda \xi + (1 - \lambda)\eta) : (x, \eta) \in S.S.(g), 0 \leq \lambda \leq 1\}.$

Proof.   Since $\mathcal{B}$ is a sheaf, it is sufficient to look at a sufficiently small neighborhood U of a point x. The first step consists in decomposing the singular spectra of f and g in such a way that we can write (with obvious meaning of the symbols)

$$f(x) = \sum_{j=1}^{m_1} F_j(x + i\Gamma_j 0),$$

$$g(x) = \sum_{k=1}^{m_2} G_k(x + i\Delta_k 0),$$

and

(2.5.4) $\Gamma_j^0 \cap -\Delta_k^0 = \emptyset$ for any pair (j, k).

On the other hand, (2.5.4) immediately implies that, for any pair (j, k), it is

$$\Gamma_j \cap \Delta_k \neq \emptyset,$$

and so a product between f and g can be defined as

(2.5.5) $f(x) \cdot g(x) := \sum_{j,k} F_j(x + i\Gamma_j 0)G_k(x + i\Delta_k 0) =$

$$= \sum_{j,k} (F_j G_k)(x + i(\Gamma_j \cap \Delta_k)0).$$

It is possible to show that the definition given by (2.5.5) is independent of the choice of the decomposition for f and g; moreover, from the right-hand side of (2.5.5) we deduce that

$$S.S.(f \cdot g) \subseteq \bigcup_{j,k} \left(U \times i(\Gamma_j \cap \Delta_k)^0 \infty\right) = \bigcup_{j,k} \left(U \times i(\Gamma_j^0 + \Delta_k^0)\infty\right),$$

which finally concludes the proof.   □

Remark 2.5.2 The right-hand side of (2.5.3) is what could be called the "convex linear combination of $S.S.(f)$ and $S.S.(g)$ with respect to fibers", i.e. at each fiber $\{x\} \times iS_\infty^{n-1}$ one takes the union of all shortest arcs connecting all pairs (x, ξ), (x, η) with $(x, \xi) \in S.S.(f)$, $(x, \eta) \in S.S.(g)$; in particular we note that such shortest arcs are uniquely defined for any pair $((x, \xi), (x, \eta))$ in view of the non antipodality of $S.S.(f)$ and $S.S.(g)$. For future reference we will denote such a set by $S.S.(f) \vee S.S.(g)$.

Proposition 2.5.2 can be easily translated in the language of microfunctions if we remember that a microfunction is an equivalence class of hyperfunctions, and that the support of a microfunction is really the singularity spectrum of any representative hyperfunction:

Corollary 2.5.1 *Let $[f]$, $[g]$ be germs of microfunctions at (x_0, ξ_0) and (x_0, η_0) respectively. If, for sufficiently small (but otherwise arbitrary) neighborhoods Γ_0 of ξ_0 and Δ_0 of η_0, the shortest arcs through ξ_0, η_0 does not intersect*

$$\{(\{x_0\} \times \partial \Gamma_0) \cap \operatorname{supp}[f]\} \vee \{(\{x_0\} \times \partial \Delta_0) \cap \operatorname{supp}[g]\},$$

then $[f] \cdot [g]$ is a well defined microfunction in a neighborhood of the set

$$\{(x_0, \xi_0) \vee (x_0, \eta_0)\} \setminus (\operatorname{supp}[f] \cup \operatorname{supp}[g]).$$

We have seen in Chapter I that (under some mild regularity condition) it is possible to define the definite integral of a hyperfunction on compact sets with smooth boundaries. However, we will now need to define microlocal operators in the same way in which pseudo-differential operators were defined in the $\mathcal{C}^\infty$ setting; this means that we need to define them as integral operators with microfunction kernels; to this purpose we need not only a notion of product for microfunctions (Corollary 2.5.1), but also a notion of integration along fibers (rather than just plain integration). We begin by quoting the following weak version of the Edge of the Wedge Theorem (see the historical appendix to Chapter I):

Proposition 2.5.3 *(Martineau). Let F_j, $j = 1, \ldots, N$, be holomorphic functions which, in a neighborhood of a compact set K, satisfy*

$$\sum_{j=1}^{N} F_j(x + i\Gamma_j 0) = 0,$$

for suitable cones $\Gamma_1, \ldots, \Gamma_N$. Then, for any proper subcone

$$\Delta_{jk} \subseteq \Gamma_j + \Gamma_k,$$

there exist

$$H_{jk} \in \mathcal{O}(K + i\Delta_{jk} 0)$$

such that $H_{jk} = -H_{kj}$ and

$$F_j(z) = \sum_{k=1}^{N} H_{jk}(z).$$

We now prove the existence of integration along fibers for hyperfunctions.

Theorem 2.5.2 *Let $f(x, y)$ be a hyperfunction of $n + m$ variables (x, y) on an open set $U \times V \subseteq \mathbb{R}^n \times \mathbb{R}^m$, and let $\pi_x : U \times V \to U$ be the canonical projection; let $K \subseteq V$ be a domain with piecewise smooth boundary and let f be real analytic on a neighborhood of $U \times \partial K$. Then one can give a good definition of*

$$\int_K f(x, y) dy$$

as an element of $\mathcal{B}(U)$. Moreover

$$\operatorname{supp}\left(\int_K f(x, y) dy\right) \subseteq \pi_x(\operatorname{supp}(f))$$

and

$$S.S.\left(\int_K f(x, y) dy\right) \subseteq \{(x, \xi) : (x, y, \xi) \in S.S.(f) \text{ for some } y\}.$$

Proof.   Once again, since $\mathcal{B}$ is a sheaf, we can proceed locally and assume U to be a sufficiently small neighborhood of a point x; we then represent f as

$$f(x, y) = \sum_{j=1}^{N} F_j\left((x, y) + i\Gamma_j 0\right),$$

with F_j analytically continuable to a neighborhood of $U \times \partial K$. Now set $\tilde{\Gamma}_j := \pi_x(\Gamma_j)$ and consider infinitesimal wedges $U + i\tilde{\Gamma}_j 0 \subseteq \pi_x((U \times K) + i\Gamma_j 0)$. In accordance to the definition of definite integral for a hyperfunction consider, for $z \in U + i\tilde{\Gamma}_j 0$, a piecewise smooth continuous map $\varphi_j(z)$ such that

$$\begin{cases} \varphi_j(t) = 0 & \text{for } t \in \partial K \\ (z, t + i\varphi_j(t)) \in (U \times K) + i\Gamma_j 0 & \text{for } t \in \overset{\circ}{K}. \end{cases}$$

Then one defines

$$G_j(z) := \int_{K + i\varphi_j} F_j(z, \eta) d\eta$$

which is holomorphic in $U \times i\tilde{\Gamma}_j 0$, and

(2.5.6) $$\int_K f(x, y) dy := \sum_{j=1}^{N} G_j(x + i\Gamma_j 0).$$

The estimates on the support and the singular spectrum of $\int_K f(x, y) dy$ are an immediate consequence of (2.5.6), while Proposition 2.5.3 allows one to show that the definition in (2.5.6) is well posed and does not depend on the various choices made during the construction of (2.5.6).   □

This result has a "microfunction translation" as follows:

Theorem 2.5.3 *Let $[f(x,y)]$ be a microfunction in (x,y) defined in a neighborhood of*

$$\{((x,y),(\xi,\eta)) : x = x_0, y \in V, \xi = \xi_0 \neq 0, \eta = 0\}$$

and with compact support with respect to y. Then the integral

$$\int f(x,y)dy$$

defines a microfunction of the variable x near (x_0,ξ_0).

This result has an immediate, and fundamental consequence (see [103], Chapter III):

Theorem 2.5.4 *Let $U \subseteq S^*\mathbb{R}^n$ be an open set, and let $K = K(x,y)$ be a microfunction defined in $U \times U^a$ such that*

$$\mathrm{supp}(K) \subseteq \{(x,y;(\xi,\eta)) : x = y, \xi = -\eta\}.$$

Then the map

$$T_K : f(y) \rightarrow \int K(x,y)f(y)dy$$

defines a sheaf endomorphism on $\mathcal{C}$ above U.

This result (see section one of this chapter) allows us to give the following definition:

Definition 2.5.2 *A sheaf homomorphism*

$$T_K : \mathcal{C} \rightarrow \mathcal{C}$$

defined as in Theorem 2.5.4 is a **microlocal operator** *and the microfunction $K(x,y)$ is its* **kernel**.

Definition 2.5.3 *If T_K is a microlocal operator such that K is a hyperfunction such that $\mathrm{supp}(K) \subseteq \{(x,y;\xi,\eta) : x = y\}$, then we say that T_K is a* **local operator**.

Remark 2.5.3 In view of Theorems 2.5.2 and 2.5.3, it is clear that a local operator actually acts as a sheaf homomorphism on the sheaf $\mathcal{B}$ of hyperfunctions.

Example 2.5.1 Even for $n = 1$, microlocal operators are more general than local ones. Consider, for example, the kernel hyperfunction

$$K(x,y) = \frac{1}{x - y + i0};$$

then the map

$$u(y) \to -\frac{1}{2\pi i} \int \frac{u(y)}{x - y + i0} dy$$

is a sheaf homomorphism of C in itself at $\{(x, i\infty)\}$, where it acts as the identity map, since

$$\delta(x) = -\frac{1}{2\pi i} \left(\frac{1}{x + i0} - \frac{1}{x - i0} \right).$$

Note that if we consider K as a hyperfunction, $\mathrm{supp}(K) = \mathbb{R}^2$, and $S.S.(K) = \{x = y, \xi = -\eta\}$.

Example 2.5.2 An important class of local operators is given by differential operators

$$P(x, D)$$

whose kernel is the hyperfunction

$$P(x, D)\delta(x - y),$$

supported at $x - y$. The same applies to infinite order differential operators.

Example 2.5.3 On the other hand, operators such as D^{-1} (in one variable) or D_j^{-1} for $j = 1, \ldots, n$ (in the case of several variables) are definitely not local, as their kernels are of the form

$$\text{const.} \quad \int \frac{dy}{y_j(x \cdot y + i0)^{n-1}},$$

where $y = (y_1, \ldots, y_n)$, $x = (x_1, \ldots, x_n)$.

We have therefore shown the following sequence of sheaf inclusions

(2.5.7) $\mathcal{P} \hookrightarrow \mathcal{P}_\infty \hookrightarrow \mathcal{L}oc \hookrightarrow \mathcal{L}$

for $\mathcal{P}$ the sheaf of (analytic coefficients) differential operators, $\mathcal{P}_\infty$ the sheaf of infinite order (analytic coefficients) differential operators, $\mathcal{L}oc$ the sheaf of local operators and $\mathcal{L}$ the sheaf of microlocal operators. The reader can easily verify that $\mathcal{P}$ and $\mathcal{P}_\infty$ are sheaves of operators; the case of $\mathcal{L}oc$ and $\mathcal{L}$ requires some more sheaf theory and the reader is referred to [123] for details.

Sequence (2.5.7), though interesting, fails to find an appropriate location for what we have already introduced, in the case of one variable, as the sheaf of microdifferential operators.

There are several reasons, both theoretical and practical, for the introduction of this special class of operators. On one hand, manipulation of microlocal operators is extremely complicated, on the other hand, microdifferential operators appear very naturally. As we shall show in Chapter VI, Sato's Fundamental

Principle claims that if the principal symbol $P_m(x,\xi)$ of a differential operator $P(x,D)$ of order m does not vanish at a point (x,ξ), then P can be inverted and its inverse is what we will call a microdifferential operator. More generally, one would like to identify a specific class of microlocal operators for which the inverse exists and for which a truly algebraic treatment is possible.

The starting point is once again John's plane wave expansion formula; if

$$P(x,D) = \sum a_\alpha(x)D^\alpha$$

is an (analytic coefficients) differential operator with $\alpha = (\alpha_1,\ldots,\alpha_n)$ a non-negative multi-index, we have already seen that $P(x,D)$ is a microlocal operator with kernel given by

$$K(x,y) = P(x,D)\delta(x-y).$$

But if we now rewrite the plane wave expansion (2.4.2) for $\delta(x-y)$ we obtain

$$(2.5.8) \qquad \delta(x-y) = \frac{(n-1)!}{(-2\pi i)^n} \int \frac{d\xi}{((x-y)\xi+i0)^n}.$$

Apply $P(x,D)$ (where D is obviously the multi-derivative with respect to x) to obtain

$$K(x,y) = \left(\frac{1}{-2\pi i}\right)^n \int \sum_j \frac{(-1)^j(j+n-1)!p_j(x,\xi)}{((x-y)\xi+i0)^{n+j}}d\xi$$

where

$$p_j(x,\xi) := \sum_{|\alpha|=j} a_\alpha(x)\xi^\alpha$$

is obviously a homogeneous polynomial of degree j in ξ.

Microdifferential operators arise when we generalize the p_j's to holomorphic functions.

Definition 2.5.4 Definition 2.5.4. *Let U be an open set of $\mathbb{C}^n \times \mathbb{C}^n$, and for any $\lambda \in \mathbb{C}$, $j \in \mathbb{Z}$ consider a holomorphic function on U*

$$p_{\lambda+j}(z,\zeta)$$

satisfying:

(i)

$$\sum_{i=1}^n \zeta_i \frac{\partial(p_{\lambda+j}(z,\zeta))}{\partial\zeta_i} = (\lambda+j)p_{\lambda+j}(z,\zeta)$$

(We say, with a language abuse, that $p_{\lambda+j}$ is homogeneous of degree $\lambda+j$ in ζ);

(ii) *for every compact set* $K \subseteq U$, *there exists a positive constant* C_K *such that, for all* $j < 0$ *and all* $(z, \zeta) \in K$,

$$|p_{\lambda+j}(z, \zeta)| \le C_K^{-j}(-j)!;$$

(iii) *for every compact set* $K \subseteq U$ *and every* $\varphi > 0$, *there exists a positive constant* $C_{K,\varphi}$ *such that, for all* $j \ge 0$ *and all* $(z, \zeta) \in K$,

$$|p_{\lambda+j}(z, \zeta)| \le \frac{1}{j!} C_{K,\varphi} \varphi^j.$$

We denote by $\mathcal{E}_{(\lambda)}^{\infty}(U)$ *the totality of sequences* $\{p_{\lambda+j}(z, \zeta)\}$ *satisfying* (i), (ii), (iii) *and we call*

$$P(z, D) := \sum_j p_{\lambda+j}(z, D)$$

a **microdifferential operator** *on* U. *When* U *ranges over a cofinal covering of* $\mathbb{C}^n \times \mathbb{C}^n$, $\{\mathcal{E}_{(\lambda)}^{\infty}(U)\}$ *originates a sheaf, denoted by* $\mathcal{E}_{(\lambda)}^{\infty}$ *and called the sheaf of* **microdifferential operators of infinite order**. *We denote by* $\mathcal{E}(\lambda)$ *the sheaf of operators for which* $p_{\lambda+j} = 0$ *if* $j > 0$ *and by* $\mathcal{E}_{(\lambda)}$ *the union* $\cup_{j \in \mathbb{Z}} \mathcal{E}(\lambda + j)$. *Finally an element of* $\mathcal{E}_{(\lambda)}$ *is a* **microdifferential operator of finite order**, *while* $\mathcal{E}(\lambda)$ *is a* **microdifferential operator of order at most** λ. *When* $\lambda = 0$, *we will simply suppress it from the notation.*

Through Definition 2.5.4, we have introduced what we have called microdifferential operators, but from the definition it is not obvious at all that the new sheaf $\mathcal{E}_{(\lambda)}^{\infty}$ is a subsheaf of $\mathcal{L}$, the sheaf of microlocal operators.

To do so, one introduces the following sequence $\{\Phi_j(\lambda)\}$ of holomorphic functions of one variable:

$$\Phi_j(\lambda) = \begin{cases} \dfrac{(-1)^j \Gamma(j)}{(-2\pi i)^n} \lambda^{-j} & \text{for } j > 0 \\[3mm] \dfrac{1}{(-2\pi i)^n \Gamma(1-j)} \lambda^{-j} \left(\log \lambda - \sum_{k=1}^{-j} \frac{1}{k} \right) & \text{for } j \le 0 \end{cases}$$

where $\Gamma(\alpha) = \int_0^{+\infty} e^{-t} t^{\alpha-1} dt$ is Euler's gamma function. Then a standard computation (see e.g. [103], [123]) shows that the kernel $K(x, y)$ of a differential operator can be rewritten as

(2.5.9)
$$K(x, y) = P(x, D)\delta(x - y) =$$
$$= \left(\frac{1}{2\pi}\right)^n \int \sum_j p_j(x, \xi) \Phi_{n+j}[(x - y)\xi + i0] d\xi.$$

The question, now, is whether the conditions of Definition 2.5.4 suffice to guarantee some sort of convergence for the right-hand side of (2.5.9) if the p_j 's are replaced by the holomorphic functions of Definition 2.5.4. It can actually be proved that this is the case (see Proposition 4.1.1 in [123]) and so we can show the following result:

Theorem 2.5.5 *Let $P(z, D)$ be a microdifferential operator in $\mathcal{E}^\infty$ over an open set $U \subseteq \mathbb{C}^n \times \mathbb{C}^n$ such that if $(z, \zeta) \in U$ and $c > 0$, then $(z, c\zeta) \in U$. Then $P(z, D)$ determines, by taking suitable boundary values, a microlocal operator on*

$$\mathcal{R}(U) := \{(x, \xi) \in S^* \mathbb{R}^n : (x, i\xi) \in U\}.$$

Proof. Indeed if one takes the boundary value of

(2.5.10) $$\sum_j p_j(z, i\zeta)\Phi_{j+n}(i\tau)$$

along $\text{Im}\,\tau > 0$, this defines a hyperfunction, in view of Proposition 4.1.1 of [123]. Set now

$$\tau = (x - y)\zeta$$

and denote by

$$K(x, y, \zeta)$$

the boundary value of (2.5.10). Then $K \in \mathcal{B}(V)$ for V an open set containing

$$\{(x, y, \zeta) : x = y, (x, i\zeta) \in U\}$$

and one can verify that

$$S.S.(K) \subseteq G := \{(x, y, \zeta; (\xi, \eta, \rho)) : (x, y, \zeta) \in V,$$

$$(x - y) \cdot \zeta = 0, \xi = -\eta = k\zeta, \ \zeta = k(x - y) \ \text{for some} \ k > 0\}.$$

Define now

$$K(x, y) = \int K(x, y, \zeta)d\zeta.$$

Then since for a point in G, $\rho = 0$ implies $x = y$, we see that $K(x, y)$ is a microfunction defined in

$$\{(x, y; (\xi, \eta)) : (x, \xi) \in \mathcal{R}(U)\}$$

with support contained in $\{x = y, \xi = -\eta\}$, and it therefore defines a microlocal operator according to Definition 2.5.2.   □

Remark 2.5.4 It can be shown that $\mathcal{L}$ is a flabby sheaf, while $\mathcal{E}$ is not.

In the next few chapters we will use the sheaf $\mathcal{E}$, and some other related sheaves of differential operators, to study systems of differential equations from an algebraic point of view.

2.6   Historical Notes

We have two appendices to this chapter. The first deals with some physical motivations for the theory of hyperfunctions and microfunctions, while the second describes Hörmander's definition of analytic wave front set, and its relationship with the singular spectrum of a hyperfunction.

2.6.1   Physical Origins for the Theory of Microfunctions

In this section we will explore some of the physical origins of the theory of microfunctions. Unfortunately, it is beyond the scope of this book, and certainly beyond our capabilities, to give a more comprehensive account, for which we refer the reader to [44], [94], [184]; it would have seemed to us unacceptable, however, to have an introductory treatise on microlocal analysis in which the reader is not alerted to the deep physical meaning that these theories have. We will restrict our attention, here, to the aspects related to the microanalyticity and the macrocausality of the S-matrix.

Even though the need for hyperfunctions (and for the study of singularities) only becomes evident when dealing with quantum field theory, we wish to begin by showing the reader how boundary values of holomorphic functions are naturally linked to macroscopic notions of causality.

Let us start with a simple example: consider the (one dimensional) motion of a harmonic dumped oscillator of mass m, acted upon by an exterior force $mf(t)$ (t being the time variable). The motion equation (for $x = x(t)$ the space variable) is

$$(2.6.1) \qquad \frac{d^2x}{dt^2} + 2\gamma\frac{dx}{dt} + \omega_0^2 = f(t),$$

where ω_0 is the natural frequency of the oscillator, and $\gamma > 0$ is the dumping constant. This physical system is known to be causal, as a consequence of the positivity of γ (in other words, as long as $f(t) = 0$, the system only admits its free oscillations); here, however, we will ignore this extra piece of information, and we derive the causality of the system from the intrinsic properties of the equation which describes it.

If now we take $f(t) \equiv 0$ (i.e. we look at the free oscillations of the system), then the general solution of (2.6.1) is

$$(2.6.2) \qquad x_0(t) = a\exp(-\omega_1 t) + b\exp(-\omega_2 t)$$

where a, b are arbitrary constants depending on the initial conditions of the system, where ω_1, ω_2 are the solutions of the characteristic equation $\omega^2 + 2i\gamma\omega - \omega_0^2$, i.e.

$$(2.6.3) \qquad \omega_{1,2} = \pm(\omega_0^2 - \gamma^2)^{1/2} - i\gamma.$$

A different situation occurs if we consider a harmonic exterior force, namely $f(t) = F_\omega \exp(-i\omega t)$. It is then immediate to verify that the solution $x(t)$ is given by

$$x(t) = X_\omega \exp(-i\omega t),$$

where, by (2.6.1) and (2.6.2),

$$X_w = G(\omega)F_\omega, \quad G(\omega) = \frac{-1}{(\omega - \omega_1)(\omega - \omega_2)}.$$

Finally, by composition of these elementary harmonic forces, one can consider the "general" case, in which $f(t)$ is assumed to be arbitrary, but represented by a Fourier integral, i.e.

$$f(t) = (2\pi)^{-1} \int_{-\infty}^{+\infty} F(\omega)\exp(-i\omega t)d\omega,$$

and, of course,

$$F(\omega) = \int_{-\infty}^{+\infty} f(s)\exp(i\omega s)ds.$$

Since we are dealing with a linear system, the solution in the general case follows, by integration, from the harmonic case, so that

$$x(t) = (2\pi)^{-1} \int_{-\infty}^{+\infty} X(\omega)\exp(-i\omega t)d\omega =$$

$$= (2\pi)^{-1} \int_{-\infty}^{+\infty} G(\omega)F(\omega)\exp(-i\omega t)d\omega =$$

$$= (2\pi)^{-1} \int_{-\infty}^{+\infty} f(s)ds \int_{-\infty}^{+\infty} G(\omega)\exp(-i\omega t)d\omega =$$

$$= g * f(t),$$

where

(2.6.4)
$$g(t) = (2\pi)^{-1} \int_{-\infty}^{+\infty} G(\omega)\exp(-i\omega t)d\omega,$$

and $*$ denotes the usual convolution product.

Remark 2.6.1 Notice that if $f(t)$ where $\delta(t)$, the Dirac distribution, then $x(t)$ would coincide with $g(t)$ which, therefore, is nothing but the fundamental solution of (2.6.1). We also see that $g(t)$ or, equivalently, its Fourier transform $G(\omega)$, contains all information necessary to the study of the system; in the terminology currently used in Quantum Physics one might say that $g(t)$ is the transition function between the input $f(t)$ and the output $x(t)$.

Let us now examine $G(\omega) = -1/(\omega - \omega_1)(\omega - \omega_2)$ more closely; its integral, which will provide the fundamental solution $g(t)$, can be easily computed with

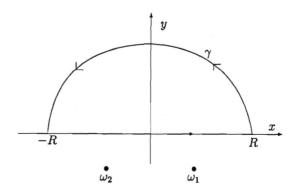

Figure 2.6.1

the help of residues theory. By (2.6.2) one sees that both ω_1 and ω_2 belong to $I_-(\omega) = \{\omega \in \mathcal{C} : \mathrm{Im}\,\omega < 0\}$, and hence $G(\omega)$ can by analytically continued in the half plane $I_+(\omega) = \{\omega \in \mathcal{C} : \mathrm{Im}\,\omega > 0\}$. Fix $t = \tau < 0$ in (2.6.4): then the integral (2.6.4) can be computed by closing the integration path in the half plane I_+, where G is holomorphic: Since the integral of G on half-circles in I_+ tends to zero as their radius R grows to infinity (indeed, $|G(\omega)|$ rapidly decreases to zero, when $|\omega| \to +\infty$), we deduce

$$g(\tau) = (2\pi)^{-1} \int_{-\infty}^{+\infty} G(\omega)\exp(-i\omega\tau)d\omega$$

$$= (2\pi)^{-1} \int_{\gamma} G(\omega)\exp(-i\omega\tau)d\omega = 0,$$

where γ is as in figure 2.6.1.

Hence

(2.6.5) $g(\tau) = 0$ for $\tau < 0$.

A similar reasoning applies if we put $\tau > 0$ in (2.6.4); this time, however, the path γ must be chosen (see figure 2.6.2) in the half-plane $I_-(\omega)$ (where G is no longer holomorphic) if we want to kill off the integral along the semi-circles. For $\tau > 0$, hence, the residues theorem gives

$$g(\tau) = -2\pi i \sum \mathrm{Res}\,\frac{G(\omega)\exp(-i\omega\tau)}{2\pi} =$$

$$= \frac{i}{(\omega_1 - \omega_2)}(\exp(-i\omega_1\tau) - \exp(-i\omega_2\tau)).$$

Finally, we have

$$x(t) = \int_{-\infty}^{+\infty} g(t-s)f(s)ds = \int_{-\infty}^{t} g(t-s)f(s)ds =$$

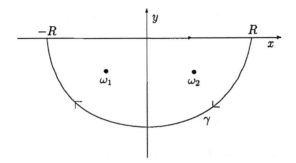

Figure 2.6.2

$$= \int_{-\infty}^{t} \exp(-\gamma(t-s)) \sin\left((\omega_0^2 - \gamma^2)^{1/2}(t-s)\right) \frac{f(s)}{(\omega_0^2 - \gamma^2)^{1/2}} ds$$

where the upper bound of integration is t instead of $+\infty$, because of (2.6.4).

Remark 2.6.2   The result just obtained can be given a quite interesting physical interpretation: the position $x(t)$ of the system at the instant t, only depends on the values $f(t)$ in instant s preceding $t(s < t)$, i.e. the system is causal. This result is of course a direct consequence of (2.6.4), which in turn follows from the fact that $G(\omega)$ has a holomorphic continuation in $I_+(\omega)$. The computations above thus show, even though in a single example, that the analyticity of $G(\omega)$ (and its behavior at infinity) plays a key role in establishing the causality of the physical system which G describes.

We now proceed to reverse the above process: more precisely we show that an abstract assumption of causality for a given system leads to obtain analyticity properties on the functions (distributions) which describe the system itself. We will adopt here an axiomatic approach: some axioms (inspired by physical considerations) are stated, and from them we deduce important properties of the objects which are to describe the system. In the sequel both the input $f(t)$ and its output $x(t)$ will be assumed to belong to a suitable class of generalized functions (e.g. the space $\mathcal{D}'$ of distributions). Three reasonable axioms will be stated:

α) linearity (superposition principle),

β) invariance with respect to time translations,

γ) macroscopic primitive causality.

The linearity axiom states that our physical system acts as a linear operator on a suitable space of distributions, hence there exists a kernel $g = g(t, s)$ such that

$$x(t) = \int_{-\infty}^{+\infty} g(t, s)f(s)ds$$

(the integrals which we write are meaningful only insofar as f, g, x are functions; in all other cases they must be looked at from a symbolic point of view, as it is customary in the theory of distributions).

Axiom β states that if the input is translated with respect to time, the same happens for the output. It is well known that the convolution operators are the most general operators which commute with translations, so that from β we deduce that $x(t)$ can be obtained from $f(t)$ via convolution with a given convolutor. In other words,

$$(2.6.6) \qquad x(t) = \int_{-\infty}^{+\infty} g(t-s)f(s)ds = g * f(t)$$

(the integral, as usual, may by symbolic, while $g * f$ is perfectly meaningful, as far as at least one of the two distributions g, f is of compact support). Finally, the primitive causality condition implies that if $f(t)$ vanishes for $t < T$, the same happens for $x(t)$ (in the example we examined before this did not occur, because the system was endowed, so to speak, with "built-in" oscillations, which should not be considered in our discussion). This, in particular, implies that $g(t) = 0$ for $t < 0$, i.e. the support of g, considered a distribution, is contained in $I\!\!R^+$ ($g(t)$ is the output $x(t)$ corresponding to the input $f(t) = \delta(t)$ and of course coincides with the kernel appearing (2.6.5)). This fact has an interesting consequence: if $G(\omega)$ is the Fourier transform of $g(t)$ (which exists if g is a compactly supported, or tempered, distribution, then the causality condition implies that

$$G(\omega) = \int_{0}^{+\infty} g(t)\exp(i\omega t)dt$$

(indeed, $g(t) = 0$ for $t < 0$). This apparently irrelevant simplification implies in particular that $G(\omega)$ admits of a holomorphic continuation in $I_+(\omega)$. Indeed, if $\omega = u + iv$, $v > 0$, then

$$G(\omega) = \int_{0}^{\infty} g(t)\exp(iut)\exp(-vt)dt$$

is still convergent, because of the factor $\exp(-vt)$ (this would not be true if the integral were extended to all of $I\!\!R$).

The preceding discussion can be made very precise, for example for $G(\omega) \in L^2(I\!\!R)$, and summarized by a result whose detailed proof the reader can find in [92]:

Theorem 2.6.1 *Let $G(\omega) \in L^2(I\!\!R)$. Then the following conditions are equivalent:*

(i) $g(t)$ vanishes for $t < 0$;

(ii) $G(\omega)$ is, almost everywhere in the sense of Lesbesgue, the limit, for $v \to 0^+$, of a function $G(u+iv)$, holomorphic in the half-plane $I_+(\omega) = I_+(u+iv)$, which is square integrable on each line parallel to the real axis, i.e.:

$$\int_{-\infty}^{+\infty} |G(u+iv)|^2 du < C \quad (v > 0);$$

(iii) $G(\omega)$ satisfies Plemelj's formula (or, in the language of physicists, the dispersion relation):

$$G(\omega) = (\pi i)^{-1} P.V. \int_{-\infty}^{+\infty} \frac{G(\omega)}{(\omega' - \omega)} d\omega',$$

where

$$P.V. \int_{-\infty}^{+\infty} \cdots := \lim_{\epsilon \to 0} \left(\int_{-\infty}^{\omega-\epsilon} \cdots + \int_{\omega+\epsilon}^{+\infty} \cdots \right)$$

denotes the Cauchy principal value of the integral.

Remark 2.6.3 The requirement that G belongs to $L^2(\mathbb{R})$ can often be translated into the physical assumption that the total energy of the system be finite.

Though this seldom occurs in practice, one can equally well handle more complicated situations (e.g. when G is of polynomial growth) with the use of theorems in the same spirit as Theorem 2.6.1.

Remark 2.6.4 A function $G(\omega)$ which satisfies any of the equivalent conditions of Theorem 2.6.1 is said to be a causal transform (an example of such a function is, of course, the transition function which arises when dealing with the harmonic oscillator:

$$G(\omega) = \frac{-1}{(\omega - \omega_1)(\omega - \omega_1)}, \quad \text{Im}\omega_{1,2} < 0.$$

Theorem 2.6.1 extends somehow to the (more interesting) case of distributions; let $\mathcal{D}'$ denote the space of Schwartz distributions on $\mathbb{R}$, $\mathcal{D}'_+$ the subspace of those distributions whose support is contained in $[0, +\infty)$, and S' the space of tempered distributions. Then the following result holds:

Theorem 2.6.2 Let $G_\omega = \hat{g}_t \in S'$. The $g_t \in \mathcal{D}_+$ if and only if:

(i) G_ω is, in the sense of distributions, the boundary value of a function $G(u+iv)$, holomorphic for $v > 0$;

(ii) for any fixed value $v > 0$, $G(u+iv)$ belongs, as a distribution in u, to S' and, in this space, converges to G_u, for $v \to 0^+$;

(iii) given $\epsilon > 0$, there exists an integer n such that

$$G(\omega) = \mathcal{O}(|\omega|^n) \quad (\mathrm{Im}\,\omega \geq \epsilon > 0).$$

Remark 2.6.5 Both Theorem 2.6.1 and Theorem 2.6.2 show that the mathematical translation of the physical assumption of causality leads to a natural introduction of the notion of boundary value of a holomorphic function satisfying suitable growth conditions. In other words, if we do not rely on the fact that every distribution can be represented as boundary value of a suitable holomorphic function (i.e. on the injection $\mathcal{D}' \to \mathcal{B}$ of the sheaf of distributions into the sheaf $\mathcal{B}$ of hyperfunctions to be thought of as the "sheaf of the boundary values of holomorphic functions"), we do not obtain a natural description of the consequences of the causality axiom. Hence, if a formal theory of boundary valued does not seem (at this stage) unavoidable, it certainly looks natural.

All the material described so far only refers to one dimensional problems. We wish to conclude this discussion with a few remarks on the situation which arises when dealing with several variables.

Let $\Gamma^+ = \{x = (x_1, \ldots, x_n) \in I\!\!R^n : x_1^2 - \sum_2^n x_i^2 > 0, x_1 > 0\}$,

$$\Gamma^- = \{x \in I\!\!R^n : x_1^2 - \sum_1^n x_i^2 > 0, x_1 < 0\};$$

correspondingly one can construct the forward tube domain $T^+ = I\!\!R^n + i\Gamma^+$, and the backward tube domain $T^- = I\!\!R^n + i\Gamma^-$. Causal distributions can then be defined as those tempered distributions (on $I\!\!R^n$) which, by convolution, describe causal systems, and it can be proved that $u \in \mathcal{S}'$ is causal if and only if $\mathrm{supp}(u) \subseteq \Gamma^+$. Theorem 2.6.2 extends to:

Theorem 2.6.3 $u \in \mathcal{S}'$ *is causal if and only if its Fourier transform $\hat{u}$ extends holomorphically in the tube T^+ to a function of which, in $\mathcal{S}'$, $\hat{u}$ is the boundary value.*

We now briefly mention a more difficult analysis of how microanlyticity is related to microcausality.

As it is well known, one of the historical interests of elementary particles physics was the study of collision processes and the theory of the scattering matrix (or S-matrix in brief) has been developed (essentially by a group of physicists [22], [43], [44], [79], [92], [93], [94], [95], [99] in the early sixties) exactly for this purpose.

Under the assumption that only strong interactions are considered, and that the particles involved are stable particles with respect to such strong interactions, one may safely assume that both incoming and outgoing particles are free, and can therefore be characterized by their impulse-energy real four-dimensional vector

$$p = (p_{(0)}, \vec{p})$$

where the component $p_{(0)} > 0$ represents the energy and where the mass of the particle is given by Minkowski's metrical relation (with the speed of light being set as $c = 1$)

$$m^2 = p_{(0)}^2 - \vec{p}^2.$$

Thus, particles can actually be represented as points in the algebraic variety (the mass-variety) given by the hyperboloid

$$M = \{p \in I\!\!R^4 : p^2 = m^2, p_{(0)} > 0\};$$

with p^2 being defined by $p_{(0)}^2 - \vec{p}^2$ and $\vec{p}^2 = p_1^2 + p_2^2 + p_3^2$.

We will not give here the details which can be found on any initial textbook in quantum physics, but we will just mention that the superposition principle allows us to say that all the information which can be derived from the collision is contained in the so called S-matrix of the process, whose entries are complex valued distributions on the algebraic variety which is the product of the mass-varieties associated to all the particles involved in the collision.

Historically, two equivalent principles have been stated in the study of the S-matrix: a mathematical one and a physical one; the reader is referred to [93], [95], [184], [185] for further details on this topic. Let us point out that when the study of the S-matrix begun, not much was known about the structure of the matrix, except that some general properties (related to its analyticity) were understood. Particularly important is the fact that strong relationships must link the physical interactions which the matrix strives to describe, and the structure of its singularities, in the sense of microlocal analysis.

In fact, it can be shown that the *physical* postulate of macrocausality, is equivalent to the *mathematical* postulate of microanalyticity. This of course, is what is known now (and it has been known for maybe 20 years), but at the origin of the theory, physicists were forced to formulate similar postulates, such as the $i\varepsilon$-postulate which can only be correctly expressed in the language of hyperfunctions in several variables.

2.6.2 Hörmander's Analytic Wave Front Set

At the beginning of this chapter we have defined the wave front set for distributions and we have used the same procedure to create an analogous concept for hyperfunctions, which we have called the singular spectrum of a hyperfunction. In the first part of this historical appendix, we have also mentioned how physicists have felt the need for the introduction of a similar notion (the microlocal essential support) in relation to the study of the S-matrix. Now we will go back to distributions to describe a notion due to Hörmander, who in [85] introduced the concept of analytic wave front set. As it will turn out, all these concepts are equivalent and demonstrate the necessity of a concept which is being born in so many different areas of study. It is also interesting the fact that all these notions were created independently and essentially at the same time.

Let therefore f be a distribution on an open set U in $\mathbb{R}^n$. The goal of Hörmander's notion is to study the open subset of U where f is not real analytic; to do so, we will once again try to use the Fourier transform: however, since no compactly supported analytic functions exist, one needs (as already pointed out in section two of this chapter) to have a more complex description. The fundamental idea is still the same; we want to say that if f is real analytic in a point x_0, then one can find compactly supported distributions which coincide with f in a neighborhood of x_0 and such that their Fourier transforms satisfy suitable bounds (a similar idea is developed also in Kaneko's [102], where a "psychological Paley-Wiener theorem" for compactly supported real analytic functions is proved).

To make the previous comments precise we follow [18] and we start by recalling the following well known characterization of real analytic functions:

Lemma 2.6.1 *A function f is real analytic in an open subset U of $\mathbb{R}^n$ if and only if it satisfies the following condition: for every compact subset K of U, there exist constants C_K and m_K such that, for all multi-indices α*

$$\sup_K |\frac{\partial^\alpha f}{\partial x^\alpha}| \leq C_K (C_K |\alpha|)^{|\alpha| + m_K}.$$

The reader will note that one might replace this Lemma with other similar results originating from the theory of quasi-analytic classes, to obtain similar notions of quasi-analytic wave-front sets.

Lemma 2.6.2 *There is a constant C_n such that if K is a compact subset of $\mathbb{R}^n$ and if K_r is the r-neighborhood of K, $K_r = \{x \in \mathbb{R}^n : d(x, K) \leq r\}$, then one can find, for every r, a sequence $\{\varphi_m\}$ of compactly supported infinitely differentiable functions, such that:*

(i) for every m, $\varphi_m = 1$ on K and $\mathrm{supp}(\varphi_m) \subseteq K_r$;

(ii) for every m and all multi-indices α of length bounded by m, it is

$$\sup |\frac{\partial^\alpha \varphi_m}{\partial x^\alpha}| \leq C_n (C_n \frac{m}{r})^{|\alpha|}.$$

The interest of this lemma, originally proved in Hörmander's [85], to which we refer the reader for the proof, is that it provides a family of cut-off functions with appropriate bounds on their derivatives. We are now ready to prove the fundamental result for the construction of the analytic wave front set:

Proposition 2.6.1 *Let f be a distribution on an open set U in $\mathbb{R}^n$, and let x_0 be a point of U. Then f is real analytic in a neighborhood of x_0 if and only if there is some bounded sequence $\{f_m\}$ in $\mathcal{E}'(\mathbb{R}^n)$, and an open neighborhood U_0 of x_0 such that the following conditions hold:*

(i) f_m coincides with f in U_0 for all values of m;

(ii) there is a constant C such that, for all m and all ξ in $\mathbb{R}^n$, it is

$$|\hat{f}_m(\xi)| \le (C)^m (1 + |\xi|)^{-m}.$$

Proof. Let us first assume that f is real analytic in a ball $B(x_0, \delta)$. Set $t = \delta/3$, $K = \overline{B(x_0, t)}$ and apply Lemma 2.6.2 (with $r = t$), to construct a sequence $\{\varphi_m\}$ satisfying the conditions of the Lemma. Now set $f_m = \varphi_m \cdot f$. It is immediate to verify that $f_m = f$ in $B(x_0, t)$ and that $\{f_m\}$ is a bounded sequence in $\mathcal{E}'(\mathbb{R}^n)$. Moreover, for a given m and a multi-index α, $|\alpha| \le m$, we have

$$D^\alpha f_m = \sum_{\beta \le \alpha} \binom{\alpha}{\beta} (D^\beta f)(D^{\alpha - \beta} f_m).$$

Since f is real analytic in $B(x_0, \delta)$, and $2t < \delta$ we use Lemma 2.6.1 to find constants C_1, m_1 such that for any α and any x in $B(x_0, 2r)$,

$$|D^\alpha f(x)| \le C_1 (C_1 |\alpha|)^{|\alpha| + m_1}.$$

Since $\text{supp}(f_m) \subseteq B(x_0, 2r)$, and since

$$\sup |D^\alpha \varphi_m(x)| \le C_n (C_n + \frac{m}{t})^{|\alpha|}, \quad \forall \alpha \text{ with } |\alpha| \le m,$$

we can use the expression for $D^\alpha f_m$ to find a constant C_2 such that

$$\sup |D^\alpha f_m| \le C_2 (C_2 m)^{|\alpha|} \quad \forall \alpha \text{ with } |\alpha| \le m.$$

(Note, in particular, that if we assume $|\alpha| \le m$ the expression $|\alpha|^{m_1}$ can be included into C_2). Standard Fourier transform properties show that

$$\xi^\alpha \hat{f}_m(\xi) = \int e^{-ix \cdot \xi} (D^\alpha f_m)(x) dx,$$

but since f_m is compactly supported in $B(x_0, 2t)$ we directly obtain

$$|\hat{f}_m(\xi)| \le C_3 (C_3 m)^m (1 + |\xi|)^{-m}, \quad m = 1, 2, \ldots,$$

with C_3 a new constant.

Vice versa, if we assume the existence of a bounded sequence $\{f_m\}$ as stated in the Proposition then, in U_0,

$$D^\alpha f(x) = (2\pi)^{-n} \int \xi^\alpha f_m(\xi) e^{ix\cdot\xi} d\xi.$$

The finiteness of the integral on the right hand side, together with the characterization given by Lemma 2.6.1, now shows that f is real analytic near x_0.

$\square$

The reader will have noticed that the condition on the boundedness of the sequence $\{f_m\}$ is crucial for the proof. This result is clearly the counterpart of the classical Paley-Wiener theorem and allows us to provide a natural definition for the analytic wave front set of a distribution; as in the cases examined earlier in this chapter, one may define such an object on the cotangent bundle of U or, and this is what we will do in this case, on its cosphere bundle:

Definition 2.6.1 *Let f be a distribution in an open set U of $\mathbb{R}^n$; then its analytic wave front set $WF_A(f)$ is the closed subset of $U \times iS_\infty^{n-1}$ whose complement consists of those points $(x_0, i\xi_0\infty)$ satisfying the following condition:*

> *There exist an open neighborhood Ω_0 of x_0, an open conic neighborhood Γ of ξ_0 and a bounded sequence $\{f_m\}$ of compactly supported distributions such that $f_m = f$ in Ω_0, and*
>
> $$|\hat{f}_m(\xi)| \le (Cm)^m (1+|\xi|)^m,$$
>
> *for all ξ in Γ and some positive constant C, which is independent of m.*

Remark 2.6.6 It is an immediate consequence of the previous Proposition that a point (x_0, ξ_0) belongs to the complement of $WF_A(f)$ if and only if there exists some $t > 0$ such that if $\{\varphi_m\}$ is the bounded sequence associated by Lemma 2.6.2 to $\overline{B(x_0, t)}$, then there exist some conic neighborhood Γ of ξ_0, a constant C, and an integer k, such that

$$|(\widehat{\varphi_m \cdot f})(\xi)| \le (Cm)^m (1+|\xi|)^{k-m}$$

for all $\xi \in \Gamma$.

It is immediate to see that if π is the canonical projection of the cosphere bundle on its basis, then $\pi(WF_A(f))$ is a closed subset of U. Moreover, an application of Proposition 2.6.1 shows that f is real analytic outside this projection and therefore the analytic wave front set satisfies the most obvious requests.

We now want to show that this new definition of analytic wave front set coincides with the notion of singular spectrum when applied to distributions; as a matter of fact, we now have two different ways of describing the singularity locus of distributions: one is given by the notion of analytic wave front set introduced just now, and one (essentially provided in Section 2.2) is based on the fact that each distribution is a hyperfunction as well. From this point of view, we know that every distribution can be seen as sum of boundary values of holomorphic functions, and as such it is a hyperfunction to which the notion of singular spectrum can be applied. If we look back at the definition of singular spectrum for a hyperfunction we readily see that if f is a distribution on U, then its singular spectrum $S.S.(f)$, which is also referred to as the analytic singular spectrum of f, can be defined as the closed subset of $U \times iS_\infty^{n-1}$ which is the complement of the set of those points $(x_0, i\xi_0\infty)$ for which the following condition holds:

> *there exists some open neighborhood Ω of x_0 and a family $\{f_\alpha\}$ of function holomorphic in $\Omega + i\Gamma_\alpha$ such that f is, in Ω, the sum of boundary values of the f_α, and for each α there exists some y_α in Γ_α for which $\langle y_\alpha, \xi_0 \rangle$ is negative.*

The proof of the equivalence of these two notions is far from being trivial, and we will follow Björk's arguments as described in [18]. We first need to notice that the distributions which arise as boundary values of holomorphic functions are not arbitrary; this was already proved in Chapter one, where a tempered growth condition was described to characterize such boundary values; we now translate that growth condition in terms of the analytic wave front set.

Proposition 2.6.2 *Let f be a holomorphic function in $\Omega + i\Gamma^\delta$, for $\Gamma^\delta = \Gamma \cap \{0 < |y| < \delta\}$, which satisfies the tempered growth condition*

$$|f(x + iy)| \leq C|y|^{-k}$$

for all point in $\Omega + i\Gamma^\delta$, and let $b(f)$ denote its boundary value (which is actually a distribution, not just a hyperfunction). If ξ_0 in S^{n-1} satisfies $\langle y, \xi_0 \rangle < 0$ for some y in Γ, then the point $(x, i\xi_0\infty)$ is outside $WF_A(b(f))$ for all x in Ω.

Proof.   Let $x_0 \in \Omega$ and let $t > 0$ be such that $\overline{B(x_0, 2t)} \subseteq \Omega$. Set, as before, $K = \overline{B(x_0, t)}$ and construct a sequence $\{\varphi_m\}$ as in Lemma 2.6.2. We now define a sequence of "almost analytic" functions as follows: for any m we set

$$\chi(\varphi_m) = g_m(x + iy) := \sum_{|\alpha| < m} \left(\frac{\partial^\alpha \varphi_m}{\partial x^\alpha}(x) \right) \frac{(iy)^\alpha}{\alpha!}.$$

Clearly, the functions g_m are not analytic but they do satisfy some kind of "limit" Cauchy-Riemann equations as $y \to 0$, since one can easily prove the existence of a positive constant $C = C(n)$ such that, for $z = x + iy$,

$$\left|\frac{\bar{\partial}(g_m)}{\partial \bar{z}}(z)\right| \leq \frac{C(C|y|)^{m-1}}{m!}\frac{(Cm)^m}{t^m} \leq C_1(C_1|y|)^{m-1},$$

with a suitable choice for $C_1 > 0$. Choose now $y_o \in \Gamma^\delta$ satisfying $\langle y_0, \xi_0 \rangle < 0$, and let us estimate the Fourier transform of $\varphi_m \cdot b(f)$ along the ray $\{s\xi_0 : s > 0\}$. If $m \geq k + 1$ we obtain

$$(\widehat{\varphi_m b(f)})(s\xi_0) = \lim_{y \to 0} \int \left(\chi(e^{-isx\cdot\xi_0}\varphi_m)(x+iy)\right) f(x+iy)dx.$$

If we set $y = ty_0$, so that $t \to 0$ when $y \to 0$, we have (in view of the existence of $b(f)$ in the sense of distributions) that

$$(\widehat{\varphi_m b(f)})(s\xi_0) = U_1(s) - U_2(s),$$

with

$$U_1(s) = \int e^{-isx\cdot\xi_0+sy_0\cdot\xi_0}(g_m(x+iy_0)) \cdot f(x+iy_0)dx$$

and

$$U_2(s) = \iint_{(x,r)\to x+iry_0} f(z)e^{-iz\cdot\xi_0}\bar{\partial}g_m \wedge dz.$$

In order to complete the proof, we only need to estimate the growth of $U_1(s)$ and $U_2(s)$, to be able to apply Remark 2.6.1.

As for $U_1(s)$, the boundedness of $f(x+iy_0)$ in Ω, together with the estimates on the derivatives of φ_m and the definition of g_m, yields

$$|U_1(s)| \leq (Cm)^m(1+s)^{-m},$$

for a suitable positive constant C.

As for $U_2(s)$ we know that for $0 < t < 1$ it is

$$\left|\frac{\partial f}{\partial \bar{z}}(x+ity_0)\right| \leq C(C|y_0|)^{m-1}t^{m-1};$$

on the other hand, the growth of $f(x+ity_0)$ is bounded by hypothesis and so we obtain the desired estimates. □

Proposition 2.6.3 *For a distribution f on an open set U, it is*

$$WF_A(f) \subseteq S.S.(f).$$

Proof. This is the easy inclusion. To prove it we assume f to be a compactly supported distribution; this does not implies any loss of generality as every distribution is a locally finite sum of compactly supported distributions to which we can apply the following arguments. Now, for a compactly supported distribution f we can always write $f = \sum b(f_\alpha)$ as a finite sum where each f_α

is holomorphic in $I\!\!R^n + i\Gamma_\alpha$. If we take $(x, i\xi_0\infty) \notin S.S.(f)$, one can choose the decomposition above in such a way that each cone Γ_α can be chosen so to contain a point y_α such that $\langle y_\alpha, \xi_0 \rangle$ is negative. Then, by Proposition 2.6.2, $(x, i\xi_0\infty) \notin WF_A(b(f_\alpha))$, for any α, and therefore $(x, i\xi_0\infty) \notin WF_A(f)$, which concludes the proof. $\qquad\square$

The inclusion of $S.S.(f)$ into $WF_A(f)$ is, on the other hand, much more complicated, and it requires some detailed analysis on the analytically uniform structure of specific spaces of distributions in the sense of Ehrenpreis [52]. We first need to introduce an auxiliary space of compactly supported infinitely differentiable functions: let Γ be an open cone in the frequency domain (i.e. the real space with variable ξ), and fix some pair (ε, δ) of positive constants; we now define the space $\mathcal{F}$ as the space of all compactly supported infinitely differentiable functions g which satisfy the following conditions:

(i) $\text{supp}(g) \subseteq \{|x| \leq \delta\}$;

(ii) $|\hat{g}(\zeta)| \leq (1 + |\zeta|)^{-2} \exp(\delta|\text{Im}\zeta|)$ if $\text{Re}(\zeta) = \xi \notin \Gamma$;

(iii) $|\hat{g}(\zeta)| \leq (1 + |\xi|)^{-2} \exp(\delta|\text{Im}\zeta| + \varepsilon|\text{Re}\zeta|)$ if $\text{Re}(\zeta) = \xi \in \Gamma$.

Note that $\mathcal{F}$ is dense in $\mathcal{D}$; we will use this fact at a later stage; the crucial lemmas which we need are given in Liess' [149]:

Lemma 2.6.3 *There are absolute positive constants A, B, C (only depending on the dimension of the space) such that if $q(\xi)$ is a non-negative real valued function which is Lipschitz continuous with norm less than or equal to one, then there exists a plurisubharmonic function $Q(\zeta)$ in $\mathbb{C}^n$ which is Lipschitz continuous with norm bounded by C and which satisfies on all of $\mathbb{C}^n$, the following chain of inequalities:*

$$q(\xi) \leq Q(\xi + i\eta) \leq 2q(\xi) + A|\eta| + B.$$

Lemma 2.6.4 *Consider a pair of cones $\Gamma \subset\subset \Gamma'$, and let $\mathcal{F}$ be defined as above, with respect to the cone Γ. Then there exist positive constants D, E, m which only depend on n and on the pair (Γ, Γ'), such that every g in $\mathcal{F}$ can be written as a finite sum of distributions supported in $\{|x| \leq \delta + E\varepsilon\}$*

$$g = g_1 + g_2 + \cdots + g_s + h,$$

where the distributions satisfy the following Fourier conditions:

(i) $|\hat{g}_j(\xi)| \leq Dj^{-2}(1 + |\xi|)^m \exp(2j\varepsilon)$ if $j/4 < |\xi| < 4j$ and $\xi \in \Gamma'$;

(ii) $|\hat{g}_j(\xi)| \leq Dj^{-2}(1 + |\xi|)^m$ otherwise;

(iii) $|\hat{h}(\xi)| \leq D(1 + s)^{-2}(1 + |\xi|)^m$ for all ξ.

We are now ready to prove the equality of $S.S.(f)$ and $WF_A(f)$.

Theorem 2.6.4 *Let f be a distribution defined on an open set U of $\mathbb{R}^n$. Then*

$$S.S.(f) = WF_A(f).$$

Proof. We have already proved that $WF_A(f) \subseteq S.S.(f)$, so we only need to prove the reverse inequality.

To do so, let us pick a point $(x_0, i\xi_0\infty)$ which does not belong to $WF_A(f)$; we will prove that it is also outside $S.S.(f)$. With no loss of generality we assume x_0 to be the origin and take $f \in \mathcal{E}'(\mathbb{R}^n)$ (just multiply f by a suitable cut-off function).

Now let us replace f by a suitable distribution g whose Fourier transform is

$$\hat{f}(\xi)(1 + |\xi|^2)^{-s},$$

where s is chosen so large enough so that

$$|\hat{g}(\xi)| \le C(1 + |\xi|)^{-m-n-1},$$

for m as in Lemma 2.6.4, and $C > 0$ a suitable constant. Notice that the location of both $S.S.(f)$ and $WF_A(f)$ are not changed as we go from f to g (just recall the definition of these sets in terms of Fourier transform).

We now take the compact $\overline{B(x_0, r)}$ and accordingly construct a sequence of cut-off functions $\{\varphi_j\}$ such that for suitable constants C_1, $C_2 > 0$, and a suitable cone $-\Gamma$ containing ξ_0 one has:

$$|\widehat{(\varphi_j g)}(\xi)| \le C_2(1 + |\xi|)^{-m-n-1} \quad \forall j;$$

$$|\widehat{(\varphi_j g)}(\xi)| \le C_1(C_1 j)^j (1 + |\xi|)^{-j} \quad \forall \xi \in -\Gamma;$$

$$\varphi_j g = g \quad \text{in } B(x_0, r).$$

Choose now an open convex cone $\Gamma_0 \subset\subset \Gamma$ such that $-\xi_0 \in \Gamma_0$, and apply Lemma 2.6.4 to the pair (Γ_0, Γ) with $\delta = r/2$ and ε small enough. It is not hard to see that for all functions $p \in \mathcal{F}$ one has $|\langle g, p \rangle| \le M$, for some positive constant M (this is essentially a consequence of the estimates in Lemma 2.6.4). By the Hahn-Banach theorem, and this last estimate, one now deduces the existence of a complex-valued Borel measure $d\lambda(\zeta)$ such that

$$\langle g, p \rangle = \int \hat{p}(\zeta) d\lambda(\zeta)$$

for all $p \in \mathcal{F}$. By the estimates which define the growth of the Fourier transform of the elements in $\mathcal{F}$, we obtain (with χ the characteristic function of Γ_0)

$$\int (1 + |\zeta|)^2 \exp(\delta|\mathrm{Im}\zeta| + \varepsilon\chi(\zeta)|\mathrm{Re}\zeta|)|d\lambda(\zeta)| < +\infty.$$

Now we note that if $|x| < \delta$, $|y| < \varepsilon$ and $\langle y, \xi \rangle \leq 0$ for all $\xi \notin \Gamma_0$, one has

$$| \exp(-i(x + iy) \cdot \zeta)| \leq \exp(\delta |\mathrm{Im}\zeta| + \varepsilon \chi(\zeta)|\mathrm{Re}\zeta|)$$

and therefore the function

$$h(x + iy) := \int \exp(-i(x + iy) \cdot \zeta) d\lambda(\zeta)$$

belongs to $\mathcal{O}(\Omega + i(\Gamma_0^*)^\varepsilon)$, where $\Omega := B(0, \delta)$ and $\Gamma_0^* := \{y : \langle y, \xi \rangle < 0$ for all $\xi \notin \Gamma_0\}$. Now it is obvious that

$$\langle g, p \rangle = \langle b(h), p \rangle$$

for all $p \in \mathcal{F}$ and so, by the density of $\mathcal{F}$ in $\mathcal{D}(\Omega)$, we deduce that $g = b(h)$ in Ω. Now it is immediate to verify (since $\xi_0 \in -\Gamma_0$) that there exists $y_0 \in \Gamma_0^*$ such that

$$\langle y_0, \xi_0 \rangle < 0,$$

which shows that

$$(x_0, i\xi_0 \infty) \notin S.S.(g) = S.S.(f),$$

thus concluding the proof. $\square$

Chapter 3

$\mathcal{D}$-Modules

3.1 Introduction

In this Chapter we will introduce the notion of a $\mathcal{D}$-module and we will provide its first fundamental properties. We will then also microlocalize such a notion, as to introduce the notion of $\mathcal{E}$-module. Since homological algebra is an important underlying notion for the theory of $\mathcal{D}$-modules, in this introduction we will provide a concise and self-contained review of the theory of derived categories and of spectral sequences for a double complex. In section 2, we discuss the foundations for $\mathcal{D}$-modules in the spirit of algebraic geometry. In section 3, we introduce the notion of a good filtration on $\mathcal{D}$-module which allows us to define the characteristic variety of a $\mathcal{D}$-module. In particular, we prove that such a characteristic variety depends only upon the module structure and not on a particularly chosen good filtration. In the last section of this Chapter we prove Sato's fundamental theorem for the category of $\mathcal{E}$-modules.

The reader who is familiar with the theory of derived categories may skip the remainder of this section and move directly to Section 3.2.

Let $\mathcal{A}$ be an abelian category. Let $X^\bullet$ be a (cochain) complex in $\mathcal{A}$, i.e. $X^\bullet = \{X^n, d_X^n\}_{n \in \mathbb{Z}}$ is a collection such that each X^n is an object in $\mathcal{A}$ and each $d_X^n : X^n \to X^{n+1}$ is a map in $\mathcal{A}$ satisfying $d_X^{n+1} \circ d_X^n = 0$. A map $f^\bullet$ from $X^\bullet$ to another (cochain) complex $Y^\bullet$, $f^\bullet : X^\bullet \to Y^\bullet$, is a sequence $f^\bullet = \{f^n\}_{n \in \mathbb{Z}}$ of maps $f^n : X^n \to Y^n$ such that the following diagram commutes:

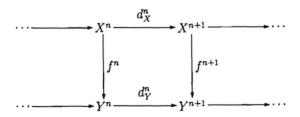

151

i.e. we have $f^{n+1} \circ d_X^n = d_Y^n \circ f^n$ for each $n \in \mathbb{Z}$.

Let $Co(\mathcal{A})$ be the abelian category of complexes in $\mathcal{A}$ with maps as defined in the above. Define the subcategory $Co^+(\mathcal{A})$ of "bounded below complexes" of $Co(\mathcal{A})$ as follows: an object $X^\bullet$ of $Co(\mathcal{A})$ belongs to $Co^+(\mathcal{A})$ if $X^n = 0$ for $n << 0$. One similarly defines the subcategory $Co^-(\mathcal{A})$ of "bounded above complexes" of $Co(\mathcal{A})$. Let $Co^b(\mathcal{A})$ be the subcategory of $Co(\mathcal{A})$ whose objects are bounded below and above, i.e., $Co^b(\mathcal{A}) = Co^+(\mathcal{A}) \cap Co^-(\mathcal{A})$.

Maps $f^\bullet$ and $g^\bullet$ in $Co(\mathcal{A})$ are said to be homotopic if there exists a sequence $s^\bullet = \{s^n\}_{n \in \mathbb{Z}}$ of maps $s^n : X^n \to Y^{n-1}$ such that

$$f^n - g^n = d_Y^{n-1} \circ s^n + s^{n+1} \circ d_X^n$$

as shown in the following diagram:

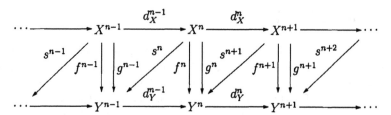

We will write $f^\bullet \sim g^\bullet$ to denote that $f^\bullet$ and $g^\bullet$ are homotopic. Notice that "$\sim$" is an equivalence relation. This relation allows us to construct a new category $K(\mathcal{A})$ whose objects are those of $Co(\mathcal{A})$, but whose maps are equivalence classes of maps under the homotopy equivalence relation. So, for any $X^\bullet$ and $Y^\bullet$ in $Co(\mathcal{A})$, it is

$$\mathcal{H}om_{K(\mathcal{A})}(X^\bullet, Y^\bullet) = \mathcal{H}om_{Co(\mathcal{A})}(X^\bullet, Y^\bullet)/\sim .$$

This in particular means that a map $f^\bullet$ in $Co(\mathcal{A})$ is the zero map in $K(\mathcal{A})$ if $f^\bullet$ is homotopic to the zero map in $Co(\mathcal{A})$. Since $\mathcal{A}$ is an abelian category, one can well-define the cohomology functor from $K(\mathcal{A})$ to $\mathcal{A}$.

We will now define the derived category of $\mathcal{A}$ as a localization of $K(\mathcal{A})$. In order to do so, we define the notion of quasi-isomorphism as follows. A map $f^\bullet : X^\bullet \to Y^\bullet$ in $K(\mathcal{A})$ is said to be a quasi-isomorphism if and only if $H^n(f^\bullet)$ is an isomorphism for all n. We denote the totality of quasi-isomorphisms by (QIS). Then the derived category of $\mathcal{A}$, denoted as $D(\mathcal{A})$, is the localization of $K(\mathcal{A})$ at (QIS)

$$D(\mathcal{A}) := K(\mathcal{A})_{(QIS)},$$

The localization functor Q from $K(\mathcal{A})$ to $K(\mathcal{A})_{(QIS)}$ has the usual universal property which allows us to characterize the process of localization. If a functor Q' carries quasi-isomorphisms in $K(\mathcal{A})$ into isomorphisms in a category $D'(\mathcal{A})$, then there exists a unique functor U from $D(\mathcal{A}) = \mathcal{K}(\mathcal{A})_{(QIS)}$ to $D'(\mathcal{A})$ such that Q' factors through $D(\mathcal{A})$. Namely, the following diagram of categories and functors commute, $Q' = U \circ Q$:

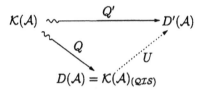

We can also provide a more explicit construction for the derived category $D(\mathcal{A})$. The objects in $D(\mathcal{A})$ are those in $Co(\mathcal{A})$ while a map in $D(\mathcal{A})$ from, say, $X^\bullet$ to $Y^\bullet$ is a suitable equivalence class of pairs $(f^\bullet, s^\bullet)$, where $f^\bullet : X^\bullet \to Y_1^\bullet$ is a map and $s^\bullet : Y^\bullet \to Y_1^\bullet$ is a quasi-isomorphism, and $X^\bullet, Y^\bullet$ and $Y_1^\bullet$ are objects of $D(\mathcal{A})$. The equivalence relation between maps $(f^\bullet, s^\bullet)$ and $(g^\bullet, t^\bullet)$ from $X^\bullet$ to $Y^\bullet$ is defined as follows: given the diagram

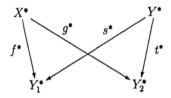

we say that $(f^\bullet, s^\bullet) \sim (g^\bullet, t^\bullet)$ if there exists an object $Y_3^\bullet$ with quasi-isomorphisms $r^\bullet$ and $u^\bullet$ satisfying $r^\bullet \circ s^\bullet = u^\bullet \circ t^\bullet$ and $r^\bullet \circ f^\bullet = u^\bullet \circ g^\bullet$ as indicated in the diagram below.

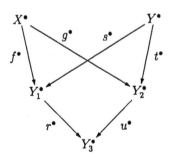

Remark 3.1.1   This is nothing but the categorical version of the familiar arithmetic property in a localized ring : $f/s = g/t$ holds if there is a common denominator $rs = ut$ such that $f/s = rf/rs = ug/ut = g/t$.

Remark 3.1.2 A map from $X^\bullet$ to $Y^\bullet$ in the derived category $D(\mathcal{A})$, therefore, may be characterized precisely as the direct limit:

$$\operatorname{Hom}_{D(\mathcal{A})}(X^\bullet, Y^\bullet) = \varinjlim_{Y'^\bullet} \left(\begin{array}{ccc} X^\bullet & & Y^\bullet \\ & \searrow^{f^\bullet} \quad \swarrow_{s^\bullet} & \\ & Y'^\bullet & \end{array} \right)$$

Then the cohomology functor H^n induced a map from $H^n(X^\bullet)$ to $H^n(Y^\bullet)$:

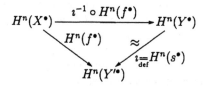

$$H^n(X^\bullet) \xrightarrow{\iota^{-1} \circ H^n(f^\bullet)} H^n(Y^\bullet)$$

with $H^n(f^\bullet)$ and $\iota \underset{\text{def}}{=} H^n(s^\bullet)$, $\approx$, and $H^n(Y'^\bullet)$.

Since spectral sequences associated with double complexes are important for what follows, we will describe these concisely here (see also Remark 3.2.1 below). A double complex in an abelian category $\mathcal{A}$ is a family of objects and maps in the category $\mathcal{A}$

$$\left(C^{p,q}, d^{p,q}_{(1,0)}, d^{p,q}_{(0,1)} \right)_{p,q \in \mathbb{Z}}$$

such that

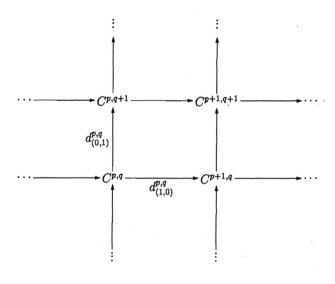

is a commutative diagram and such that $d_{(1,0)}^{p+1,q} \circ d_{(1,0)}^{p,q} = 0$ and $d_{(0,1)}^{p,q+1} \circ d_{(0,1)}^{p,q} = 0$ hold. Then let $C^n = \bigoplus_{p+q=n} C^{p,q}$ and define $d^n : C^n \to C^{n+1}$ by

$$d^n|_{C^{p,q}} = (-1)^p d_{(1,0)}^{p,q} + (-1)^q d_{(0,1)}^{p,q}.$$

It is immediate to verify that $(C^\bullet, d^\bullet)$ is a complex. We can define filtrations on C^n as follows:

$$F^p C^n = \bigoplus_{\substack{p' + q = n \\ p' \geq p}} C^{p',q}$$

and

$$'F^p C^n = \bigoplus_{\substack{q + p' = n \\ p' \geq p}} C^{q,p'}.$$

From those filtrations, the following spectral sequences are induced

$$E_0^{p,q} = C^{p,q}, \quad 'E_0^{p,q} = C^{q,p}$$

$$E_1^{p,q} =_{(0,1)} H^q(C^{p,\bullet}), \quad 'E_1^{p,q} =_{(1,0)} H^q(C^{\bullet,p})$$

$$E_2^{p,q} =_{(1,0)} H^p({}_{(0,1)}H^q(C^{\bullet\bullet})), \quad 'E_2^{p,q} =_{(0,1)} H^p({}_{(1,0)}H^q(C^{\bullet\bullet}))$$

which abut to $E^n = H^n(C^\bullet)$, the n-th cohomology of the associated complex $C^\bullet$ defined above.

For example, let us consider a double complex $(C^{p,q})_{p,q\in\mathbb{Z}}$ in the first quadrant. Namely, $C^{p,q} = 0$ unless $p \geq 0$ and $q \geq 0$. Assume that the double complex is vertically exact, i.e.

$$E_1^{p,q} =_{(0,1)} H^q(C^{p,\bullet}) = 0 \text{ for } q > 0.$$

Then the complex $E_1^{\bullet,0}$ is quasi-isomorphic to $C^\bullet$. This is because we have $E_2^{n,0} =_{(1,0)} H^n(E_1^{\bullet,0})$ and

$$0 = E_2^{n-2,1} \longrightarrow E_2^{n,0} \longrightarrow E_2^{n+2,-1} = 0.$$

Therefore, we have the isomorphisms among their cohomologies:

$$E_2^{n,0} \approx E_3^{n,0} \approx \ldots E_\infty^{n,0} \approx E^n = H^n(C^\bullet).$$

We refer the reader interested in more details to [35].

3.2 Algebraic Geometry and Algebraic Analysis

In this section we will give the reader a first taste of algebraic analysis. Our goal here is to introduce the reader to some basic material, as well as to show the philosophical links which exist between algebraic analysis and algebraic geometry. We will, therefore, begin with some very basics notion from classical algebraic geometry.

One of the most fundamental objects to be studied in algebraic geometry is the zero set of a system given by finitely many relations among finitely many generators for an algebra B over a field A. Let us study this in some detail. Let $x_1, x_2, \ldots, x_n$ be a set of generators for the algebra B over A. Then if we consider the polynomial ring $A[X_1, \ldots, X_n]$ in n variables, we have an exact sequence

$$A[X_1, X_2, \ldots, X_n] \xrightarrow{\varphi} B \longrightarrow 0,$$

where the epimorphism φ is defined by $\varphi(X_i) = x_i$ for $i = 1, 2, \ldots, n$. The kernel of the map φ, denoted as ker φ, is an ideal in $A[X_1, X_2, \ldots X_n]$. Since A is a field (hence noetherian), the Hilbert Basis Theorem implies that the ideal ker φ is finitely generated. Let $f_1(X_1, \ldots, X_n), f_2(X_1, \ldots, X_n), \ldots, f_\ell(X_1, \ldots, X_n)$ be a set of generators for ker φ. Then we can write the A-algebra B as follows:

$$B = A[x_1, x_2, \ldots, x_n],$$

where the generators $x_1, x_2, \ldots, x_n$ satisfy the polynomial relations

(3.2.1)   (P.R.) $\begin{cases} f_1(x_1, \ldots, x_n) = 0 \\ f_2(x_1, \ldots, x_n) = 0 \\ \\ f_\ell(x_1, \ldots, x_n) = 0 \end{cases}$

That is, we have the isomorphism

$$A[X_1, \ldots, X_n]/(f_1, \ldots, f_\ell) \xrightarrow{\approx} B = A[x_1, \ldots, x_n].$$

Let now K' be an extension field of A. Let us try to find a solution in K' for the system (3.2.1); this is equivalent to finding a K'-rational point $(x_1, \ldots, x_n)$ on the algebraic variety consisting of the common zeros of the system (3.2.1), and this is a process which can be described as follows.

To find a K'-rational point on the variety embedded in the Euclidean space $K' \times \ldots \times K'$ means to find a map Ψ from B to K' so that the diagram

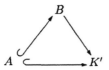

commutes, or, from the point of view of affine schemes (see e.g. Hartshorne [74] for the notions and notations in algebraic geometry).

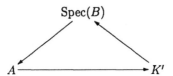

In terms of the relations (3.2.1), the above A-algebra homomorphism Ψ operates as follows:

$$\Psi\left(f_i(x_1,\ldots,x_n)\right) = f_i\left(\Psi(x_1),\ldots\Psi(x_n)\right) = f_i(c_1,\ldots,c_n) = 0,$$

where c_j is the assigned value in K', i.e., $c_j = \Psi(x_j)$ for $j = 1, 2, \ldots, n$. This is exactly what a solution of the system (3.2.1) means.

In algebraic analysis, the above coefficient field A is replaced by the sheaf $\mathcal{D}$ of non-commutative rings of differential operators over $X = \mathbb{C}^n$, whose precise definition will be given in the next section. We also replace the A-algebra B by a $\mathcal{D}$-Module $\mathcal{M}$. Note that we write "$\mathcal{D}$-Modules" instead of "sheaves of germs of $\mathcal{D}$-modules" since no confusion can arise. Then a system of linear partial differential equations will be described as a $\mathcal{D}$-Module $\mathcal{M}$.

Let $\mathcal{M}$ be finitely generated over $\mathcal{D}$, and let $u_1, u_2, \ldots, u_m$ be generators for $\mathcal{M}$ over $\mathcal{D}$. Then we have an exact sequence of $\mathcal{D}$-Modules

$$\mathcal{D}^m \xrightarrow{\bullet u} \mathcal{M} \longrightarrow 0,$$

where the augmentation map $\bullet u$ from the direct sum $\mathcal{D}^m$ to $\mathcal{M}$ is defined by

$$(A_1 U_1 \oplus \ldots \oplus A_m U_m) \bullet u = A_1 u_1 + \ldots + A_m u_m,$$

where $U_i = [0, \ldots, 0, \overset{i}{1}, 0, \ldots, 0], i = 1, \ldots, m$, are the elements of the canonical basis for the free Module $\mathcal{D}^m$. When $\mathcal{D}$ is left noetherian, the submodule $\ker \bullet u$ of $\mathcal{D}^m$ is finitely generated over $\mathcal{D}$. Hence, for a set of generators $v_1, v_2, \ldots, v_\ell$ for the submodule $\ker \bullet u$, there exists an epimorphism

$$\mathcal{D}^\ell \xrightarrow{\bullet v} \ker u \longrightarrow 0,$$

where each v_j can be written as

$$v_j = P_{j1}U_1 + P_{j2}U_2 + \ldots + P_{jm}U_m.$$

Then the map $\bullet v$ is given as

$$(B_1V_1 \oplus \ldots \oplus B_\ell V_\ell) \bullet v = B_1 v_1 + \ldots + B_\ell v_\ell,$$

where the vectors $V_k = [0,\ldots,0,\overset{k}{1},0,\ldots,0], k = 1,\ldots,\ell$, form a canonical basis for the free Module $\mathcal{D}^\ell$. Therefore, we obtain the following commutative diagram:

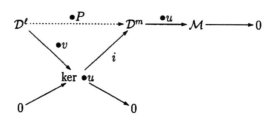

Note that the above map P from $\mathcal{D}^\ell$ to $\mathcal{D}^m$ is the composition of the inclusion $\ker \bullet u \overset{i}{\hookrightarrow} \mathcal{D}^m$ and the epimorphism $\bullet v : \mathcal{D}^\ell \to \ker \bullet u$. Then observe that for any element $[B_1, B_2, \ldots, B_\ell]$ of $\mathcal{D}^\ell$

$$([B_1, \ldots, B_\ell] \bullet P) \bullet u = ([B_1, \ldots, B_\ell]v)i) u = 0,$$

since the image of v is contained in $\ker u$. Then the $\mathcal{D}$-Module homomorphism P may be written as an $\ell \times m$-matrix $[P_{i,j}]$ with entries in $\mathcal{D}$:

$$[P_{i,j}] = \begin{bmatrix} P_{11} & P_{12} & \cdots & P_{1m} \\ P_{21} & P_{22} & \cdots & P_{2m} \\ \vdots & \vdots & \ddots & \vdots \\ P_{\ell 1} & P_{\ell 2} & \cdots & P_{\ell m} \end{bmatrix} = \begin{bmatrix} v_1 \\ v_2 \\ \vdots \\ v_\ell \end{bmatrix}$$

By writing the augmentation map $\bullet u$ in vector form

$$u = \begin{bmatrix} u_1 \\ \vdots \\ u_m \end{bmatrix}$$

the composition map $P \circ u = 0$ becomes a system of $\mathcal{D}_X$-linear relations among the generators $u_1, \ldots, u_m$ for the $\mathcal{D}$-Module $\mathcal{M}$:

$$(3.2.2) \qquad (D.R.) \qquad \begin{cases} P_{11}u_1 + P_{12}u_2 + \ldots + P_{1m}u_m = 0 \\ P_{21}u_1 + P_{22}u_2 + \ldots + P_{2m}u_m = 0 \\ \quad\vdots = \vdots \\ P_{\ell 1}u_1 + P_{\ell 2}u_2 + \ldots + P_{\ell m}u_m = 0 \end{cases}$$

that is, a system of partial differential equations.

Furthermore, the exact sequence

$$\mathcal{D}^\ell \xrightarrow{\ \bullet P\ } \mathcal{D}^m \xrightarrow{\ \bullet u\ } \mathcal{M} \longrightarrow 0,$$

can be extended as

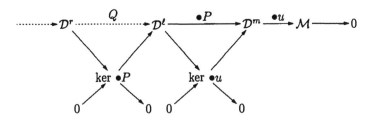

obtaining a free (projective) resolution of $\mathcal{M}$. One may like to write this resolution of $\mathcal{M}$ as follows:

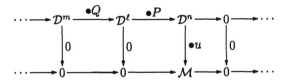

regarding the two row sequences as quasi-isomorphic objects in a suitable derived category.

It is important to regard a system like (3.2.2) as a free resolution of a $\mathcal{D}$-Module $\mathcal{M}$. That is, the systems equivalent to (3.2.2), i.e., isomorphic systems as $\mathcal{D}$-Modules, provide various free resolutions for $\mathcal{M}$.

Let now $\mathcal{O}$ be the sheaf of germs of holomorphic functions over X. Then a local solution of the $\mathcal{D}$-Module $\mathcal{M}$ in $\mathcal{O}$ is a $\mathcal{D}$-homomorphism f from $\mathcal{M}$ to $\mathcal{O}$. Namely, f assigns each generator u_i to a holomorphic function

$$f(u_i) = f_i \text{ for } i = 1, 2, \ldots, n.$$

Then, given the $\mathcal{D}$-homomorphism f, the system (3.2.2)

$$f\left(\sum_{j=1}^{n} P_{ij}u_j\right) = 0$$

implies

$$\sum f(P_{ij}u_j) = \sum P_{ij}f(u_j) = \sum P_{ij}f_j = 0,$$

indicating that $(f_1, \ldots, f_n)$ is a holomorphic solution in the most orthodox sense for the system of partial differential equations. Since we denote the sheaf of $\mathcal{D}$-homomorphisms from $\mathcal{M}$ to $\mathcal{O}$ by $\mathcal{H}om_{\mathcal{D}}(\mathcal{M}, \mathcal{O})$ we now can see that the sheaf $\mathcal{H}om_{\mathcal{D}}(\mathcal{M}, \mathcal{O})$ is in fact the sheaf of holomorphic solutions to the system $\mathcal{M}$. Later on in this chapter, we will consider the higher derived functors $\mathcal{E}xt_{\mathcal{D}}^h(\mathcal{M}, \mathcal{O})$ of $\mathcal{H}om_{\mathcal{D}}(\mathcal{M}, \mathcal{O})$ and we will provide an analytic interpretation for them.

This short discussion should have been sufficient to justify the strong links which exist between algebraic geometry and algebraic analysis. Such a correspondence can be quickly summarized as follows:

Algebraic Geometry

1) A scheme over a base ring

2) A map between schemes

$$X \to Y$$

3) To find K-rational points on a scheme X over A is to construct an A-algebra homomorphism from K to X such that

commutes.

Algebraic Analysis

1') An analytic $\mathcal{D}$-module

2') A $\mathcal{D}$-linear map between $\mathcal{D}$-Modules

$$\mathcal{M} \to \mathcal{M}'$$

3') To find solutions of a system of partial differential equations in a sheaf S of certain functions, i.e., a "known" $\mathcal{D}$-Module, is to construct a $\mathcal{D}$-linear map from $\mathcal{M}$ to S, namely an element of $\mathcal{H}om_{\mathcal{D}}(\mathcal{M}, S)$.

We are now ready to give some significant examples of $\mathcal{D}$-Modules.

Example 3.2.1 The sheaf $\mathcal{O}$ is a $\mathcal{D}$-Module, called de Rham Module. For germs $P = \sum_{\alpha} f_{\alpha}(z)\partial^{\alpha}$ in $\mathcal{D}$ and $h(z) \in \mathcal{O}$, the structure of the $\mathcal{D}$-module is given by $Ph(z) = \sum_{\alpha} f_{\alpha}(z)\partial^{\alpha}h(z) \in \mathcal{O}$, where $\partial^{\alpha} = (\partial/\partial_{z_1})^{\alpha_1} \ldots (\partial/\partial_{z_n})^{\alpha_n}$.

Let us now give a projective resolution for the de Rham Module $\mathcal{O}$. Since one can easily see that

$$\mathcal{O} \approx \mathcal{D}/\mathcal{D}\partial_1 + \ldots + \mathcal{D}\partial_n,$$

where $\partial_i = (\partial/\partial z_i), i = 1, 2, \ldots, n$, we obtain

$$\mathcal{D}^n \xrightarrow{\ \bullet\partial_{[n,1]}\ } \mathcal{D} \xrightarrow{\ \bullet u\ } \mathcal{O} \longrightarrow 0,$$

where $\partial_{[n,1]} = [\partial_1, \partial_2, \ldots, \partial_n]^t$. As a system of relations, we have

(3.2.3) $\partial_i u = 0$ for $i = 1, 2, \ldots, n.$

As we will see in the section on the de Rham Functor in the next chapter, the de Rham functor $\mathcal{H}om_{\mathcal{D}}(\mathcal{O}, -)$ evaluated at $\mathcal{O}$ can be computed as follows:

$$\mathbb{R}\mathcal{H}om_{\mathcal{D}}(\mathcal{O}, \mathcal{O}) \approx \Omega^{\bullet} \bigotimes_{\mathcal{O}} \mathcal{O} \approx \Omega^{\bullet},$$

where $\Omega^{\bullet}$ is the complex of sheaves of differential forms whose coefficients are holomorphic functions. Then the Poincaré lemma states that $\mathcal{H}^q(\Omega^{\bullet}) = 0$ for $q \neq 0$, and $\mathcal{H}^0(\Omega^{\bullet}) \approx \mathbb{C}$. The Poincaré lemma implies simply that $\mathbb{C}$ and $\Omega^{\bullet}$ are quasi-isomorphic.

Hence, it is clear that the sheaf $\mathcal{H}om_{\mathcal{D}}(\mathcal{O}, \mathcal{O})$ of germs of holomorphic solutions is the sheaf $\mathbb{C}$. Hence the derived functor $\mathbb{R}\mathcal{H}om_{\mathcal{D}}(\mathcal{O}, \mathcal{O}) \approx \Omega^{\bullet}$ is a resolution of $\mathbb{C}$.

The left exact functor $\mathcal{H}om_{\mathcal{D}}(-, \mathcal{O})$, on the other hand, takes the above projective resolution for $\mathcal{O}$

$$0 \longrightarrow \mathcal{H}om_{\mathcal{D}}(\mathcal{O}, \mathcal{O}) \longrightarrow \mathcal{H}om_{\mathcal{D}}(\mathcal{D}, \mathcal{O}) \xrightarrow{\ \tilde{\partial}_{[1,n]}\ } \mathcal{H}om_{\mathcal{D}}(\mathcal{D}^n, \mathcal{O})$$

$$0 \longrightarrow \mathbb{C} \longrightarrow \mathcal{O} \xrightarrow{\ [\partial_1, \ldots, \partial_n]\ } \mathcal{O}^n$$

The second and third vertical isomorphisms in this diagram are induced by the $\mathcal{D}$-linear maps which assign the canonical generators to the elements of $\mathcal{O}$.

The $\mathcal{D}$-Module $\mathcal{O}$ is a holonomic $\mathcal{D}$-Module, in the sense of Chapter 5. (A maximally overdetermined system is the older terminology for a holonomic system or a holonomic $\mathcal{D}$-Module.)

Example 3.2.2   Another extreme case of a $\mathcal{D}$-Module contrasting the $\mathcal{D}$-Module $\mathcal{O}$, is the $\mathcal{D}$-Module $\mathcal{D}$ itself. That is to say the $\mathcal{D}$-module for which there are no relations. Then, since there is no restriction, the sheaf of local solutions of the $\mathcal{D}$-Module $\mathcal{D}$ in the sheaf $\mathcal{O}$, will be the whole $\mathcal{O}$. This is what the canonical isomorphism $\mathcal{H}om_{\mathcal{D}}(\mathcal{D}, \mathcal{O}) \xrightarrow{\approx} \mathcal{O}$ means, i.e., each element $f \in \mathcal{H}om_{\mathcal{D}}(\mathcal{D}, \mathcal{O})$ is mapped to $f(1) \in \mathcal{O}$.

In general, for a projective resolution

$$\cdots \to \mathcal{D}^n \xrightarrow{Q} \mathcal{D}^{\ell} \xrightarrow{P} \mathcal{D}^m \xrightarrow{u} \mathcal{M} \to 0$$

of a $\mathcal{D}$-Module $\mathcal{M}$, the left exact contravariant functor $\mathcal{H}om_{\mathcal{D}}(-, \mathcal{O})$ induces the following complex:

$$(3.2.4) \qquad 0 \longrightarrow \mathcal{H}om_{\mathcal{D}}(\mathcal{M}, \mathcal{O}) \xrightarrow{\tilde{u}^{\bullet}} \mathcal{O}^m \xrightarrow{\tilde{P}^{\bullet}} \mathcal{O}^{\ell} \xrightarrow{\tilde{Q}^{\bullet}} \mathcal{O}^n \longrightarrow \ldots,$$

where $\tilde{Q} \circ \tilde{P} = \mathcal{H}om_{\mathcal{D}}(Q, \mathcal{D}) \circ \mathcal{H}om_{\mathcal{D}}(P, \mathcal{D}) = \mathcal{H}om_{\mathcal{D}}(P \circ Q, \mathcal{D}) = 0$. Then observe that $\mathcal{H}om_{\mathcal{D}}(\mathcal{M}, \mathcal{O}) \cong \ker \tilde{P}$ is the solution sheaf of the partial differential operators $\tilde{P}u = 0$. Therefore, the 0-th cohomology of the complex (3.2.4) is isomorphic to the solution sheaf $\mathcal{H}om_{\mathcal{D}}(\mathcal{M}, \mathcal{O})$, and the first cohomology of (3.2.4), i.e. $\ker \tilde{Q}/\mathrm{im}\tilde{P}$, is the first derived functor $\mathcal{E}xt^1_{\mathcal{D}}(\mathcal{M}, \mathcal{O}) = R^1\mathcal{H}om_{\mathcal{D}}(-, \mathcal{O})(\mathcal{M})$. In the sense of derived categories, the complex (3.2.4) is precisely $\mathbb{R}\mathcal{H}om_{\mathcal{D}}(\mathcal{M}, \mathcal{O})$, whose h-th cohomology is given by $\mathcal{H}^h\left(\mathbb{R}\mathcal{H}om_{\mathcal{D}}(\mathcal{M}, \mathcal{O})\right) \overset{\mathrm{def}}{=} \mathbb{R}^h\mathcal{H}om_{\mathcal{D}}(\mathcal{M}, \mathcal{O}) \approx \mathcal{E}xt^h_{\mathcal{D}}(\mathcal{M}, \mathcal{O})$.

For example, $\mathcal{E}xt^1_{\mathcal{D}}(\mathcal{M}, \mathcal{O}) = 0$ means $\ker \tilde{Q}/\mathrm{im}\tilde{P} = 0$, namely if $\tilde{Q}f = 0$ for $f \in \mathcal{O}^{\ell}$, then there exists u in $\mathcal{O}^m$ satisfying $\tilde{P}u = f$.

When the sequence is exact, that is $\mathcal{H}om_{\mathcal{D}}(-, \mathcal{O})$ is an exact functor, all the "higher" solution sheaves $\mathcal{E}xt^q_{\mathcal{D}}(\mathcal{M}, \mathcal{O})$ vanish. Then there is an induced doubly indexed spectral sequence $E^{p,q}_2$ with the property

$$E^{p,q}_2 = H^p(X, \mathcal{E}xt^q_{\mathcal{D}}(\mathcal{M}, \mathcal{O})) = 0 \text{ for } q \neq 0.$$

The sheaf of higher global solutions over X, $Ext_D^n(X, \mathcal{M}, \mathcal{O})$ can now be computed as the abutment from the above spectral sequence:

$$0 = E_2^{p,1} \longrightarrow E_2^{p,0} \longrightarrow E_2^{p+2,-1} = 0.$$

Namely, $E_2^{p,0} = H^p(X, \mathcal{H}om_D(\mathcal{M}, \mathcal{O})) \cong E_3^{p,0} \cong \ldots \cong E_\infty^{p,0} \cong E^p$, and $Ext_D^p(X, \mathcal{M}, \mathcal{O}) = H^p(X, \mathcal{N}^\bullet)$, for $\mathcal{N}^\bullet$ the complex in (3.2.4).

Remark 3.2.1 Let us provide a little more background on the notion of abutment and on the relationships between spectral sequences and derived categories. Consider a doubly indexed spectral sequence in an abelian category $\mathcal{A}$ i.e. a family of objects $\{E_r^{p,q}, E^n; p, q, r, n \in \mathbb{Z}\}$ such that there exists a map

$$d_r^{p,q} : E_r^{p,q} \longrightarrow E_r^{p+r,q-r+1}$$

satisfying $d_r^{p+r,q-r+1} \circ d_r^{p,q} = 0$. Furthermore there exists an isomorphism

$$d_r^{p,q} : Ker d_r^{p,q} / Im d_r^{p-r,q+r-1} \xrightarrow{\approx} E_{r+1}^{p,q}.$$

An object E^n is said to be the abutment of the spectral sequence when E^n is a filtered object: $E^n \supset \ldots \supset F^p(E^n) \supset F^{p+1}(E^n) \supset \ldots$ and these inclusions satisfy $F^p(E^n)/F^{p+1}(E^n) \xrightarrow{\approx} E_\infty^{p,q}$ and $E^n = \bigoplus_{p+q=n} E_\infty^{p,q}$.

The following spectral sequence, called the spectral sequence of a composite functor, is also useful. Let $\mathcal{A}$ and $\mathcal{B}$ be abelian categories having enough injectives. Let F be a left exact functor from $\mathcal{A}$ to $\mathcal{B}$ and let G be a left exact functor from $\mathcal{B}$ to an abelian category $\mathcal{C}$. We assume that F takes injective objects in $\mathcal{A}$ into G-acyclic objects in $\mathcal{B}$, i.e., the higher derived functors $R^i G(F(I))$ vanish for any injective object I in $\mathcal{A}$. This induces a doubly indexed cohomological spectral sequence in $\mathcal{C}$ beginning with

$$E_2^{p,q} = R^p G(R^q F(A))$$

such that the derived functor of the composite functor $G \circ F$ $E^n = R^n(G \circ F(A))$ is its abutment.

One might ask how the theory of the spectral sequence of a composite functor can be rephrased in the language of derived categories. The answer is quite simple, as follows. With the same assumptions given above, let $X^\bullet$ be an object in $D^+(\mathcal{A}) = \{X^\bullet \in D(\mathcal{A}); H^k(X^\bullet) = 0$ for sufficiently small $k\}$, i.e. $D^+(\mathcal{A}) = K^+(\mathcal{A})_{(QIS)}$. Let $I^+(\mathcal{A})$ be a full subcategory of $K^+(\mathcal{A})$ such that every object in $I^+(\mathcal{A})$ is a complex of injective objects of $\mathcal{A}$. Then for each object $X^\bullet$ in $K^+(\mathcal{A})$ there exists a unique complex $I^\bullet$ in $I(\mathcal{A})$ such that $X^\bullet$ and $I^\bullet$ are quasi-isomorphic. Namely, we have a functor

$$Q' : K(\mathcal{A}) \rightsquigarrow I(\mathcal{A})$$

sending quasi-isomorphisms in $K^+(\mathcal{A})$ to isomorphisms in $I^+(\mathcal{A})$. By the universal mapping property of the localization, there exists a unique functor U such that the diagram

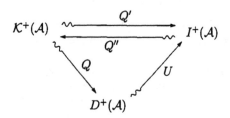

commutes. We also have an embedding functor Q'' from $I^+(\mathcal{A})$ to $K^+(\mathcal{A})$. Then its composition with the localization Q, $Q \circ Q''$ is a functor from $I^+(\mathcal{A})$ to $D^+(\mathcal{A})$. Now the uniqueness of an object in $I(\mathcal{A})$ corresponding to an object in $K^+(\mathcal{A})$, i.e., the definition of Q', implies that $Q \circ Q''$ and U are inverse to each other. That is $I^+(\mathcal{A}) \approx D^+(\mathcal{A})$.

Let now F be a left exact functor from $\mathcal{A}$ to $\mathcal{B}$; we can define its right derived functor $\mathbb{R}F$ from $D^+(\mathcal{A})$ to $D^+(\mathcal{B})$ by setting

$$\mathbb{R}F(X^\bullet) = F(I^\bullet),$$

where $I^\bullet = Q'X^\bullet$ is an object in $I(\mathcal{A})$. In order to compute the right derived functor of $F \circ G$, we note that

$$(\mathbb{R}G \circ \mathbb{R}F)X^\bullet = \mathbb{R}G(F(I^\bullet)).$$

Since FI^k is a G-cyclic object in $\mathcal{B}$, we now obtain that

$$\mathbb{R}G(F(I^\bullet)) = G(F(I^\bullet)) = (G \circ F)I^\bullet = \mathbb{R}(G \circ F)X^\bullet.$$

Therefore we have shown that the derived category version of the spectral sequence of the composite functor is simply

$$\mathbb{R}G \circ \mathbb{R}F = \mathbb{R}(G \circ F).$$

As an application of this concept, we will now discuss the notion of hyperderived functor (hypercohomology) of a left exact functor. Let F be a left exact functor from an abelian category $\mathcal{A}$ with enough injectives to another abelian category $\mathcal{B}$. Then F induces a functor Co F from the abelian category of positively indexed complexes in $\mathrm{Co}^+(\mathcal{A})$ of $\mathcal{A}$ to the abelian category of positively indexed complexes $\mathrm{Co}^+(\mathcal{B})$ of $\mathcal{B}$ such that

$$\mathrm{Co}F(X^\bullet) = F(X^\bullet)$$

for any $X^\bullet \in \mathrm{Co}^+(\mathcal{A})$. We therefore obtain the following commutative diagram of categories and functors:

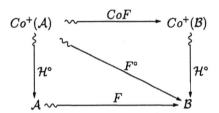

where $\mathcal{H}^0(X^\bullet) = \mathrm{Ker}\,(d^0 : X^0 \to X^1)$ is the zero-th cohomology of $X^\bullet$. Then, for the left exact functor F, we can define a functor F^0 from $\mathrm{Co}^+(\mathcal{A})$ to $\mathcal{B}$ by setting

$$
\begin{aligned}
F^0 := \mathcal{H}^0 \circ \mathrm{Co}F &= \mathrm{Ker}\,(Fd^0 : FX^0 \to FX') \\
&= F(\mathrm{Ker}(d^0 : X^0 \to X')) \\
&= F \circ \mathcal{H}^0.
\end{aligned}
$$

Consequently, we obtain two spectral sequences

$$
'E_2^{p,q} = R^p F(\mathcal{H}^q(X^\bullet))
$$

and

$$
E_1^{p,q} = R^q F(X^p)
$$

which abut to the hyperderived functor $F^n X^\bullet$ of F.

Let us now consider a complex non-singular analytic variety V. Let $\mathcal{O}_V$ be the sheaf of germs of holomorphic functions on V, and let $\Omega_V^\bullet$ be the sheaf of exterior algebra of $\mathbb{C}$-differentials on V. Since Poincaré lemma implies that $\mathcal{H}^q(\Omega_V^\bullet) = 0$ for $q \neq 0$, we see that the complex $\Omega_V^\bullet$ is acyclic. Therefore, the spectral sequence corresponding to the above $'E_2^{p,q}$ becomes

$$
'E_2^{p,q} = H^p(V, \mathcal{H}^q(\Omega_V^\bullet)) = 0,
$$

for any $q \neq 0$. The only non-vanishing term is given by

$$
'E_2^{p,0} = H^p(V, \mathbb{C}),
$$

since $\mathbb{C} = Ker(d^0 : \mathcal{O}_V \to \Omega_V^1)$.

Consider now the following sequence:

$$0 = {}'E_2^{p-2,1} \xrightarrow{d_2^{p-2,1}} {}'E_2^{p,0} \xrightarrow{d_2^{p,0}} {}'E_2^{p+2,0-2+1} = 0.$$

From it, we deduce

$${}'E_2^{p,0} \approx {}'E_3^{p,0} \approx \ldots \approx {}'E_\infty^{p,o}$$

which is isomorphic to the abutment $E^p = \bigoplus_{p+0=p} E_\infty^{p,0} = E_\infty^{p,0}$, i.e.,

$${}'E_2^{p,0} = H^p(V,\mathbb{C}) \approx E^p = H^p(V,\Omega_V^\bullet).$$

Furthermore, if V is Stein, the first spectral sequence becomes

$$E_1^{p,q} = H^q(V,\Omega_V^p) = 0 \text{ for } q \neq 0.$$

Consider the sequence

$$
\begin{array}{ccccccc}
\cdots \longrightarrow & E_1^{p-1,0} & \xrightarrow{d_1^{p-1,0}} & E_1^{p,0} & \xrightarrow{d_1^{p,0}} & E_1^{p+1,0} & \longrightarrow \cdots \\
& \| & & \| & & \| & \\
\cdots \longrightarrow & \Gamma(V,\Omega_V^{p-1}) & \xrightarrow{\Gamma(V,d_V^{p-1})} & \Gamma(V,\Omega_V^p) & \xrightarrow{\Gamma(V,d_V^p)} & \Gamma(V,\Omega_V^{p+1}) & \longrightarrow \cdots
\end{array}
$$

Then $E_2^{p,0} = Ker\ d_1^{p,0}/Im\ d_1^{p-1,0}$ is nothing but the p-th cohomology of the de Rham complex of V. Since we have, as before, $E_2^{p,0} \approx E_3^{p,0} \approx \ldots \approx E_\infty^{p,0}$, we can compute the hypercohomology by $H^p(V,\Omega_V^\bullet) \cong E^p \approx E_\infty^{p,0} \approx E_2^{p,0}$ which is $H^p(\Gamma(V,\Omega_V^\bullet))$. Summarizing the above discussion, we have:

$$
\begin{array}{ccccc}
H^p(V,\mathbb{C}) & \cong & H^p(V,\Omega_V^\bullet) & \cong & H^p(\Gamma(V,\Omega_V^\bullet)) \\
\| & & \| & & \| \\
{}'E_2^{p,0} & \xrightarrow{\approx} & E^p & \xleftarrow{\approx} & E_2^{p,0}
\end{array}
$$

Remark 3.2.2 Let $\mathcal{M}'$ be a finitely generated $\mathcal{D}$-Module with generators $v_1,\ldots,v_{m'}$, i.e. $\mathcal{M}' = \mathcal{D}v_1 + \ldots + \mathcal{D}v_{m'}$. Let $\bullet v : \mathcal{D}^{m'} \to \mathcal{M}'$ be the augmentation map induced by the generators.

Choose $\bullet Q : D^{\ell'} \to \mathcal{D}^{m'}$ in such a way that $Ker(\bullet v) = Im(\bullet Q)$.

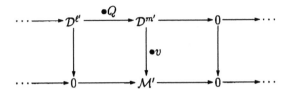

Then the map $\bullet v$ induces an isomorphism from the 0-th cohomology $\mathcal{D}^{m'}/Im(\bullet Q)$ of the top row of the diagram to the 0-th cohomology of $\mathcal{M}'$ in the bottom row. If the $\mathcal{D}$-Module $\mathcal{M}'$ is isomorphic to a $\mathcal{D}$-Module $\mathcal{M}$ determined by a projective resolution

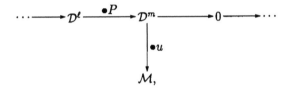

then the complexes $\ldots \to \mathcal{D}^{\ell'} \xrightarrow{\bullet Q} \mathcal{D}^{m'} \to 0 \to \ldots$ and $\ldots \to \mathcal{D}^{\ell} \xrightarrow{\bullet P} \mathcal{D}^{m} \to 0 \to$ $\ldots$ are exact, except at the 0-th spot. Their 0-th cohomologies are moreover isomorphic to each other. This shows that the notion of a $\mathcal{D}$-Module is an intrinsic characterization of a system of partial differential equations.

In the next chapter, a single $\mathcal{D}$-Module M will be generalized to a complex of $\mathcal{D}$-Modules $\mathcal{M}^{\bullet}$ replacing a projective resolution by a quasi-isomorphic complex of objects.

Example 3.2.3 Let $\mathcal{M}$ and $\mathcal{M}'$ be cyclic $\mathcal{D}$-Modules generated by u and v satisfying the ordinary differential equations $(xD-\lambda)u = 0$ and $(xD-\lambda-1)v = 0$ respectively, where $\lambda \neq 1$. This can be rewritten as

$$\begin{cases} \mathcal{M} = \mathcal{D}u \\ (xD - \lambda)u = 0, \ \lambda \neq 1, \end{cases}$$

and

$$\begin{cases} \mathcal{M}' = \mathcal{D}v \\ (xD - \lambda - 1)v = 0, \ \lambda \neq 1. \end{cases}$$

In terms of free resolutions, we have

$$\cdots \longrightarrow \mathcal{D} \xrightarrow{\ \bullet(xD - \lambda)\ } \mathcal{D} \xrightarrow{\ \bullet u\ } \mathcal{M} \longrightarrow 0$$

and

$$\cdots \longrightarrow \mathcal{D} \xrightarrow{\ \bullet(xD - \lambda - 1)\ } \mathcal{D} \xrightarrow{\ \bullet v\ } \mathcal{M}' \longrightarrow 0.$$

We claim that the cyclic $\mathcal{D}$-Modules $\mathcal{M} = \mathcal{D}u = \mathcal{D}/\mathcal{D}(xD - \lambda)$ and $\mathcal{M}' = \mathcal{D}v = \mathcal{D}/\mathcal{D}(xD - \lambda - 1)$ are isomorphic. First note that $Dx - xD = 1$ holds in $\mathcal{D}$. Now, if u is a "solution" of the ordinary differential equation $(xD - \lambda)u = 0$, put $v = xu$ so to have

$$(xD - \lambda - 1)v = (xD - \lambda - 1)xu = (xDx - \lambda x - x)u$$

$$= (x(xD + 1) - \lambda x - x)u = (x^2 D + x - \lambda x - x)u = x(xD - \lambda)u = 0.$$

On the other hand

$$Dv = Dxu = (xD+1)u = (xD-\lambda+\lambda+1)u = (xD-\lambda)u+(\lambda+1)u = (\lambda+1)u.$$

Let now

$$u = \left(\frac{1}{\lambda+1}D\right)v.$$

Then notice that we obtain

$$xu = x\frac{1}{\lambda+1}Dv = \frac{1}{\lambda+1}xDv = \frac{1}{\lambda+1}(\lambda+1)v = v$$

and also

$$\frac{1}{\lambda+1}Dxu = \frac{1}{\lambda+1}(xD+1)u = \frac{1}{\lambda+1}(xD-\lambda+\lambda+1)u$$

$$= \frac{1}{\lambda+1}\left((xD-\lambda)u+(\lambda+1)u\right) = u.$$

Therefore, we obtain an isomorphism between $\mathcal{M}$ and $\mathcal{M}'$ given by

$$\mathcal{D}/\mathcal{D}(xD - \lambda) \longleftrightarrow \mathcal{D}/\mathcal{D}(xD - \lambda - 1)$$

$$\bar{1} \longmapsto \overline{\tfrac{1}{\lambda+1}D}$$

$$\bar{x} \longleftarrow\mapsto \bar{1},$$

where $\bar{}$ indicates the equivalence class modulo an ideal.

3.3 Filtrations and Characteristic Varieties

Let X be a complex manifold of dimension n. We will denote by $\mathcal{D}$ the sheaf of germs of linear partial differential operators with holomorphic coefficients of finite order over X. The stalk $\mathcal{D}_z$ at z consists of all finite sums

$$(3.3.1) \qquad P(z, \partial) = \sum_\alpha a_\alpha(z) \partial^\alpha,$$

where $\alpha = (\alpha_1, \ldots, \alpha_n)$ takes on a finite number of n-tuples with $\alpha_i \geq 0, i = 1, \ldots, n$, $\partial^\alpha = (\partial/\partial z_1)^{\alpha_1} \ldots (\partial/\partial z_n)^{\alpha_n}$, and the coefficients $a_\alpha(z)$ are germs of holomorphic functions at z. Then $\mathcal{D}$ has locally the following increasing filtration

$$\mathcal{O} = \mathcal{D}^{(0)} \subset \mathcal{D}^{(1)} \subset \ldots \subset \mathcal{D}^{(k)} \subset \ldots \subset \mathcal{D}, \mathcal{D}^{(-1)} = \{0\},$$

where the $\mathcal{O}$-Submodule $\mathcal{D}^{(k)}$ is defined at each stalk by

$$\mathcal{D}_z^{(k)} = \{P(z, \partial) \in \mathcal{D}_z \mid P(z, \partial) = \sum_{|\alpha| \leq k} a_\alpha(z) \partial^\alpha, |\alpha| = \alpha_1 + \ldots a + \alpha_n\}.$$

Notice that we have $\mathcal{D}_z = \bigcup_{k=0}^\infty \mathcal{D}_z^{(k)}$. The increasing filtration $\{\mathcal{D}^{(k)}\}_{k \geq 0}$ at each stalk has the following fundamental properties:

$$\mathcal{D}^{(k)} \mathcal{D}^{(k')} \subset \mathcal{D}^{(k+k')},$$

and

$$\text{for } P \in \mathcal{D}^{(k)} \text{ and } Q \in \mathcal{D}^{(k')}$$

we have that the Lie bracket $[P, Q] = PQ - QP$ belongs to $\mathcal{D}^{(k+k'-1)}$, that is $[\mathcal{D}^{(k)}, \mathcal{D}^{(k')}] \subset \mathcal{D}^{(k+k'-1)}$ holds. This second property follows from the well-known fact that the principal parts of P and Q commute.

Remark 3.3.1 Locally speaking, the non-commutative Ring $\mathcal{D}$ can be written as

$$\left\{ \sum_{|\alpha_1 + \ldots + \alpha_n| \leq k} \left(\sum a_{\alpha_1' \ldots \alpha_n'} z_1^{\alpha_1'} z_2^{\alpha_2'} \ldots z_n^{\alpha_n'} \right) \partial_1^{\alpha_1} \partial_2^{\alpha_2} \ldots \partial_n^{\alpha_n}, k \geq 0, a_{\alpha_1' \ldots \alpha_n'} \in \mathcal{C} \right\}$$

using a local coordinate system $(z_1, \ldots, z_n, \partial_1, \ldots, \partial_n)$, and where the series $\sum a_{\alpha_1' \ldots \alpha_n'} z_1^{\alpha_1'} \ldots z_n^{\alpha_n'}$ belongs to $\mathcal{O}$. Namely, for a small neighborhood U of a point, we have

$$\mathcal{D}(U) = \mathcal{O}(U) \bigotimes_{\mathcal{C}} \mathcal{C}[\partial_1, \ldots, \partial_n].$$

In other words,

$$\mathcal{D}(U) = \{P \in \text{Hom}_{\mathbb{C}}(\mathcal{O}(U), \mathcal{O}(U)) : [\ldots[[P, f_1]f_2]\ldots]f_k] \in \mathcal{O} \text{ for some } k\}.$$

In particular, for $Q = f_1$ in $\mathcal{D}^{(0)} = \mathcal{O}$, we have $[P, f_1] \in \mathcal{D}^{(k-1)}$; conversely, if $[P, f_1] \in \mathcal{D}^{(k-1)}$ for all $f \in \mathcal{D}^{(0)}$, then P must be in $\mathcal{D}^{(k)}$. Therefore, we can also define $\mathcal{D}^{(k)}$ as follows:

$$\mathcal{D}^{(k)} = \{P \in \mathcal{D} | [\ldots[[P, f_1]f_2]\ldots]f_k] \in \mathcal{D}^{(0)} = \mathcal{O} \text{ for all } f_1, f_2, \ldots, f_k \text{ in } \mathcal{O}\}.$$

A differential operator P in $\mathcal{D}$ is said to be of order k if $P \in \mathcal{D}^{(k)}$ and $P \notin \mathcal{D}^{(k-1)}$. Then the canonical image of P in $\mathcal{D}^{(k)}/\mathcal{D}^{(k-1)}$ is the principal symbol of P. Namely, if $P = \sum_{|\alpha| \le k} a_\alpha(z)\partial^\alpha$ is of order k, then the principal symbol of P is $\sigma_k(P) = \sum_{|\alpha|=k} a_\alpha(z)\xi^\alpha$, which is a homogeneous polynomial of degree k in the cotangent coordinates $\xi = (\xi_1, \ldots, \xi_n)$ and a holomorphic function in $z = (z_1, \ldots, z_n)$. We have the following ring isomorphism $\oplus \sigma_k$ from the associated graded sheaf of rings to the sheaf of graded rings of holomorphic functions that are homogeneous with respect to the cotangent coordinates:

$$(3.3.2) \qquad \oplus \sigma_k : \bar{\mathcal{D}} = \bigoplus_{k \ge 0} \left(\mathcal{D}^{(k)}/\mathcal{D}^{(k-1)}\right) \xrightarrow{\approx} \bigoplus_{k \ge 0} \mathcal{O}_{T^* X}(k);$$

this homomorphism satisfies

$$(3.3.3) \qquad \sigma_{k+k'}(PQ) = \sigma_k(P) \cdot \sigma_{k'}(Q)$$

for any $P \in \mathcal{D}^{(k)}$ and $Q \in \mathcal{D}^{(k')}$. Note that locally $\bigoplus_{k \ge 0} \mathcal{O}_{T^* X}(k)$ is just $\mathcal{O}[\xi_1, \ldots, \xi_n]$, the ring of polynomials with coefficient in $\mathcal{O}$, where $\xi_i = \sigma_1(\frac{\partial}{\partial z_i})$. This shows that the associated graded sheaf of rings $\bar{\mathcal{D}}$ is canonically isomorphic to the symmetric algebra of the holomorphic tangent sheaf. Thus, through the notion of a filtration, we have obtained the commutative object $\bar{\mathcal{D}}$ from the non-commutative object $\mathcal{D}$.

We will now define a filtration on the direct sum $\mathcal{D}^m$, as follows. Let $U_i = [0, 0, \ldots, 0, \overset{i}{\overbrace{1}}, 0, \ldots, 0], i = 1, 2, \ldots, m$, be the canonical generators for $\mathcal{D}^m$: define

$$(3.3.4) \qquad (\mathcal{D}^m)^{(k)} = \bigoplus_{i=1}^{m} \mathcal{D}^{(k-l_i)} U_i \underset{\text{def}}{\overrightarrow{=}} \bigoplus \mathcal{D}^{(k)}(-l_i)U_i$$

where l_i are integers. It is easily seen that $\{(\mathcal{D}^m)^{(k)}\}_{k=0,1,\ldots}$ is an increasing filtration on $\mathcal{D}^{(m)}$.

Let now $\mathcal{M}$ be a finitely generated $\mathcal{D}$-Module $\mathcal{M}$: we need to introduce a special kind of filtration which we will call a "good filtration": such a filtration is locally defined by

(3.3.5) $$\mathcal{M}^{(k)} = \mathcal{D}^{(k-l_1)}u_1 + \mathcal{D}^{(k-l_2)}u_2 + \ldots + \mathcal{D}^{(k-l_m)}u_m,$$

where $l_i \in \mathbb{Z}, i = 1, 2, \ldots, m$, and $u_1, u_2, \ldots, u_m$ are generators for $\mathcal{M}$ over $\mathcal{D}$. Notice that we have

(3.3.6) $$\mathcal{M}^{(k)} \subset \mathcal{M}^{(k+1)}, \mathcal{M} = \bigcup_k \mathcal{M}^{(k)},$$

and

$$\mathcal{D}^{(k)}\mathcal{M}^{(h)} \subset \mathcal{M}^{(k+h)} \text{ for integers } k \text{ and } h.$$

Now that we have a good filtration on a finitely generated $\mathcal{D}$-module $\mathcal{M}$, we can define the $\bar{\mathcal{D}}$-Module $\bar{\mathcal{M}}$ by setting

$$\bar{\mathcal{M}} = \bigoplus \mathcal{M}^{(k)}/\mathcal{M}^{(k-1)},$$

where the $\bar{\mathcal{D}}$-Module structure on $\bar{\mathcal{M}}$ is given by

(3.3.7) $$\bar{P} \cdot \bar{v} = \overline{Pv},$$

where $\bar{P} = \bigoplus_{k\geq 0} \overline{P^{(k)}} \in \bar{\mathcal{D}}, \overline{P^{(k)}} \in \mathcal{D}^{(k)}/\mathcal{D}^{(k-1)}, P^{(k)} \in \mathcal{D}^{(k)}$, and $\bar{v} = \bigoplus_h \overline{v^{(h)}} \in \overline{\mathcal{M}}, \overline{v^{(h)}} \in \mathcal{M}^{(h)}/\mathcal{M}^{(h-1)}, v^{(h)} \in \mathcal{M}^{(h)}$, and the product $\overline{Pv}$ is defined by $\overline{Pv} = \bigoplus \overline{P^{(k)}v^{(h)}} \in \bar{\mathcal{M}}, \overline{P^{(k)}v^{(h)}} \in \mathcal{M}^{(k+h)}/\mathcal{M}^{(k+h-1)}$.

Let, for another set of generators $u'_1, \ldots, u_{m'}$ of $\mathcal{M}$,

$$\mathcal{M}_0^{(k)} = \mathcal{D}^{(k-l'_1)}u'_1 + \ldots + \mathcal{D}^{(k-l'_m)}u_{m'}$$

be another good filtration for the $\mathcal{D}$-Module $\mathcal{M}$ (in particular $\mathcal{M} = \bigcup_k \mathcal{M}_0^{(k)}$). Then, for any k, one can find k' large enough to have $\mathcal{M}_0^{(k)} \subset \mathcal{M}^{(k')}$ and for any k' one can find k small enough to have $\mathcal{M}^{(k)} \subset \mathcal{M}_0^{(k')}$, i.e., the topologies induced by the filtrations $\{\mathcal{M}^{(k)}\}$ and $\{\mathcal{M}_0^{(k)}\}$ are equivalent.

Let $\mathcal{N}$ be a $\mathcal{D}$-submodule of $\mathcal{M}$: one can obtain an induced filtration by simply putting $\mathcal{N}^{(k)} = \mathcal{M}^{(k)} \cap \mathcal{N}$. Furthermore, for an epimorphism φ of $\mathcal{D}$-Modules

$$\mathcal{M} \xrightarrow{\varphi} \mathcal{M}'' \longrightarrow 0,$$

one can define a filtration on $\mathcal{M}''$ as

$$\mathcal{M}''^{(k)} = \varphi\left(\mathcal{M}^{(k)}\right).$$

Obviously, for a finitely generated $\mathcal{D}$-Module $\mathcal{M}$ represented by

$$\mathcal{D}^m \xrightarrow{\bullet u} \mathcal{M} \longrightarrow 0,$$

the good filtration (3.3.5) is given by the epimorphism $\bullet u$ applied to the filtration (3.3.4) on $\mathcal{D}^m$.

Consider now a cyclic $\mathcal{D}$-Module $\mathcal{M}$ with one relation, i.e.

$$\begin{cases} \mathcal{M} = \mathcal{D}u \\ Pu = 0, \quad P \in \mathcal{D}. \end{cases}$$

Assume that P is of order k, i.e., $P \in \mathcal{D}^{(k)}$ but $P \notin \mathcal{D}^{(k-1)}$, and let P^k be the principal part of P, i.e. $P = P^k + P^{(k-1)}, P^{(k-1)} \in \mathcal{D}^{(k-1)}$. A good filtration on $\mathcal{M}$ is of the form

$$\mathcal{M}^{(\ell)} = \mathcal{D}^{(\ell-i)}u.$$

Let $\bar{v} \in \bar{\mathcal{M}}$ be arbitrary, $\bar{v} = \oplus \overline{v^{(h)}}, \overline{v^{(h)}} \in \mathcal{M}^{(h)}/\mathcal{M}^{(h-1)}$. Then the principal part $P^k \in \mathcal{D}^{(k)}/\mathcal{D}^{(k-1)} \subset \bar{\mathcal{D}}$, annihilates $\bar{v}$ as follows.
Since $v^{(h)}$ is in $\mathcal{M}^{(h)}, v^{(h)}$ can be written as

$$v^{(h)} = Q^{(h-i)}u, \text{ where } Q^{(h-i)} \in \mathcal{D}^{(h-i)}.$$

Hence $\overline{v^{(h)}}$ is written as $Q^{h-i}u$, where Q^{h-i} is the principal part of $Q^{(h-i)}$. Then $P^k\overline{v^{(h)}} = P^kQ^{h-i}u = Q^{h-i}P^ku = Q^{h-i}(-P^{(k-1)}u)$ holds, since $Pu = (P^k + P^{(k-1)})u = 0$ implies $P^ku = -P^{(k-1)}u$. As $Q^{h-i}P^{(k-1)}u$ belongs to $\mathcal{D}^{(h-i+k-1)}u = \mathcal{M}^{(h+k-1)}$, we have $P^k\overline{v^{(h)}} = 0$ in $\mathcal{M}^{(h+k)}/\mathcal{M}^{(h+k-1)}$ for all h.

In general, for a finitely generated $\mathcal{D}$-Module $\mathcal{M}$, let $\mathcal{J}(\mathcal{M})$ be the annihilator ideal of the $\bar{\mathcal{D}}$-Module $\bar{\mathcal{M}}$ that is $\mathcal{J}(\mathcal{M}) = \{\bar{P} \in \bar{\mathcal{D}}; \bar{P}\bar{\mathcal{M}} = \bar{0}\}$. For another good filtration $\{\mathcal{M}_0^{(k)}\}$, we also obtain the associated annihilator ideal $\mathcal{J}_0(\mathcal{M})$. Then let $\bar{P}$ be an arbitrary element of the radical ideal $\sqrt{\mathcal{J}(\mathcal{M})}$ of $\mathcal{J}(\mathcal{M})$. For some k, $\bar{P}$ is the principal symbol P^k of some element $P^{(k)}$ in $\mathcal{D}^{(k)}$. Then there exists t so that $\bar{P}^t = (P^k)^t$ belongs to $\mathcal{J}(\mathcal{M})$. Therefore, $(P^k)^t \cdot \bar{v}^{(h)}$ is in $\mathcal{M}^{(kt+h-1)}$ for any $\bar{v}^{(h)}$ in $\mathcal{M}^{(h)}$ and all h. Hence, for any $\bar{v}^{(kt+h-1)}$ in $\mathcal{M}^{(kt+h-1)}$ we have

$$(P^k)^t \cdot \bar{v}^{(kt+h-1)} \in \mathcal{M}^{(kt+h-1+kt-1)} = \mathcal{M}^{(2kt-2+h)}.$$

Consequently, we have that $(P^k)^{2t} \cdot \bar{v}^{(h)}$ belongs to $\mathcal{M}^{(2kt-2+h)}$.
Repeating this process s times, we get

(3.3.8) $(P^k)^{st}\bar{v}^{(h)} \in \mathcal{M}^{(skt-s+h)}.$

As we noted earlier, for another good filtration $\{\mathcal{M}_0^{(h)}\}$ the relations

(3.3.9) $\mathcal{M}^{(h')} \subset \mathcal{M}_0^{(h)}$ and

(3.3.10) $\mathcal{M}_0^{(h)} \subset \mathcal{M}^{(h'')}$

hold for some h' and h''. Let $\bar{v}_0^{(h)}$ be an arbitrary element in $\mathcal{M}_0^{(h)}$. Then from (3.3.10) we have

$$(P^k)^{s't} \cdot \bar{v}_0^{(h)} \in (P^k)^{s't} \mathcal{M}^{(h'')}.$$

By (3.3.8), for any $\bar{v}^{(h'')}$ in $\mathcal{M}^{(h'')}$ we have

$$(P^k)^{s't} \cdot \bar{v}^{(h'')} \in \mathcal{M}^{(s'kt-s'+h'')}.$$

Then we can find s' large enough to have

$$\mathcal{M}^{(s'kt-s'+h'')} \subset \mathcal{M}_0^{(s'kt+h-1)}.$$

This shows that $\bar{P}^{s't}$ is in $\sqrt{\mathcal{J}_0(\mathcal{M})}$, and therefore we obtain the inclusion

$$\sqrt{\mathcal{J}(\mathcal{M})} \subset \sqrt{\mathcal{J}_0(\mathcal{M})}.$$

Reversing the role of the two filtrations, we conclude that the radical $\sqrt{\mathcal{J}(\mathcal{M})}$ of $\mathcal{J}(\mathcal{M})$ is determined only by the $\mathcal{D}$-Module $\mathcal{M}$ independently of the choice of a good filtration. That is, we have obtained an ideal $\sqrt{\mathcal{J}(\mathcal{M})}$ globally defined on T^*X.

Proposition 3.3.1 *For an exact sequence of finitely generated $\mathcal{D}$-Modules*

$$0 \longrightarrow \mathcal{M}' \longrightarrow \mathcal{M} \longrightarrow \mathcal{M}'' \longrightarrow 0,$$

we have

$$\sqrt{\mathcal{J}(\mathcal{M})} = \sqrt{\mathcal{J}(\mathcal{M}')} \cap \sqrt{\mathcal{J}(\mathcal{M}'')}.$$

Proof. Let $\{\mathcal{M}^{(k)}\}$ be a good filtration on $\mathcal{M}$. This induces two filtrations on $\mathcal{M}'$ and $\mathcal{M}''$ which make the following sequence

$$0 \longrightarrow \mathcal{M}'^{(k)} \longrightarrow \mathcal{M}^{(k)} \longrightarrow \mathcal{M}''^{(k)} \longrightarrow 0$$

exact. Consider now the commutative diagram

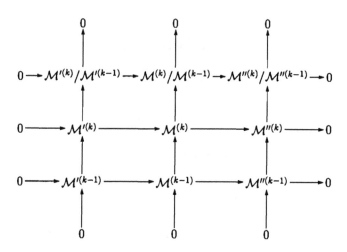

The nine lemma implies that the top row is exact. Therefore, we obtain the exact sequence of $\bar{\mathcal{D}}$-Modules

$$0 \longrightarrow \bar{\mathcal{M}}' \longrightarrow \bar{\mathcal{M}} \longrightarrow \bar{\mathcal{M}}'' \longrightarrow 0.$$

This implies that $\mathcal{J}(\mathcal{M}) = \mathcal{J}(\mathcal{M}') \cap \mathcal{J}(\mathcal{M}'')$ holds, and therefore we have $\sqrt{\mathcal{J}(\mathcal{M})} = \sqrt{\mathcal{J}(\mathcal{M}')} \cap \sqrt{\mathcal{J}(\mathcal{M}'')}$.   □

Definition 3.3.1 *Let U be an open subset of $X = \mathbb{C}^n$. Define $V(\mathcal{M}) = V\left(\sqrt{\mathcal{J}(\mathcal{M})}\right) = \{(z, \xi) \in T^*U; \bar{P}(z, \xi) = 0 \text{ for all } \bar{P} \in \sqrt{\mathcal{J}(\mathcal{M})}\}$. This complex analytic variety in T^*U is called the characteristic variety of the $\mathcal{D}$-Module $\mathcal{M}$.*

According to this definition, the characteristic variety of $\mathcal{M}$ is the closed reduced complex analytic variety in T^*X determined by the homogeneous radical Ideal $\sqrt{\mathcal{J}(\mathcal{M})}$ of $\bar{\mathcal{D}}$: in abstract terms

$$V(\mathcal{M}) = \text{Specan}\left(\bar{\mathcal{D}}/\sqrt{\mathcal{J}(\mathcal{M})}\right).$$

In particular, the characteristic variety $V(\mathcal{M})$ is a conic subset since for $(z, \xi) \in V(\mathcal{M})$ and $\lambda \in \mathbb{C}\backslash\{0\}$ we always have $(z, \lambda\xi) \in V(\mathcal{M})$.

Definition 3.3.1 is somewhat abstract, so we now proceed to give a more direct construction for the characteristic variety of a $\mathcal{D}$-Module. Let $\mathcal{M}$ be a $\mathcal{D}$-Module over X, and let $\bar{\mathcal{M}}$ be the $\bar{\mathcal{D}}$-Module defined by $\bar{\mathcal{M}} = \bigoplus_k \mathcal{M}^{(k)}/\mathcal{M}^{(k-1)}$; we can then define the characteristic variety of $\mathcal{M}$ as the support of $\mathcal{O}_{T^*X} \otimes_{\bar{\mathcal{D}}} \bar{\mathcal{M}}$. Then, our earlier definition of the characteristic variety $V(\mathcal{M})$ coincides with the above definition. Indeed, those holomorphic functions on T^*X that are in the ideal $\mathcal{O}_{T^*X} \otimes_{\bar{\mathcal{D}}} \sqrt{\mathcal{J}(\mathcal{M})}$ are precisely those that vanish at each point in

$Supp(\mathcal{O}_{T^{\cdot}X} \otimes_{\bar{\mathcal{D}}} \bar{\mathcal{M}})$. Hence, the NullstellenSatz of Hilbert for conic complex analytic sets implies that $V(\mathcal{M}) = V(\sqrt{\mathcal{J}(\mathcal{M})}) = V(\mathcal{J}(Supp(\mathcal{O}_{T^{\cdot}X} \otimes_{\bar{\mathcal{D}}} \bar{\mathcal{M}}))) = Supp(\mathcal{O}_{T^{\cdot}X} \otimes_{\bar{\mathcal{D}}} \bar{\mathcal{M}})$.

As an example, let us consider a cyclic $\mathcal{D}$-Module $\mathcal{M}$, of the form

$$\begin{cases} \mathcal{M} = \mathcal{D}u, \\ P_1 u = 0 \\ P_2 u = 0 \\ \cdots \\ \cdots \\ P_\ell u = 0, \quad P_j \in \mathcal{D}. \end{cases}$$

such a $\mathcal{D}$-module is represented by the following exact sequence.

$$\mathcal{D}^\ell \xrightarrow{\bullet \begin{bmatrix} P_1 \\ \vdots \\ P_\ell \end{bmatrix}} \mathcal{D} \xrightarrow{\bullet u} \mathcal{M} \longrightarrow 0$$

Let $\mathcal{J} = (P_1, \ldots, P_\ell)$ be the ideal in $\mathcal{D}$ generated by those differential operators. Since the following short sequence is exact

$$0 \longrightarrow \mathcal{J} \hookrightarrow \mathcal{D} \xrightarrow{\bullet u} \mathcal{M} \longrightarrow 0,$$

the canonical filtration on $\mathcal{D}$ induces good filtrations on $\mathcal{M}$ and $\mathcal{J}$. We, therefore, obtain the exact sequence

$$0 \longrightarrow \bar{\mathcal{J}} \longrightarrow \bar{\mathcal{D}} \longrightarrow \bar{\mathcal{M}} \longrightarrow 0.$$

Apply now the exact functor $\mathcal{O}_{T^{\cdot}X} \overline{\otimes_{\bar{\mathcal{D}}}}$——, (see Chapter IV, Section 3): we obtain the following exact sequence

$$0 \longrightarrow \mathcal{O}_{T^{\cdot}X} \underset{\bar{\mathcal{D}}}{\otimes} \bar{\mathcal{J}} \longrightarrow \mathcal{O}_{T^{\cdot}X} \longrightarrow \mathcal{O}_{T^{\cdot}X} \underset{\bar{\mathcal{D}}}{\otimes} \bar{\mathcal{M}} \longrightarrow 0.$$

This implies

$$\mathcal{O}_{T^{\cdot}X} \underset{\bar{\mathcal{D}}}{\otimes} \bar{\mathcal{M}} \overset{\approx}{\leftarrow} \mathcal{O}_{T^{\cdot}X} / \mathcal{O}_{T^{\cdot}X} \underset{\bar{\mathcal{D}}}{\otimes} \bar{\mathcal{J}}.$$

Since $\mathcal{O}_{T^{\cdot}X} \otimes_{\bar{\mathcal{D}}} \bar{\mathcal{J}}$ is the ideal in $\mathcal{O}_{T^{\cdot}X}$ generated by the symbols

$$\sigma(P_1), \sigma(P_2), \ldots, \sigma(P_\ell),$$

the support of $\mathcal{O}_{T^{\cdot}X} \otimes_{\bar{\mathcal{D}}} \bar{\mathcal{M}}$ is the zero set of the ideal

$$(\sigma(P_1), \ldots, \sigma(P_\ell)).$$

In particular, for $\mathcal{M} = \mathcal{D}_X u$, i.e.

$$\mathcal{M} = \mathcal{D}_X / \{P \in \mathcal{D}_X : Pu = 0\},$$

we have

$$V(\mathcal{M}) = \{(z, \xi) \in T^*X, \sigma(P)(z, \xi) = 0, Pu = 0\}.$$

We conclude this section with the following result:

Proposition 3.3.2 *For an exact sequence*

$$0 \longrightarrow \mathcal{M}' \longrightarrow \mathcal{M} \longrightarrow \mathcal{M}'' \longrightarrow 0,$$

of finitely generated $\mathcal{D}$-Modules, we have

$$V(\mathcal{M}) = V(\mathcal{M}') \cup V(\mathcal{M}'').$$

Proof.   For $0 \to \mathcal{M}' \to \mathcal{M} \to \mathcal{M}'' \to 0$, the NullstellenSatz of Hilbert implies that $\sqrt{\mathcal{J}(\mathcal{M})} = \sqrt{\mathcal{J}(\mathcal{M}')} \cap \sqrt{\mathcal{J}(\mathcal{M}'')}$ if and only if $V(\sqrt{\mathcal{J}(\mathcal{M})}) = V(\sqrt{\mathcal{J}(\mathcal{M}')}) \cap \sqrt{\mathcal{J}(\mathcal{M}'')})$. Since $V(\sqrt{\mathcal{J}(\mathcal{M}')} \cap \sqrt{\mathcal{J}(\mathcal{M}'')}) = V(\sqrt{\mathcal{J}(\mathcal{M}')}) \cup V(\sqrt{\mathcal{J}(\mathcal{M}'')})$, and in view of Proposition 3.3.1, we have concluded the proof. $\square$

3.4 $\mathcal{E}$-Modules

In this section, we microlocalize the sheaf $\mathcal{D}_X$ and introduce the sheaf $\mathcal{E}_X$. We then prove, in this setting, a version of Sato's fundamental theorem for infinite order differential operators. This is intended to prepare the reader for the more complete treatment which will be given in Chapter VI.

The microlocalization, namely the process which will bring us from $\mathcal{D}_X$ to $\mathcal{E}_X$, or from $\mathcal{D}_X^\infty$ to $\mathcal{E}_X^\infty$, is the non-commutative version of the localization for a commutative ring.

Let X be an open subset of $\mathbb{C}^n$, and let T^*X be the cotangent bundle. Let us recall that for a partial differential operator $P = \sum_{|j| \le k} a_j(z) \partial^j$ in $\mathcal{D}_X$, the symbols $\{\sigma_j(P)\} = \{P_j(z, \xi)\}$ are homogeneous polynomials in ξ and holomorphic functions in z.

Let Ω be an open conic subset in $T^*X \cong X \times \mathbb{C}^n$. We will define the sheaf $\mathcal{E}_X^\infty$ of germs of microdifferential operators of infinite order via the sequence of symbols that are holomorphic in both ξ and z. Namely, the totality $\mathcal{E}_X^\infty(\Omega)$ of sections over $\Omega \subset T^*X$ is isomorphic to the totality of sequences of holomorphic functions $\{p_j(z, \xi)\}_{j \in \mathbb{Z}}$ in Ω satisfying the growth conditions in the following definition.

Definition 3.4.1 *A sequence of holomorphic functions defined on an open conic subset Ω of T^*X, denoted as $\{p_j(z,\xi)\}_{j\in\mathbb{Z}}$, is said to be a symbol sequence if the following conditions are satisfied.*

(a) Each $p_j(z,\xi)$ is homogeneous of degree j with respect to ξ, i.e.

$$\sum_{1\leq\ell\leq n} \xi_\ell \frac{\partial p_j(z,\xi)}{\partial \xi_\ell} = j p_j(z,\xi).$$

(b) For any $\epsilon > 0$ and any compact set K in Ω, there exists a positive constant $C_{\epsilon,K}$ such that for any $j \geq 0$

$$\sup_K |p_j(z,\xi)| \leq C_{\epsilon,K}\frac{\epsilon^j}{j!}.$$

(c) For any compact set K in Ω, there exists a positive constant R_K such that for any $j < 0$

$$\sup_K |p_j(z,\xi)| \leq R_K^{-j}(-j)!.$$

The sheaf $\mathcal{E}_X^\infty$ is defined as the sheaf associated to the presheaf whose sections on an open set Ω are given by $P(z,\partial_z) = \sum_{j\in\mathbb{Z}} p_j(z,\partial_z) = \sum_{j\in\mathbb{Z}} p_j(z,\xi)$. For $P(z,\partial_z) = \sum_{j\in\mathbb{Z}} p_j(z,\xi)$ and $Q(z,\partial_z) = \sum_{j\in\mathbb{Z}} q_i(z,\xi)$ in $\mathcal{E}_X^\infty(\Omega)$, the product $(P \circ Q)(z,\partial_z) = \sum_k r_k(z,\xi)$ is defined by

$$r_k(z,\xi) = \sum_{\substack{k = j+i-|\alpha| \\ j,i\in\mathbb{Z}, \alpha\in\mathbb{N}^n}} \frac{1}{\alpha!}\partial_\xi^\alpha p_j(z,\xi)\cdot\partial_z^\alpha q_i(z,\xi),$$

where $\alpha! = \alpha_1!\ldots\alpha_n!$, $|\alpha| = \alpha_1 + \ldots + \alpha_n$ and $\partial_\xi^\alpha = \partial^{|\alpha|}/\partial\xi_1^{\alpha_1}\ldots\partial\xi_n^{\alpha_n}$. The sum $(P+Q)(z,\partial_z) = \sum s_\ell(z,\xi)$ is given by

$$s_\ell(z,\xi) = p_\ell(z,\xi) + q_\ell(z,\xi).$$

We can also define the subsheaf $\mathcal{E}_X^{(m)}$ of microdifferential operators of order m defining its sections as follows:

$$\mathcal{E}_X^{(m)}(\Omega) = \{P(z,\partial_z) = \sum p_j(z,\xi) \in \mathcal{E}_X^\infty(\Omega) | p_j(z,\xi) = 0 \text{ for } j > m\}.$$

Then one has

$$\mathcal{E}_X = \bigcup_{m\in\mathbb{Z}} \mathcal{E}_X^{(m)}.$$

The sheaf $\mathcal{E}_X$ is called the sheaf of microdifferential operators of finite order. Notice that we have

$$\mathcal{E}_X|_{T_X^* X} = \mathcal{D}_X$$

and

$$\mathcal{E}_X^{(m)}|_{T_X^* X} = \mathcal{D}_X^{(m)}.$$

We also have an isomorphism from $\mathcal{E}_X^{(m)}/\mathcal{E}_X^{(m-1)}$ to $\mathcal{O}_{T^*X}(m)$ obtained by assigning the principal symbol $\sigma_m(P)$ to every element $P \in \mathcal{E}_X^{(m)}$.

Remark 3.4.1 Let $\pi : T^*X \to X$ be the projection defined by $\pi(z, \xi) = z$. The non-commutative sheaves $\mathcal{E}_X^{\infty}$ and $\mathcal{E}_X$ have the following algebraic properties (see [19] for proofs):

(1) $\mathcal{E}_X$ is a coherent left $\mathcal{E}_X$-Module.

(2) $\mathcal{E}_X^{\infty}$ is $\mathcal{E}_X$-faithfully flat.

(3) $\mathcal{D}_X^{\infty}$ is $\mathcal{D}_X$-faithfully flat, where $\mathcal{D}_X^{\infty}$ is the ring of differential operators of infinite order.

(4) $\mathcal{E}_X$ is $\mathcal{E}_X(0)$-flat.

(5) $\mathcal{E}_X$ is $\pi^{-1}\mathcal{D}_X$-flat.

Remark 3.4.2 We also have the following natural properties:

$$\mathcal{E}_X^{(k)}\mathcal{E}_X^{(k')} \subset \mathcal{E}_X^{(k+k')},$$

$$\sigma_{k+k'}(P \circ Q) = \sigma_k(P)\sigma_{k'}(Q),$$

and

$$[P, Q] = PQ - QP \in \mathcal{E}_X^{(k+k'-1)},$$

$\sigma_{k+k'-1}([P, Q]) = \{\sigma_k(P), \sigma_{k'}(Q)\} := \sum_{1 \leq j \leq n} \left(\frac{\partial \sigma_k(P)}{\partial \xi_j} \cdot \frac{\partial \sigma_{k'}(Q)}{\partial z_j} - \frac{\partial \sigma_{k'}(Q)}{\partial \xi_j} \cdot \frac{\partial \sigma_k(P)}{\partial z_j} \right)$
for any $P \in \mathcal{E}_X^{(k)}$ and $Q \in \mathcal{E}_X^{(k')}$.

The most fundamental result for $\mathcal{E}_X$ is the analogue of Sato's fundamental theorem, a cornerstone of algebraic analysis, which will be discussed in Chapter VI.

Theorem 3.4.1 *Let Ω be an open subset of T^*X and let $P(z, \partial_z)$ be an element of $\mathcal{E}_X^{(m)}$. If $\sigma_m(P) \neq 0$ in Ω, then there exists $Q(z, \partial_z)$ in $\mathcal{E}_X^{(-m)}(\Omega)$ such that*

$$P \circ Q = Q \circ P = I.$$

Proof. Let $P = \sum_{j \leq m} p_j$. Then we have

$$\sigma_o(\frac{1}{p_m} \circ P) = \sigma_o(\frac{1}{p_m}) \circ \sigma(P) = 1.$$

Hence, $\frac{1}{p_m} \circ P$ can be written as $I - R$, where $R \in \mathcal{E}_X^{(-1)}(\Omega)$. Since the inverse of $I - R$ is the formal sum $\sum_{k \geq 0} R^k$, one needs to show that $\sum_{k \geq 0} R^k$ belongs to $\mathcal{E}_X(U)$. Consider the element $Q = \sum R^k \circ \frac{1}{p_m}$ in $\mathcal{E}_X^{(-m)}(\Omega)$. In order to show the convergence of $\sum_{k \geq 0} R^k$ in $\mathcal{E}_X(U)$, one introduces the following quasi-formal norm

$$N_m(P, K, t) = \sum_{\substack{k \geq 0 \\ \alpha, \beta \in \mathbb{N}^n}} \frac{2 \cdot (2n)^{-k} \cdot k!}{(|\alpha| + k)!(|\beta| + k)!} \sup_K \left|\partial_z^\alpha \partial_\xi^\beta p_{m-k}\right| t^{2k+|\alpha|+|\beta|},$$

which (for any compact set K in Ω) belongs to the formal power ring $\mathcal{C}[[t]]$. Then, $P \in \mathcal{E}_X^{(m)}(\Omega)$ if, for any arbitrary compact set K in Ω there exists $\epsilon_K > 0$ such that $N_m(P, K, \epsilon_K)$ is finite. We also have the following property:

$$N_m(P + Q, K, t) \prec\prec N_m(P, K, t) + N_m(Q, K, t)$$

for P and Q in $\mathcal{E}_X^{(m)}(\Omega)$, where $A(t) \prec\prec B(t), A(t), B(t) \in \mathcal{C}[[t]]$, means that $B(t)$ is a majorant series of $A(t)$. In our case, the positive real coefficients of $N_m(P + Q, K, t)$ are less than or equal to the corresponding positive real coefficients of $N_m(P, K, t) + N_m(Q, K, t)$. See [18] for details. From these general properties of $N_m(P, K, t)$, we obtain

$$N_0\left(\sum_{k \geq 0} R^k, K, t\right) \prec\prec \sum_{k \geq 0} N_0(R, K, t)^k.$$

But, since $R \in \mathcal{E}_X^{(-1)}(\Omega)$, there exist positive constants $\mathcal{E}_K, C$ so that $N_0(R, K, \epsilon_K) \leq C\epsilon_K^2 < 1$. Hence, we have $\sum_{k \geq 0} N_0(R, K, t)^k \leq \sum_{k \geq 0} (C\epsilon_K^2)^k$. Consequently, $\sum_{k \geq 0} R^k$ belongs to $\mathcal{E}_X^{(0)}(\Omega)$. $\square$

3.5 Historical Notes

The rather short history of $\mathcal{D}$-module theory goes back to the late 1950's when Mikio Sato began to develop his philosophy of algebraic analysis. We will make a short comment on the role of homological algebra and sheaf theory, essential for the theory of 𝒟-modules as shown in Chapter III.

The role of homological algebra, or cohomological methods, (sometimes called "General Nonsense") has been very important for algebraic analysis since the appearance of hyperfunctions in several variables, as was mentioned in Chapter I. For example, the first chapter of Seminar Notes, No.22 (in Japanese, 1968), University of Tokyo, by Hikosaburo Komatsu, begins with spectral sequences. This series of seminars began in April 1967, and continued through February 1968, for undergraduate seniors and first year graduate students at the University of Tokyo. At that time, homological algebra was not common knowledge among analysis students who came to listen to Komatsu's seminars. Among students in the audience were Kawai and Kashiwara. Sato was encouraged to become serious about "his old theory" of hyperfunctions by this renewed movement by Komatsu. As we mentioned in Chapter I, even though Sato had a plan to write Part III of hyperfunction theory in which his theory of skew categories would have been exposed, Part III was never written. His skew category theory, which is now called derived category, was developed by Grothendieck and Verdier in France. Hartshorne's Lecture Notes, Springer-Verlag, 1966, on this subject has been a constant reference for algebraic analysis since SKK. It is worth noticing that Sato did not receive any encouragement in Japan and at Princeton to write Part III of his hyperfunction theory, possibly because such an abstract approach by Sato to hyperfunction theory and analysis might have been too radical at that time. In fact, as he mentioned to us, his ideas were not too warmly received in Princeton, when he presented them to A. Weil.

In the 1950's, sheaf cohomology theory was employed by H. Cartan to clarify some of Kiyoshi Oka's fundamental results in several complex variables, e.g, the now famous Theorems A and B of Oka-Cartan-Serre. In algebraic geometry in the late 1950's, it was homological algebra which not only rephrased old notions, but also became a powerful device which allowed Grothendieck to develop a new algebraic geometry. Sato's algebraic analysis, an algebraic geometry-type analysis, naturally demanded homological algebra even to define the sheaf of hyperfunctions in several variables.

Around 1957, Sato began the cohomological study of systems of partial differential equations. The first occasion in which he publicly announced his work on this topic was the series of talks which he delivered in the Spring of 1960 at the Kawada Friday-Seminar just before his departure for the Institute for Advanced Study. During his talks, Sato put emphasis on the importance of cohomological treatments of systems of linear and non-linear partial differential equations. We can say that Sato's algebraic analysis began in that moment. The most spectacular successes of his program are microlocal analysis and 𝒟-module theory, the topics of this monograph. We will make no extended comments on Sato's number theoretic work done around 1960, other than to mention Sato Conjecture on L-functions, the connection between Ramanujan and Weil Conjectures, and the theory of prehomogenous vector spaces. The theory of holonomic quantum fields and soliton equations on Grassmann manifolds are also some of Sato's work in mathematics-physics. Note that these physics fields are rooted in his original program outlined in 1960 on the general theory of non-linear partial differential equations.

The arrival of Masaki Kashiwara established another main character on the $\mathcal{D}$-module theory stage. It was Kashiwara who established almost all that we have in the theory of $\mathcal{D}$-modules. In the case of partial differential operators with constant coefficients, the straightforward applications of various finiteness properties of noetherian rings can be used. In Kashiwara's thesis, the notion of a filtration was developed to overcome the variable coefficients case. Namely, a filtration is used as a bridge between commutative objects and non-commutative ones. Essentially all the mathematical devices in Chapter III came from Kashiwara. From his very first publication on $\mathcal{D}$-modules, namely his astonishing Master's Thesis of 1970, Kashiwara established the methods and the fundamental results for further study of $\mathcal{D}$-modules. Within a few years, Kashiwara obtained the most crucial theorem in $\mathcal{D}$-module theory, known as the Constructibility Theorem for a holonomic $\mathcal{D}$-module. One of the climaxes, if not the climax, is Kashiwara's Riemann-Hilbert Correspondence Theorem.

Chapter 4

Functors Associated with $\mathcal{D}$-modules

4.1 Introduction and Preliminary Material

In this chapter we study some fundamental functors associated with the $\mathcal{D}$-modules which we have introduced in Chapter III. Before we get to the de Rham functor in Section 4.2, we remind the reader the conditions which an $\mathcal{O}_X$-Module $\mathcal{M}$ must satisfy in order to be a $\mathcal{D}_X$-module. In particular, we will show that $\mathcal{H}om_{\mathcal{O}_X}(\Omega_X^n, -)$ and $-\otimes \Omega_X^n$ are one the inverse of the other when acting on the categories of left and right $\mathcal{D}_X$-modules.

The de Rham complex is a well known resolution for $\mathcal{O}_X$. In Section 4.2 we construct and study the de Rham functor which allows a generalization to an arbitrary left $\mathcal{D}_X$-module. In particular we discuss the important result of Mebkhout who has proved that the de Rham functor $\mathcal{H}om_{\mathcal{D}_X}(\mathcal{O}_X, -)$ and the solution functor $\mathcal{H}om_{\mathcal{D}_X}(-, \mathcal{O}_X)$ are in global duality, and we show that both Poincaré and Serre duality are just special cases of Mebkhout's theorem.

After a quick review of algebraic local cohomology (which we introduce in Section 4.3 as a natural extension of the space of hyperfunctions in one variable supported at the origin) we prove, in Section 4.4, two cohomology vanishing theorems for $\mathcal{D}_X$-modules.

Let X be a complex manifold of dimension n, $\mathcal{O}_X$ the sheaf of germs of holomorphic functions, and $\mathcal{D}_X$ the sheaf of germs of partial differential operators with coefficients in holomorphic functions. Let us denote by Θ_X the sheaf of germs of holomorphic tangent vector fields.

If a and b are holomorphic functions, the chain rule $\partial(ab) = \partial a \cdot b + a \cdot \partial b$ gives the formula $\partial \cdot a = \partial a + a \cdot \partial$ which induces a non-commutative ring structure on $\mathcal{D}_X$; as a first example, $\partial_i \cdot z_i = 1 + z_i \cdot \partial_i$. The repeated use of this formula provides Leibniz's rule, i.e., for $a \in \mathcal{O}_X$ and for ∂^α the differential operator

$\partial^\alpha = \left(\dfrac{\partial^{\alpha_1}}{\partial z_1^{\alpha_1}}, \ldots, \dfrac{\partial^{\alpha_n}}{\partial z_n^{\alpha_n}} \right)$ we have that $\partial^\alpha \cdot a = \sum\limits_{\beta \leq \alpha} \binom{\alpha}{\beta} \partial^\beta a \cdot \partial^{\alpha-\beta}$. The adjoint operator of $P(z, \partial) = \sum f_\alpha(z) \partial^\alpha$ is defined as $\sum (-1)^{|\alpha|} \partial^\alpha \cdot f_\alpha(z)$ so that, for example, the adjoint of ∂_i is $-\partial_i$. See also Lemma 6.2.3. Since $\mathcal{D}_X$ is the sheaf of rings (Ring in short) generated by Θ_X over $\mathcal{O}_X$, the action of Θ_X determines a $\mathcal{D}_X$-Module structure.

Definition 4.1.1 *Let $\mathcal{M}$ be an $\mathcal{O}_X$-Module, and let ψ be the $\mathbb{C}$-linear map from $\Theta_X \otimes_{\mathcal{O}_X} \mathcal{M}$ to $\mathcal{M}$ defined by $\psi(\theta \otimes m) = \theta \cdot m$ for all $\theta \otimes m \in \Theta_X \otimes_{\mathcal{O}_X} \mathcal{M}$. If for all $a \in \mathcal{O}_X, \theta, \theta' \in \Theta_X, m \in \mathcal{M}$, ψ satisfies*

(1) $\psi(a\theta \otimes m) = a\psi(\theta \otimes m)$, i.e. $(a\theta) \cdot m = a(\theta \cdot m)$

(2) $\psi(\theta \otimes am) = a\psi(\theta \otimes m) + \theta(a)m$, i.e., $\theta \cdot (am) = a(\theta m) + \theta(a) \cdot m$

(3) $\psi([\theta, \theta'] \otimes m) = \psi(\theta \otimes \psi(\theta' \otimes m)) - \psi(\theta' \otimes \psi(\theta \otimes m))$, i.e., $[\theta, \theta'] \cdot m = \theta \cdot (\theta' \cdot m) - \theta'(\theta \cdot m)$, then $\mathcal{M}$ is a left $\mathcal{D}_X$-Module.

The corresponding definition for a right $\mathcal{D}_X$-Module is the following.

Definition 4.1.2 *Let $\mathcal{N}$ be an $\mathcal{O}_X$-Module, and let φ be the $\mathbb{C}_X$-linear map from $\Theta_X \otimes_{\mathcal{O}_X} \mathcal{N}$ to $\mathcal{N}$ defined by $\varphi(\theta \otimes n) = n \cdot \theta$ for all $\theta \otimes n \in \Theta_X \otimes_{\mathcal{O}_X} \mathcal{N}$. If φ, for all $a \in \mathcal{O}_X, \theta, \theta' \in \Theta_X, n \in \mathcal{N}$, satisfies*

(1) $\varphi(a\theta \otimes n) = \varphi(\theta \otimes an)$, i.e. $n \cdot (a\theta) = (an) \cdot \theta$*

(2) $\varphi(\theta \otimes an) = -\theta(a) \cdot n + a\varphi(\theta \otimes n)$, i.e., $(an) \cdot \theta = -\theta(a) \cdot n - a(n\theta)$*

(3) $\varphi([\theta, \theta'] \otimes n) = \varphi(\theta' \otimes \varphi(\theta \otimes n)) - \varphi(\theta \otimes \varphi(\theta' \otimes n))$, i.e., $n \cdot [\theta', \theta] = (n \cdot \theta) \cdot \theta' - (n \cdot \theta') \cdot \theta$, then $\mathcal{N}$ is a right $\mathcal{D}_X$-Module.*

Note that the induced $\mathcal{O}_X$-Module structure of the left $\mathcal{D}_X$-Module $\mathcal{M}$ coincides with the original $\mathcal{O}_X$-Module structure. See [19].

Since n is the dimension of the complex manifold X, the sheaf Ω_X^n of germs of holomorphic n-forms is locally free of rank one as an $\mathcal{O}_X$-Module. For $\theta_1 \wedge \ldots \wedge \theta_n \in \wedge^n \Theta_X$, an n-form $\omega \in \Omega_X^n$ determines an element $\omega(\theta_1 \wedge \ldots \wedge \theta_n)$ in $\mathcal{O}_X$ as an $\mathcal{O}_X$-linear map from $\wedge^n \Theta_X$ to $\mathcal{O}_X$. We will define a right $\mathcal{D}_X$-Module structure on Ω_X^n via Lie derivatives. Namely we define

$$((\text{Lie }(\theta)\omega)(\theta_1 \wedge \ldots \wedge \theta_n) :=$$

$$:= \theta(\omega(\theta_1 \wedge \ldots \wedge \theta_n)) - \sum_{i=1}^n \omega(\theta_1 \wedge \ldots \wedge [\theta, \theta_i] \wedge \ldots \wedge \theta_n).$$

More explicitly, this means that the tangent vector fields $\theta \in \Theta_X$ act on Ω_X^n by $\omega\theta \overset{\text{def}}{=} -(\text{Lie }\theta)\omega$. Thus Ω_X^n has the structure of a right $\mathcal{D}_X$-Module. In terms of local coordinates, if $\omega = g dz_1 \wedge \ldots \wedge dz_n \in \Omega_X^n$, then $\omega \partial_i = -(\text{Lie } \partial_i)\omega = -(\partial_i g) dz_1 \wedge \ldots \wedge dz_n$. More generally, the extended right action of $\mathcal{D}_X$ on Ω_X^n is given by

$$\omega \cdot P(z, \partial) = (P(z, \partial)^* g) dz_1 \wedge \ldots \wedge dz_n,$$

where $P(z, \partial)^*$ is the adjoint differential operator of $P(z, \partial)$. Note that the right module axioms are satisfied because of the property of adjointness which gives $(P(z, \partial) Q(z, \partial))^* = Q(z, \partial)^* P(z, \partial)^*$.

The right $\mathcal{D}_X$-Module Ω_X^n, a free $\mathcal{O}_X$-Module of rank one, induces an exact covariant functor from the category of left $\mathcal{D}_X$-Modules to the category of right $\mathcal{D}_X$-Modules as follows. Given a left $\mathcal{D}_X$-Module $\mathcal{M}$, one can construct the right $\mathcal{D}_X$-Module $\mathcal{M} \otimes_{\mathcal{O}_X} \Omega_X^n$, where the right module structure is induced by

$$(u \otimes \omega) \cdot \theta = -(\theta u) \otimes \omega + u \otimes \omega \theta$$

for $\theta \in \Theta$ and $u \otimes \omega \in \mathcal{M} \otimes_{\mathcal{O}_X} \Omega_X^n$. We leave it to the reader to verify that this definition make $\mathcal{M} \otimes_{\mathcal{O}_X} \Omega_X^n$ a right $\mathcal{D}_X$-module. Notice also that, more generally, for any right $\mathcal{D}_X$-Module $\mathcal{N}$ and any left $\mathcal{D}_X$-Module $\mathcal{M}$ one can construct a right $\mathcal{D}_X$-Module $\mathcal{M} \otimes_{\mathcal{O}_X} \mathcal{N}$ where the right module structure is defined by

$$(u \otimes v)\theta = -(\theta u) \otimes v + u \otimes v\theta.$$

Proposition 4.1.1 *Let $\mathcal{M}$ and $\mathcal{N}$ be, respectively, a left $\mathcal{D}_X$-Module and a right $\mathcal{D}_X$-Module. Then*

$$\mathcal{H}om_{\mathcal{O}_X}(\Omega_X^n, \mathcal{M} \otimes_{\mathcal{O}_X} \Omega_X^n) \cong \mathcal{M}$$

and

$$\mathcal{H}om_{\mathcal{O}_X}(\Omega_X^n, \mathcal{N}) \otimes_{\mathcal{O}_X} \Omega_X^n \cong \mathcal{N}.$$

Proof. Just note that $\mathcal{H}om_{\mathcal{O}_X}(\Omega_X^n, \mathcal{N})$ becomes a left $\mathcal{D}_X$-Module if, for $\varphi \in \mathcal{H}om_{\mathcal{O}_X}(\Omega_X^n, \mathcal{N})$, the action of Θ_X is defined by $(\theta \varphi)\omega = -(\varphi \omega)\theta + \varphi(\omega \theta)$. The isomorphism follows since Ω_X^n is isomorphic to $\mathcal{O}_X$ as an $\mathcal{O}_X$-module. □

The above proposition shows us that the compositions of functors $\mathcal{H}om_{\mathcal{O}_X}(\Omega_X^n, -)$ and $- \otimes \Omega_X^n$ and the identity functors are isomorphic on objects in both the categories of right $\mathcal{D}_X$-Modules and of left $\mathcal{D}_X$-Modules. Hence, $- \otimes_{\mathcal{O}_X} \Omega_X^n$ is an equivalence from the category of left $\mathcal{D}_X$-Modules to the category of right $\mathcal{D}_X$-Modules, and vice versa $\mathcal{H}om_{\mathcal{O}_X}(\Omega_X^n, -)$ is an equivalence from the category of right $\mathcal{D}_X$-modules to the category of left $\mathcal{D}_X$ modules.

This result can be stated in a more general setting as follows:

Proposition 4.1.2 *Let $\mathcal{M}$ and $\mathcal{M}'$ be left $\mathcal{D}_X$-Modules and let $\mathcal{N}$ and $\mathcal{N}'$ be right $\mathcal{D}_X$-Modules. Then:*

(4.1.1)
$$\mathcal{M} \otimes_{\mathcal{O}_X} \mathcal{M}'$$
is a left $\mathcal{D}_X$-Module where the action of Θ_X is defined by

$$\theta(m \otimes m') = (\theta m) \otimes m' + m \otimes (\theta m');$$

(4.1.2)
$$\mathcal{M} \otimes_{\mathcal{O}_X} \mathcal{N}$$
is a right $\mathcal{D}_X$-Module where the action of Θ_X is defined by

$$(m \otimes n)\theta = -(\theta m) \otimes n + m \otimes (n\theta);$$

(4.1.3)
$$\mathcal{H}om_{\mathcal{O}_X}(\mathcal{M}, \mathcal{M}')$$
is a left $\mathcal{D}_X$-Module where the action of Θ_X is defined by

$$(\theta\varphi)(m) = \theta(\varphi(m)) - \varphi(\theta(m));$$

(4.1.4)
$$\mathcal{H}om_{\mathcal{O}_X}(\mathcal{N}, \mathcal{N}')$$
is a left $\mathcal{D}_X$-Module where the action of Θ_X is defined by

$$(\theta\varphi)(n) = -\varphi(n)\theta + \varphi(n\theta);$$

and finally

(4.1.5)
$$\mathcal{H}om_{\mathcal{O}_X}(\mathcal{M}, \mathcal{N})$$
is a right $\mathcal{D}_X$-Module where the action of Θ_X is defined by

$$(\varphi\theta)(m) = \varphi(m)\theta + \varphi(\theta m).$$

Proof.   We will prove (4.1.1), i.e. that, $\mathcal{M} \otimes_{\mathcal{O}_X} \mathcal{M}'$ is a left $\mathcal{D}_X$-Module. We need to verify (1), (2) and (3) of Definition 4.1.1. To prove (1) consider $a \in \mathcal{O}_X, \theta \in \Theta_X, m, m' \in \mathcal{M}$

$$\begin{aligned}
(a\theta)m \otimes m' &= ((a\theta) \cdot m) \otimes m' + m \otimes ((a\theta)m') \\
&= a(\theta m) \otimes m' + m \otimes a(\theta m) = a((\theta m) \otimes m') + a(m \otimes (\theta m)) \\
&= a(\theta m \otimes m' + m \otimes \theta m') = a(\theta(m \otimes m')).
\end{aligned}$$

Next, (2) is established by the following sequence of equalities:

$$\theta \cdot (a(m \otimes m')) = \theta((am) \otimes m') = \theta(am) \otimes m' + am \otimes \theta m'$$
$$= (a(\theta m) + \theta(a) \cdot m) \otimes m' + am \otimes \theta' m'$$
$$= a(\theta m) \otimes m' + \theta(a) \cdot m \otimes m' + am \otimes \theta' m'$$
$$= a((\theta m \otimes m' + m \otimes \theta' m') + \theta(a) \cdot m \otimes m'$$
$$= a(\theta(m \otimes m')) + \theta(a) \cdot m \otimes m'.$$

Finally let us prove (3), here θ' is another element of Θ_X:

$$[\theta, \theta'](m \otimes m')$$
$$= ([\theta, \theta']m) \otimes m' + m \otimes ([\theta, \theta']m')$$
$$= (\theta \cdot (\theta'm) - \theta'(\theta m)) \otimes m' + m \otimes (\theta(\theta'm') - \theta'(\theta m'))$$
$$= \theta \cdot (\theta'm) \otimes m' - \theta'(\theta m) \otimes m' + m \otimes \theta(\theta'm') - m \otimes \theta'(\theta m')$$

By adding

$$\theta'm \otimes \theta m' + \theta m \otimes \theta'm' - \theta m \otimes \theta'm' - \theta'm \otimes \theta m'$$

to the above equation we obtain

$$\theta(\theta'm \otimes m') + \theta(m \otimes \theta'm') - \theta'(\theta m \otimes m') - \theta'(m \otimes \theta m')$$
$$= \theta(\theta'm \otimes m' + m \otimes \theta'm') - \theta'(\theta m \otimes m' + m \otimes \theta m') \quad ,$$
$$= \theta(\theta'(m \otimes m')) - \theta'(\theta(m \otimes m'))$$

which concludes the proof of (4.1.1).

The other cases are left to the reader. $\qquad\qquad\square$

4.2 The de Rham Functor

The structure of a $\mathcal{D}_X$-Module $\mathcal{M}$ is determined by a sheaf homomorphism of the $\mathcal{O}_X$-Algebra $\mathcal{D}_X$ to $\mathcal{E}nd_{\mathcal{C}_X}(\mathcal{M})$. Then the restriction of this sheaf homomorphism to the sheaf of tangent vector fields Θ_X (considered as a subsheaf of $\mathcal{D}_X$) satisfies properties (1), (2) and (3) mentioned in the Section 4.1. Therefore this induces an integrable connection ∇ on the $\mathcal{O}_X$-Module $\mathcal{M}$. Namely we can define an integrable connection ∇ on $\mathcal{M}$ as a $\mathcal{O}_X$-linear map from the tangent sheaf Θ_X to $\mathcal{E}nd_{\mathcal{C}_X}(\mathcal{M})$ such that for $\theta \in \Theta_X$ we have that the element $\nabla_\theta = \nabla(\theta) \in \mathcal{E}nd_{\mathcal{C}_X}(\mathcal{M})$ satisfies

$$\nabla_\theta(am) = \theta(a) \cdot m + a \cdot \nabla_\theta(m),$$

and

$$\nabla_{[\theta,\theta']}(m) = \nabla_\theta(\nabla_{\theta'}(m)) - \nabla_{\theta'}(\nabla_\theta(m)).$$

Note that these two properties are, respectively, properties (2) and (3) of Definition 4.1.1.

The integrable connection ∇ on $\mathcal{M}$ can now be used to obtain the de Rham complex associated to $\mathcal{M}$ as the complex

(4.2.1) $\qquad 0 \to \mathcal{M} \xrightarrow{\tilde{d}^0} \Omega^1_X \otimes_{\mathcal{O}_X} \mathcal{M} \to \dots \xrightarrow{\tilde{d}^{n-1}} \Omega^n_X \otimes_{\mathcal{O}_X} \mathcal{M} \to 0$

where

$$\tilde{d}^0(m) = \sum_{j=1}^n dz_j \otimes \nabla_{\theta_j}(m) = \sum_{j=1}^n dz_j \otimes \partial_j m,$$

and

$$\tilde{d}^s(\omega \otimes m) = d^s(\omega) \otimes m + (-1)^s \omega \wedge \tilde{d}^0(m) = d^s(\omega) \otimes m + \sum_{j=1}^n (dz_j \wedge \omega) \otimes \partial_j m;$$

in this formula, the operators d^s are the coboundary operators of the complex $\Omega^\bullet_X$.

Notice that $\omega \wedge \tilde{d}^0(m)$ is the corresponding element in $\Omega^{s+1}_X \otimes_{\mathcal{O}_X} \mathcal{M}$ when we take $\omega \otimes \tilde{d}^0(m)$ in $\Omega^s_X \otimes_{\mathcal{O}_X} (\Omega^1_X \otimes_{\mathcal{O}_X} \mathcal{M})$. In particular, since $\tilde{d}^1(\tilde{d}^0(m)) = \sum_{j=1}^n (\sum_{k=1}^n (dz_k \wedge dz_j) \otimes \partial_k(\partial_j m))$ and ∂_k and ∂_j commute, we have $\tilde{d}^1 \circ \tilde{d}^0 = 0$. Then since $(\tilde{d}^{s+1} \circ \tilde{d}^s)(\omega \otimes m) = \omega \wedge (\tilde{d}^1 \circ \tilde{d}^0)(m)$ holds, we obtain $\tilde{d}^{s+1}\tilde{d}^s = 0$. When $\mathcal{M}$ is replaced by $\mathcal{O}_X$, one gets the usual de Rham complex.

The right $\mathcal{D}_X$-Module structure of Ω^n_X described in Section 4.1, allows us to obtain the augmentation map ε' from $\Omega^n_X \otimes_{\mathcal{O}_X} \mathcal{D}_X$ to Ω^n_X. By replacing $\mathcal{M}$ in (4.2.1) by $\mathcal{D}_X$, regarded as left $\mathcal{O}_X$-Modules, we obtain that the following (de Rham) complex is a free resolution of the right $\mathcal{D}_X$-Module Ω^n_X:

(4.2.2)

$$0 \to \mathcal{D}_X \xrightarrow{\tilde{d}^0} \Omega^1_X \otimes \mathcal{D}_X \xrightarrow{\tilde{d}^1} \dots \to \Omega^n_X \otimes_{\mathcal{O}_X} \mathcal{D}_X \to 0$$
$$\downarrow{\scriptstyle \varepsilon'}$$
$$\Omega^n_X$$

where

$$\tilde{d}^0(1_{\mathcal{D}}) = \sum_{j=1}^{n} dz_j \otimes \nabla_{\partial_j}(1_{\mathcal{D}}) = \sum_{j=1}^{n} dz_j \otimes \partial_j,$$

$$\tilde{d}^1(\omega \wedge 1_{\mathcal{D}}) = d^1(\omega) \otimes 1_{\mathcal{D}} - \omega \wedge \tilde{d}^0(1_{\mathcal{D}}),$$

and in general

$$\tilde{d}^s(\omega \wedge 1_{\mathcal{D}}) = d^s(\omega) \otimes 1_{\mathcal{D}} + (-1)^s \omega \wedge \tilde{d}^0(1_{\mathcal{D}}).$$

On the other hand, since Θ_X is a free $\mathcal{O}_X$-Module, these complexes induce the usual Koszul complex associated to the elements $\partial_1, \ldots, \partial_n$ in Θ_X:

(4.2.3)

$$\mathcal{D}_X \xleftarrow{\delta_1} \mathcal{D}_X \otimes_{\mathcal{O}_X} \wedge^1 \Theta_X \xleftarrow{\delta_2} \mathcal{D}_X \otimes_{\mathcal{O}_X} \wedge^2 \Theta_X \xleftarrow{\delta_3} \ldots \longleftarrow \mathcal{D}_X \otimes_{\mathcal{O}_X} \wedge^n \Theta_X$$
$$\downarrow{\scriptstyle \varepsilon''}$$
$$\mathcal{O}_X$$

where the augmentation map ε'' in (4.2.3) is defined by $\varepsilon''(1_{\mathcal{D}}) = 1_{\mathcal{D}} \in \mathcal{O}_X$ and δ_1 is induced by right multiplication by $\partial_1, \ldots, \partial_n$.

This procedure allows us, in particular, to obtain a free resolution of $\mathcal{O}_X$. Note that $\mathcal{D}_X \otimes_{\mathcal{O}_X} \wedge^i \Theta_X$ is at the $-i$-th place of the resolution for $\mathcal{O}_X$. The i-th boundary map of (4.2.3)

$$\mathcal{D}_X \otimes_{\mathcal{O}_X} \wedge^{i-1} \Theta_X \xleftarrow{\delta_i} \mathcal{D}_X \otimes_{\mathcal{O}_X} \wedge^i \Theta_X$$

is defined by

$$\delta_i(P \otimes (\theta_1 \wedge \ldots \wedge \theta_i)) = \sum_{j=1}^{i}(-1)^{j-1} P \theta_j \otimes (\theta_1 \wedge \ldots \wedge \hat{\theta}_j \wedge \ldots \wedge \theta_i)$$

$$+ \sum_{1 \le j < k \le i}(-1)^{j+k} P \otimes ([\theta_j, \theta_k] \wedge \theta_1 \wedge \ldots \wedge \hat{\theta}_j \wedge \ldots \wedge \hat{\theta}_k \wedge \ldots \wedge \theta_i),$$

where $P \in \mathcal{D}_X$ and $\theta_j \in \Theta_X$ for $j = 1, 2, \ldots, n$. In particular, $\delta_1(P \otimes \theta) = P\theta$. Hence, we have $\operatorname{Im} \delta_1 = \sum_{j=1}^{n} \mathcal{D}_X \partial_j$. Then the augmentation map ε'' in (4.2.3) induces the isomorphisms

$$\mathcal{H}_0(\mathcal{D}_X \otimes_{\mathcal{O}_X} \wedge^{-\bullet} \Theta_X) \approx \mathcal{D}_X / \sum_{j=1}^{n} \mathcal{D}_X \partial_j \approx \mathcal{O}_X.$$

In terms of local coordinates $(z_1, \ldots, z_n, \partial_1, \ldots, \partial_n)$ for $\mathcal{D}_X$, we obtain the vector space spanned by the $\mathbb{C}$-regular sequence $(\partial_1, \ldots, \partial_n)$. Then (4.2.3) becomes

$$\mathcal{D}_X \leftarrow \mathcal{D}_X \bigotimes_{\mathbb{C}} \wedge^1 (\oplus_{j=1}^n \mathbb{C}\partial_j) \leftarrow \ldots \overset{\delta_i'}{\leftarrow} \mathcal{D}_X \bigotimes_{\mathbb{C}} \wedge^i (\oplus_{j=1}^n \mathbb{C}\partial_j) \leftarrow \ldots$$

$$\ldots \leftarrow \mathcal{D}_X \bigotimes_{\mathbb{C}} \wedge^n (\oplus_{j=1}^n \mathbb{C}\partial_j),$$

which is the familiar Koszul complex in which

$$\delta_i'(P \bigotimes \partial_{j_1} \wedge \ldots \wedge \partial_{j_i}) = \sum_{k=1}^i (-1)^{k-1} P\partial_{j_k} \bigotimes \partial_{j_k} \wedge \ldots \wedge \hat{\partial}_{j_k} \wedge \ldots \wedge \partial_{j_i}).$$

Therefore we have obtained a locally free resolution of

$$\mathcal{O}_X \overset{\cong}{\leftarrow} \mathcal{D}_X/(\partial_1, \ldots, \partial_n).$$

Remark 4.2.1 For a vector bundle $\mathcal{M}$ with an integrable connection, (namely, if $\mathcal{M}$ is free of finite rank as an $\mathcal{O}_X$-Module), sequence (4.2.3) allows us to obtain a free resolution of $\mathcal{M}$ as a $\mathcal{D}_X$-Module:

$$\mathcal{M} \leftarrow \mathcal{D}_X \bigotimes_{\mathcal{O}_X} \wedge^\bullet \Theta_X \bigotimes_{\mathcal{O}_X} \mathcal{M}.$$

For a left $\mathcal{D}_X$-Module with a good filtration, we need on the other hand the Spencer sequence in order to construct a free resolution. Let $\{\mathcal{M}^{(k)}\}$ be a good filtration of $\mathcal{M}$. We will show that for any $k \geq 0$ the sequence

$$(4.2.4) \quad 0 \leftarrow \mathcal{M} \overset{\varepsilon}{\leftarrow} \mathcal{D}_X \bigotimes_{\mathcal{O}_X} \mathcal{M}^{(k)} \overset{\delta_1}{\leftarrow} \mathcal{D}_X \bigotimes_{\mathcal{O}_X} \Theta_X \bigotimes_{\mathcal{O}_X} \mathcal{M}^{(k-1)} \overset{\delta_2}{\leftarrow} \ldots$$

$$\ldots \leftarrow \mathcal{D}_X \bigotimes_{\mathcal{O}_X} \wedge^{i-1} \Theta_X \bigotimes_{\mathcal{O}_X} \mathcal{M}^{(k-i+1)} \overset{\delta_i}{\leftarrow} \mathcal{D}_X \bigotimes_{\mathcal{O}_X} \wedge^i \Theta_X \bigotimes_{\mathcal{O}_X} \mathcal{M}^{(k-i)} \overset{\delta_{i+1}}{\leftarrow} \ldots$$

$$\ldots \leftarrow \mathcal{D}_X \bigotimes_{\mathcal{O}_X} \wedge^n \Theta_X \bigotimes_{\mathcal{O}_X} \mathcal{M}^{(k-n)} \leftarrow 0$$

is locally a $\mathcal{D}_X$-free resolution of $\mathcal{M}$. The augmentation map ε in (4.2.4) is defined by $\varepsilon(P \otimes u) = Pu$, and

$$\delta_i(P \bigotimes (\theta_1 \wedge \ldots \wedge \theta_i) \bigotimes u) =$$

$$= \sum_{j=1}^i (-1)^{j-1} P\theta_j \bigotimes (\theta_1 \wedge \ldots \wedge \hat{\theta}_j \wedge \ldots \wedge \theta_i) \bigotimes u$$

$$- \sum_{j=1}^i (-1)^{j-1} P \bigotimes (\theta_1 \wedge \ldots \wedge \hat{\theta}_j \wedge \ldots \wedge \theta_i) \bigotimes \theta_j u$$

$$+ \sum_{1 \leq j \leq k \leq i} (-1)^{j+k} P \bigotimes ([\theta_j, \theta_k] \wedge \theta_1 \wedge \ldots \wedge \hat{\theta}_j \wedge \ldots \wedge \hat{\theta}_k \wedge \ldots \wedge \theta_i) \bigotimes u.$$

Before we explain the interplay among sequences (4.2.1), (4.2.2), (4.2.3) and (4.2.4), we will prove the exactness of Spencer sequence (4.2.4).

Proposition 4.2.1 *Let $\mathcal{M}$ be a left $\mathcal{D}_X$-Module with a good filtration $\{\mathcal{M}^{(k)}\}$. Then the Spencer sequence (4.2.4) associated to $\mathcal{M}$ is exact.*

Proof. The following proof is based on Kashiwara's ideas from [108]. First we consider the case when $\mathcal{M} = \mathcal{D}_X$ with the canonical filtration on $\mathcal{D}_X$ as defined in Chapter III. For $k = 0$, we have

$$0 \leftarrow \mathcal{D}_X \overset{\cong}{\leftarrow} \mathcal{D}_X \leftarrow 0$$

since $\mathcal{D}^{(m)} = 0$ for $m < 0$. So now let $k > 0$ and proceed by induction. Consider the following short exact sequences of complexes where $\overline{\mathcal{D}_X^{(k-n)}}$ denotes $\mathcal{D}_X^{(k-n)}/\mathcal{D}_X^{(k-n-1)}$.

$$
\begin{array}{ccccccccc}
 & & 0 & & & & 0 & & \\
 & & \uparrow & & & & \uparrow & & \\
0 & \to & \mathcal{D}_X \otimes \wedge^n \Theta_X \otimes \overline{\mathcal{D}^{(k-n)}} & \to & \cdots & \to & \mathcal{D}_X \otimes \Theta_X \otimes \overline{\mathcal{D}^{(k-1)}} & \to & (1) \\
 & & \uparrow & & & & \uparrow & & \\
0 & \to & \mathcal{D}_X \otimes \wedge^n \Theta_X \otimes \mathcal{D}^{(k-n)} & \to & \cdots & \to & \mathcal{D}_X \otimes \Theta_X \otimes \mathcal{D}^{(k-1)} & \to & (2) \\
 & & \uparrow & & & & \uparrow & & \\
0 & \to & \mathcal{D}_X \otimes \wedge^n \Theta_X \otimes \mathcal{D}^{(k-n-1)} & \to & \cdots & \to & \mathcal{D}_X \otimes \Theta_X \otimes \mathcal{D}^{(k-2)} & \to & (3) \\
 & & \uparrow & & & & \uparrow & & \\
 & & 0 & & & & 0 & &
\end{array}
$$

$$
\begin{array}{ccccc}
 & & 0 & & \\
 & & \uparrow & & \\
(1) & \longrightarrow & \mathcal{D}_X \otimes \overline{\mathcal{D}^{(k)}} & \longrightarrow & 0 \\
 & & \uparrow & & \uparrow \\
(2) & \longrightarrow & \mathcal{D}_X \otimes \mathcal{D}_X^{(k)} & \longrightarrow & \mathcal{D}_X \longrightarrow 0 \\
 & & \uparrow & & \uparrow \\
(3) & \longrightarrow & \mathcal{D}_X \otimes \mathcal{D}_X^{(k-1)} & \longrightarrow & \mathcal{D}_X \longrightarrow 0 \\
 & & \uparrow & & \uparrow \\
 & & 0 & & 0
\end{array}
$$

In view of the inductive assumption and of the induced long exact sequence of cohomologies, it is sufficient to prove the exactness of the top row of the diagram. On the other hand,

$$\wedge^n \Theta_X \otimes_{o_X} \overline{\mathcal{D}_X^{(k-n)}} \longrightarrow \dots \xrightarrow{\delta_2} \Theta_X \otimes \overline{\mathcal{D}_X^{(k-1)}} \xrightarrow{\delta_1} \overline{\mathcal{D}_X^{(k)}} \longrightarrow 0,$$

where

$$\delta_i((\theta_1 \wedge \dots \wedge \theta_i) \otimes p) = \sum_{j=1}^{i} (-1)^j (\theta_1 \wedge \dots \wedge \hat{\theta}_j \wedge \dots \wedge \theta_i) \otimes \theta_j p,$$

is the Koszul complex associated to the regular sequence $(\partial_1, \partial_2, \dots, \partial_n)$. Therefore, the above sequence is exact. Consequently, the exact functor $\mathcal{D}_X \otimes_{o_X} -$
induces the exactness of the top sequence in the above diagram.

We will now prove the case of a general $\mathcal{D}_X$-module $\mathcal{M}$. By the definition
of a good filtration there exists locally an epimorphism φ of filtered objects
$\varphi : \mathcal{D}_X^m \longrightarrow \mathcal{M}$ which induces the short exact sequence

$$0 \leftarrow \mathcal{M} \xleftarrow{\varphi} \mathcal{D}_X^m \leftarrow \mathcal{N} \leftarrow 0$$

where $\mathcal{N} = \ker\varphi$. By what we proved in the first part of this proof, the Spencer
sequence for the free Module $\mathcal{D}_X^m$ is exact. The long exact sequence of cohomologies associated with Spencer sequences of $\mathcal{M}$, $\mathcal{D}_X^m$ and $\mathcal{N}$ is

$$\mathcal{H}_0(\mathcal{N}) \longrightarrow \mathcal{H}_0(\mathcal{D}_X^m) \longrightarrow \mathcal{H}_0(\mathcal{M}) \longrightarrow$$

$$\longrightarrow \mathcal{H}_1(\mathcal{N}) \longrightarrow \mathcal{H}_1(\mathcal{D}_X^m) \longrightarrow \mathcal{H}_1(\mathcal{M}) \longrightarrow$$

$$\longrightarrow \mathcal{H}_2(\mathcal{N}) \longrightarrow \mathcal{H}_2(\mathcal{D}_X^m) \longrightarrow \mathcal{H}_2(\mathcal{M}) \longrightarrow \dots$$

Proceeding as we did in the first part of the proof, the middle column vanishes.
The inductive assumption that $\mathcal{H}_{j-1}(\mathcal{N}) = 0$ for $k \gg 0$ and for any left $\mathcal{D}_X$-
Module with a good filtration implies the exactness of the sequence

$$0 = \mathcal{H}_j(\mathcal{D}_X^m) \longrightarrow \mathcal{H}_j(\mathcal{M}) \longrightarrow \mathcal{H}_{j-1}(\mathcal{N}) = 0.$$

Therefore we have $\mathcal{H}_j(\mathcal{M}) = 0$ for $k \gg 0$ and all $j = 2, 3, \dots$. Since $\mathcal{H}_{-1}(\mathcal{N})$ is
the (-1)-th cohomology at $\mathcal{N}$ of

$$0 \leftarrow \mathcal{N} \leftarrow \mathcal{D}_X \otimes_{o_X} \mathcal{N}^{(k)} \leftarrow \dots,$$

the fact that $\mathcal{N} = \mathcal{D}_X \otimes_{o_X} \mathcal{N}^{(k)}$ for $k \gg 0$ implies $\mathcal{H}_{-1}(\mathcal{N}) = 0$. Hence, we
have $\mathcal{H}_0(\mathcal{M}) = 0$ for any $\mathcal{M}$. Consequently, $\mathcal{H}_0(\mathcal{N}) = 0$. From the above long
exact sequence, we finally obtain $\mathcal{H}_1(\mathcal{M}) = 0$.

Now we are ready to describe the interplay among (4.2.1), (4.2.2), (4.2.3) and
(4.2.4). Replace $\mathcal{M}$ by $\mathcal{O}_X$ with its trivial good filtration in (4.2.4). Namely set

$\mathcal{O}_X^{(k)} = \mathcal{O}_X$ for all $k \geq 0$ and $\mathcal{O}_X^{(k)} = 0$ for $k < 0$. Then we regain (4.2.3) as a special case of the Spencer sequence for a left $\mathcal{D}_X$-Module.

We will compute the derived functor $\mathbb{R}\mathcal{H}om_{\mathcal{D}_X}(\mathcal{O}_X, \mathcal{M})$ using (4.2.3) as follows.

$$\mathbb{R}\mathcal{H}om_{\mathcal{D}_X}(\mathcal{O}_X, \mathcal{M})$$
$$\cong \mathcal{H}om_{\mathcal{D}_X}(\mathcal{D}_X \otimes_{\mathcal{O}_X} \wedge^\bullet \Theta_X, \mathcal{M})$$
$$\cong \mathcal{H}om_{\mathcal{O}_X}(\wedge^\bullet \Theta_X, \mathcal{M})$$
$$\cong \mathcal{H}om_{\mathcal{O}_X}(\mathcal{O}_X, \Omega_X^\bullet \otimes_{\mathcal{O}_X} \mathcal{M})$$
$$\cong \Omega_X^\bullet \otimes_{\mathcal{O}_X} \mathcal{M}$$

That is, we have proved the isomorphism

$$\mathbb{R}\mathcal{H}om_{\mathcal{D}_X}(\mathcal{O}_X, \mathcal{M}) \cong \Omega_X^\bullet \otimes_{\mathcal{O}_X} \mathcal{M}.$$

$\square$

Definition 4.2.1 *The functor* $\mathcal{H}om_{\mathcal{D}_X}(\mathcal{O}_X, -)$ *that takes a left* $\mathcal{D}_X$-*Module* $\mathcal{M}$ *to the de Rham complex* $\Omega_X^\bullet \otimes_{\mathcal{O}_X} \mathcal{M}$ *is said to be the de Rham functor.*

Remark 4.2.2 As a consequence of our arguments above, the h-th cohomology sheaf of the de Rham complex $\Omega_X^\bullet \otimes_{\mathcal{O}_X} \mathcal{M}$ is the h-th extension sheaf. Namely,

$$\mathcal{H}^h(\Omega_X^\bullet \otimes_{\mathcal{O}_X} \mathcal{M}) \approx \mathcal{E}xt_{\mathcal{D}_X}^h(\mathcal{O}_X, \mathcal{M}),$$

where the h-th differential of the de Rham complex

$$0 \to \mathcal{M} \to \Omega_X^1 \otimes_{\mathcal{O}_X} \mathcal{M} \xrightarrow{d_\mathcal{M}^1} \dots \longrightarrow \Omega_X^h \otimes_{\mathcal{O}_X} \mathcal{M} \xrightarrow{d_\mathcal{M}^h} \dots \longrightarrow \Omega_X^n \otimes_{\mathcal{O}_X} \mathcal{M} \to 0$$

is given by

$$d_\mathcal{M}^h(\omega \otimes u) = d^h \omega \otimes u + \sum_{j=1}^n (dz_j \wedge \omega) \otimes \partial_j u.$$

Notice also that $\mathbb{R}\mathcal{H}om_{\mathcal{D}_X}(\mathcal{O}_X, \mathcal{D}_X) \approx \Omega_X^\bullet \otimes_{\mathcal{O}_X} \mathcal{D}_X$ is a free resolution of the right $\mathcal{D}_X$-Module Ω_X^n, i.e., (4.2.2).

Let us now compute

$$\mathbb{R}\mathcal{H}om_{\mathcal{D}_X}(\mathcal{O}_X, \mathcal{D}_X)$$

via (4.2.3). By what we have just seen, this complex is isomorphic to the complex

$$\mathcal{H}om_{\mathcal{D}_X}(\mathcal{D}_X \otimes_{\mathcal{O}_X} \wedge^{\bullet} \Theta_X, \mathcal{D}_X),$$

which is the complex of right $\mathcal{D}_X$-Modules $\Omega_X^{\bullet} \otimes_{\mathcal{O}_X} \mathcal{D}_X$, i.e., (4.2.2). The holonomicity of $\mathcal{O}_X$, see Chapter V, implies the pure dimensionality of $I\!RHom_{\mathcal{D}_X}(\mathcal{O}_X, \mathcal{D}_X)$, i.e.

$$I\!R^h\mathcal{H}om_{\mathcal{D}_X}(\mathcal{O}_X, \mathcal{D}_X) \overset{\text{def}}{=} \mathcal{E}xt_{\mathcal{D}_X}^h(\mathcal{O}_X, \mathcal{D}_X) \approx \mathcal{H}^h(\Omega_X^{\bullet} \otimes_{\mathcal{O}_X} \mathcal{D}_X) = 0$$
unless $h = n$,

and $\mathcal{H}^n(\Omega_X^{\bullet} \otimes_{\mathcal{O}_X} \mathcal{D}_X) \approx \Omega_X^n$. This shows that

$$I\!RHom_{\mathcal{D}_X}(\mathcal{O}_X, \mathcal{D}_X)[n] = \Omega_X^n.$$

On the other hand, from (4.2.2) we have

$$\mathcal{E}xt_{\mathcal{D}_X}^h(\Omega_X^n, \mathcal{D}_X) \cong \mathcal{H}^h(\mathcal{H}om_{\mathcal{D}_X}(\Omega_X^{\bullet} \otimes_{\mathcal{O}_X} \mathcal{D}_X, \mathcal{D}_X)) \approx \mathcal{H}^h(\mathcal{D}_X \otimes_{\mathcal{O}_X} \wedge^{\bullet} \Theta_X)$$

the h-th cohomology of (4.2.3). Therefore, we obtain

$$I\!RHom_{\mathcal{D}_X}(\Omega_X^n, \mathcal{D}_X)[n] = \mathcal{O}_X.$$

Consequently, we have

$$I\!RHom_{\mathcal{D}_X}(I\!RHom_{\mathcal{D}_X}(\mathcal{O}_X, \mathcal{D}_X)[n], \mathcal{D}_X)[n]$$
$$= I\!RHom_{\mathcal{D}_X}(\Omega_X^n.\mathcal{D}_X)[n]$$
$$= \mathcal{O}_X,$$

i.e., $\mathcal{E}xt_{\mathcal{D}_X}^n(\mathcal{E}xt_{\mathcal{D}_X}^n(\mathcal{O}_X, \mathcal{D}_X), \mathcal{D}_X) \approx \mathcal{O}_X$. We will return to this isomorphism in the section on holonomic $\mathcal{D}_X$-Modules in the next chapter.

Remark 4.2.3 In Chapter III, we constructed a free resolution of $\mathcal{O}_X$ as follows:

$$\mathcal{D}_X^n \overset{\begin{bmatrix} \partial_1 \\ \vdots \\ \partial_n \end{bmatrix}}{\longrightarrow} \mathcal{D}_X \longrightarrow \mathcal{O}_X \longrightarrow 0.$$

From this resolution, we obtained $I\!RHom_{\mathcal{D}_X}(\mathcal{O}_X, \mathcal{O}_X) = \mathbb{C}_X$. The de Rham functor $\mathcal{H}om_{\mathcal{D}_X}(\mathcal{O}_X, -)$ evaluated at $\mathcal{O}_X$ gives the usual de Rham complex $\Omega_X^{\bullet} \otimes_{\mathcal{O}_X} \mathcal{O}_X$, i.e.,

$$0 \longrightarrow \mathcal{O}_X \longrightarrow \Omega_X^1 \longrightarrow \ldots \longrightarrow \Omega_X^n \longrightarrow 0.$$

The Poincaré lemma then implies that

$$\mathcal{E}xt_{\mathcal{D}_X}^h(\mathcal{O}_X, \mathcal{O}_X) = 0 \quad \text{unless} \quad h = 0,$$

and $\mathcal{E}xt_{\mathcal{D}_X}^0(\mathcal{O}_X, \mathcal{O}_X) \approx \mathbb{C}_X$. That is, $I\!RHom_{\mathcal{D}_X}(\mathcal{O}_X, \mathcal{O}_X) = \mathbb{C}_X$.

Proposition 4.2.2 *For a left $\mathcal{D}_X$-Module $\mathcal{M}$, we have an isomorphism*

$$\mathcal{E}xt^h_{\mathcal{D}_X}(\mathcal{O}_X, \mathcal{M}) \cong \mathcal{T}or^{\mathcal{D}_X}_{n-h}(\Omega^n_X, \mathcal{M}).$$

Proof. By the pure dimensionality of $\mathcal{O}_X$ or equivalently the holonomicity of $\mathcal{O}_X$ (see Chapter V), we have

$$\mathcal{E}xt^h_{\mathcal{D}_X}(\mathcal{O}_X, \mathcal{D}_X) = 0 \quad \text{unless} \quad h = n.$$

Hence (see Remark 4.2.5 below), there is a spectral sequence such that

$$E_2^{p,q} = \mathcal{T}or^{\mathcal{D}_X}_p(\mathcal{E}xt^q_{\mathcal{D}_X}(\mathcal{O}_X, \mathcal{D}_X), \mathcal{M})$$

abuts to $\mathcal{E}xt^{q-p}_{\mathcal{D}_X}(\mathcal{O}_X, \mathcal{M})$. Since $E_2^{p,q} = 0$ unless $q = n$, we have

$$
\begin{array}{ccccc}
E_2^{p-2,n+1} & \longrightarrow & E_2^{p,n} & \longrightarrow & E_2^{p+2,n-1} \\
\| & & & & \| \\
0 & & & & 0
\end{array}
$$

i.e., we see that the abutment is given by

$$E_2^{p,n} \cong E_3^{p,n} \cong \ldots \cong E_\infty^{p,n} \cong E^{n-p} = \mathcal{E}xt^{n-p}_{\mathcal{D}_X}(\mathcal{O}_X, \mathcal{M}).$$

In particular, for $p = n - h$

$$E_2^{n-h,n} = \mathcal{T}or^{\mathcal{D}_X}_{n-h}(\mathcal{E}xt^n_{\mathcal{D}_X}(\mathcal{O}_X, \mathcal{D}_X), \mathcal{M}) \cong \mathcal{E}xt^h_{\mathcal{D}_X}(\mathcal{O}_X, \mathcal{M}).$$

$\square$

Remark 4.2.4 We can also compute $\Omega^n_X \otimes^{\mathbb{L}}_{\mathcal{D}_X} \mathcal{M}$ using (4.2.2). Namely, if we consider the free resolution of Ω^n_X

$$0 \longrightarrow \mathcal{D}_X \longrightarrow \Omega^1_X \otimes_{\mathcal{O}_X} \mathcal{D}_X \longrightarrow \ldots \longrightarrow \Omega^n_X \otimes_{\mathcal{O}_X} \mathcal{D}_X \xrightarrow{\epsilon'} \Omega^n_X \longrightarrow 0$$

we get a complex

$$(\Omega^\bullet_X \otimes_{\mathcal{O}_X} \mathcal{D}_X[n]) \otimes_{\mathcal{D}_X} \mathcal{M},$$

i.e., we obtain the de Rham complex

$$\Omega^\bullet_X \otimes_{\mathcal{O}_X} \mathcal{M}[n].$$

Therefore

$$\Omega^n_X \otimes^{\mathbb{L}}_{\mathcal{D}_X} \mathcal{M}[-n] = \Omega^\bullet_X \otimes_{\mathcal{O}_X} \mathcal{M}.$$

We have shown before that the right-hand side is $\mathbb{R}\mathcal{H}om_{\mathcal{D}_X}(\mathcal{O}_X, \mathcal{M})$. Hence

$$\Omega_X^n \otimes_{\mathcal{D}_X}^{I\!L} \mathcal{M}[-n] = I\!\!R\mathcal{H}om_{\mathcal{D}_X}(\mathcal{O}_X, \mathcal{M}),$$

namely, once again we have

$$\mathcal{T}or_{n-h}^{\mathcal{D}_X}(\Omega_X^n, \mathcal{M}) \approx \mathcal{E}xt_{\mathcal{D}_X}^h(\mathcal{O}_X, \mathcal{M}).$$

Remark 4.2.5 In this lengthy remark, we will discuss the spectral sequences associated with $I\!\!R\mathcal{H}om_{\mathcal{D}_X}(\mathcal{N}, \mathcal{D}_X) \otimes^{I\!L} \mathcal{M}$, where $\mathcal{M}$ is a left $\mathcal{D}_X$-Module and $\mathcal{N}$ is a right $\mathcal{D}_X$-Module.

Let $\mathcal{A}$ and $\mathcal{B}$ be abelian categories, and let $\mathcal{A}$ be a category with enough injectives. Let G be a right exact functor from $\mathcal{A}$ to $\mathcal{B}$, and let $C_0(\mathcal{A})$ and $C_0(\mathcal{B})$ indicate the categories of complexes of objects in $\mathcal{A}$ and $\mathcal{B}$, respectively. Define a functor $C_0(G)$ from $C_0(\mathcal{A})$ to $C_0(\mathcal{B})$ as follows.

$$(\ldots \longrightarrow A_j \longrightarrow A_{j-1} \longrightarrow \ldots) \xrightarrow{C_0(G)} (\ldots \longrightarrow GA_j \longrightarrow GA_{j-1} \longrightarrow \ldots)$$

and consider the diagram

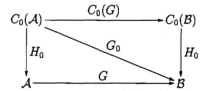

In this diagram, we define G_0 as $G \circ H_0 = H_0 \circ C_0(G)$, i.e., G_0 is a functor from $C_0(\mathcal{A})$ to $\mathcal{B}$ such that

$$\begin{aligned} G_0(A_\bullet) &= G(\ker(A_1 \longrightarrow A_0)) \\ &= \ker(GA_1 \longrightarrow GA_0). \end{aligned}$$

Then the sequence $\{G_n A_\bullet\}$ gives the derived functors of G_0, that is, the functors $\{G_n A_\bullet\}$ are the hyperderived functors of G. This induces spectral sequences

$$\begin{aligned} 'E_{p,q}^2 &= L_p G(H_q(A_\bullet)) \\ E_{p,q}^1 &= L_q G A_p \end{aligned}$$

which abuts to the hyperderived functor $G_n A_\bullet$ of G.

Now let $'\mathcal{N}^\bullet$ be a complex of right $\mathcal{D}_X$-Modules, and let $\mathcal{M}$ be a left $\mathcal{D}_X$-Module. Then there is induced a spectral sequence

(4.2.5) $$E_2^{p,q} = \mathcal{H}^p(\mathcal{T}or_{-q}^{\mathcal{D}_X}('\mathcal{N}^*, \mathcal{M}))$$

abutting to the hypertor $\mathcal{T}or_n^{\mathcal{D}_X}('\mathcal{N}^\bullet, \mathcal{M})$. Notice that $E_2^{p,q} = 0$ for all $q > 0$. One can begin at

(4.2.6) $E_{p,q}^1 = \mathcal{T}or_q^{\mathcal{D}_X}('\mathcal{N}_p, \mathcal{M})$,

where $'\mathcal{N}_p = '\mathcal{N}^{-p}$. From the above $E_{p,q}^1$ we get

$$E_{p,q}^2 = \mathcal{H}_p \mathcal{T}or_q^{\mathcal{D}_X}('\mathcal{N}^\bullet, \mathcal{M})),$$

i.e., cohomologically

$$E_2^{p,q} = \mathcal{H}^p \mathcal{T}or_{-q}^{\mathcal{D}_X}('\mathcal{N}^\bullet, \mathcal{M})).$$

Next, replace the above complex $'\mathcal{N}^\bullet$ of right $\mathcal{D}_X$-Modules by $I\!R\mathcal{H}om_{\mathcal{D}_X}(\mathcal{N}, \mathcal{D}_X)$, where $\mathcal{N}$ is a left $\mathcal{D}_X$-Module. Namely, we will consider

$$I\!R\mathcal{H}om_{\mathcal{D}_X}(\mathcal{N}, \mathcal{D}_X) \otimes_{\mathcal{D}_X}^{I\!L} \mathcal{M},$$

where $\mathcal{M}$ is a left $\mathcal{D}_X$-Module, not necessarily finitely presented as a $\mathcal{D}_X$-Module, i.e. not necessarily $\mathcal{D}_X$-coherent. In the case where $\mathcal{M}$ is of finite presentation the cohomology sheaf of $I\!R\mathcal{H}om_{\mathcal{D}_X}(\mathcal{N}, \mathcal{D}_X) \otimes_{\mathcal{D}_X}^{I\!L} \mathcal{M}$ is nothing but the usual extension sheaf $\mathcal{E}xt_{\mathcal{D}_X}^h(\mathcal{N}, \mathcal{M})$. This is because we have $\mathcal{H}om_{\mathcal{D}_X}(-, \mathcal{D}_X) \otimes_{\mathcal{D}_X} \mathcal{F} = \mathcal{H}om_{\mathcal{D}_X}(-, \mathcal{F})$ for a finitely generated projective $\mathcal{D}_X$-Module $\mathcal{F}$. First take a projective resolution of $\mathcal{N}$:

$$\mathcal{P}_\bullet \longrightarrow \mathcal{N}.$$

Then, let $'\mathcal{R}^\bullet \overset{\text{def}}{=} \mathcal{H}om_{\mathcal{D}_X}(\mathcal{P}_\bullet, \mathcal{D}_X)$. By the definition,

$$\mathcal{E}xt_{\mathcal{D}_X}^h(\mathcal{N}, \mathcal{D}_X) = \mathcal{H}^h('\mathcal{R}^\bullet).$$

Next, let $'\mathcal{P}^\bullet$ be a complex of flat right $\mathcal{D}_X$-Modules satisfying $\mathcal{H}^h('\mathcal{P}^\bullet) \overset{\approx}{\longrightarrow} \mathcal{H}^h('\mathcal{R}^\bullet)$, i.e., $'\mathcal{P}^\bullet$ and $'\mathcal{R}^\bullet$ are quasi-isomorphic. Then

$$\mathcal{H}^h('\mathcal{P}^\bullet \otimes_{\mathcal{D}_X} \mathcal{M})$$

is an invariant of two variables, contravariant in $\mathcal{N}$ and covariant in $\mathcal{M}$. Denote this functor by $\mathcal{C}^h(\mathcal{N}, \mathcal{M})$. Then $\mathcal{C}^h(\mathcal{N}, \mathcal{M})$ is an exact connected sequence of functors. (See [35] or [150] for further details). This induces a second quadrant spectral sequence such that

(4.2.7) $E_2^{p,q} = \mathcal{T}or_{-p}^{\mathcal{D}_X}(\mathcal{E}xt_{\mathcal{D}_X}^q(\mathcal{N}, \mathcal{D}_X), \mathcal{M})$;

this sequence abuts to $\mathcal{C}^n(\mathcal{N}, \mathcal{M})$.

Finally, let $\mathcal{Q}_\bullet \longrightarrow \mathcal{M}$ be an acyclic flat resolution of $\mathcal{M}$. Then we have that $I\!R\mathcal{H}om_{\mathcal{D}_X}(\mathcal{N}, \mathcal{D}_X) \otimes_{\mathcal{D}_X}^{I\!L} \mathcal{M}$ can be computed as follows:

$$\mathbb{R}\mathcal{H}om_{\mathcal{D}_X}(\mathcal{N},\mathcal{D}_X)\otimes\tfrac{\mathbb{L}}{\mathcal{D}_X}\mathcal{M}$$

$$=\mathcal{H}om_{\mathcal{D}_X}(\mathcal{P}_\bullet,\mathcal{D}_X)\otimes\tfrac{\mathbb{L}}{\mathcal{D}_X}\mathcal{M}\overset{\text{def}}{=}{}'\mathcal{R}^\bullet\otimes\tfrac{\mathbb{L}}{\mathcal{D}_X}\mathcal{M}$$

$$={}'\mathcal{P}^\bullet\otimes_{\mathcal{D}_X}\mathcal{M}$$

$$={}'\mathcal{P}^\bullet\underset{\mathcal{D}_X}{\otimes}\mathcal{Q}^\bullet.$$

As an application of equation (4.2.7), let $\mathcal{N}=\mathcal{O}_X$ and let $\mathcal{M}$ be a coherent left $\mathcal{D}_X$-Module. Since, as we saw earlier, we have:

$$\mathcal{E}xt^q_{\mathcal{D}_X}(\mathcal{O}_X,\mathcal{D}_X)=\begin{cases}\Omega^n_X & \text{for }q=n\\0 & \text{for }q\neq n\end{cases}$$

we obtain that

$$E_2^{h-n,n}=\mathcal{T}or^{\mathcal{D}_X}_{n-h}(\mathcal{E}xt^n_{\mathcal{D}_X}(\mathcal{O}_X,\mathcal{D}_X),\mathcal{M})$$

is isomorphic to the abutment $\mathcal{E}xt^h_{\mathcal{D}_X}(\mathcal{O}_X,\mathcal{M})$. That is,

$$\mathcal{T}or^{\mathcal{D}_X}_{n-h}(\Omega^n_X,\mathcal{M})\approx\mathcal{E}xt^h_{\mathcal{D}_X}(\mathcal{O}_X,\mathcal{M}),$$

which is Proposition 4.2.2.

Remark 4.2.6 We have introduced the de Rham functor $\mathcal{H}om_{\mathcal{D}_X}(\mathcal{O}_X,-)$ and the solution functor $\mathcal{H}om_{\mathcal{D}_X}(-,\mathcal{O}_X)$ in Chapter III. In his recent work, Mebkhout proved an important global duality theorem for these functors. We will now quickly describe Mebkhout's result and we will show how both Poincaré duality and Serre duality follow from his theorem.

Let $\Gamma_c(X,\mathcal{S})$ denote the set of global sections of a sheaf $\mathcal{S}$ over X with compact supports, and let $\Gamma(X,\mathcal{S})$ denote the usual set of global sections of $\mathcal{S}$ over X. Note that $\Gamma(X,-)$ and $\Gamma_c(X,-)$ are left exact functors from the category of sheaves over X to the category of abelian groups. Using the left exact functors $\mathcal{H}om_{\mathcal{D}_X}(\mathcal{O}_X,-)$ and $\Gamma_c(X,-)$, we can define a left exact functor $Hom_{\mathcal{D}_X,c}(\mathcal{O}_X,-)$ by setting, for any left $\mathcal{D}_X$-module $\mathcal{M}$,

$$Hom_{\mathcal{D}_X,c}(\mathcal{O}_X,\mathcal{M})=\Gamma_c(X,\mathcal{H}om_{\mathcal{D}_X}(\mathcal{O}_X,\mathcal{M})).$$

Similarly, we can define

$$Hom_{\mathcal{D}_X}(\mathcal{M},\mathcal{O}_X)=\Gamma(X,\mathcal{H}om_{\mathcal{D}_X}(\mathcal{M},\mathcal{O}_X)).$$

Then, there are induced spectral sequences of composite functors such that

$$E_{2,c}^{p,q}=H^p_c(X,\mathcal{E}xt^q_{\mathcal{D}_X}(\mathcal{O}_X,\mathcal{M}))$$

abutting to $Ext_{\mathcal{D}_X,c}^n(\mathcal{O}_X, \mathcal{M}) \overset{\text{def}}{=} \mathbb{R}^n Hom_{\mathcal{D}_X,c}(\mathcal{O}_X, \mathcal{M})$, and

$$E_2^{p,q} = H^p(X, \mathcal{E}xt_{\mathcal{D}_X}^q(\mathcal{M}, \mathcal{O}_X))$$

abutting to

$$Ext_{\mathcal{D}_X}^n(\mathcal{M}, \mathcal{O}_X) = \mathbb{R}^n Hom_{\mathcal{D}_X}(\mathcal{M}, \mathcal{O}_X).$$

In terms of the theory of derived categories, these can be written as

$$\mathbb{R}Hom_{\mathcal{D}_X,c}(\mathcal{O}_X, \mathcal{M}) = \mathbb{R}\Gamma_c(X, \mathbb{R}\mathcal{H}om_{\mathcal{D}_X}(\mathcal{O}_X, \mathcal{M})),$$

and

$$\mathbb{R}Hom_{\mathcal{D}_X}(\mathcal{M}, \mathcal{O}_X) = \mathbb{R}\Gamma(X, \mathbb{R}\mathcal{H}om_{\mathcal{D}_X}(\mathcal{M}, \mathcal{O}_X)),$$

respectively. Then Mebkhout's global duality states that the Yoneda pairing

$$Ext_{\mathcal{D}_X,c}^{2n-j}(\mathcal{O}_X, \mathcal{M}) \times Ext_{\mathcal{D}_X}^j(\mathcal{M}, \mathcal{O}_X) \longrightarrow Ext_{\mathcal{D}_X,c}^{2n}(\mathcal{O}_X, \mathcal{O}_X),$$

composed with the trace map

$$Ext_{\mathcal{D}_X,c}^{2n}(\mathcal{O}_X, \mathcal{O}_X) \longrightarrow \mathbb{C},$$

induces a topological duality between $Ext_{\mathcal{D}_X,c}^{2n-j}(\mathcal{O}_X, \mathcal{M})$ and $Ext_{\mathcal{D}_X}^j(\mathcal{M}, \mathcal{O}_X)$.

We will show that the above assertion is a generalization of both Poincaré and Serre duality. In order to get Poincaré duality, we choose the holonomic $\mathcal{D}$-Module $\mathcal{O}_X$. Then the above spectral sequence $E_{2,c}^{p,q}$ becomes:

$$E_{2,c}^{p,q} = H_c^p(X, \mathcal{E}xt_{\mathcal{D}_X}^q(\mathcal{O}_X, \mathcal{O}_X)).$$

Poincaré lemma implies that $\mathcal{E}xt_{\mathcal{D}_X}^q(\mathcal{O}_X, \mathcal{O}_X) = 0$ for $q \neq 0$, and $\mathcal{E}xt_{\mathcal{D}_X}^0(\mathcal{O}_X, \mathcal{O}_X) \cong \mathbb{C}_X$, as it was observed in Remark 4.2.3. Hence, we have $E_{2,c}^{p,q} = 0$ for $q \neq 0$, and $E_{2,c}^{p,0} \cong H_c^p(X, \mathbb{C}_X)$. Similarly, we also obtain $E_2^{p,0} \cong H^p(X, \mathbb{C}_X)$. From these collapsing spectral sequences, we can compute their abutments as follows:

$$\begin{cases} Ext_{\mathcal{D}_X,c}^{2n-j}(\mathcal{O}_X, \mathcal{O}_X) \cong E_{2,c}^{2n-p,0} \cong H_c^{2n-p}(X, \mathbb{C}_X) \cong H_p^c(X, \mathbb{C}_X) \\ Ext_{\mathcal{D}_X}^j(\mathcal{O}_X, \mathcal{O}_X) \cong E_2^{p,0} \cong H^p(X, \mathbb{C}_X). \end{cases}$$

We then obtain the pairing

$$H_p^c(X, \mathbb{C}_X) \times H^p(X, \mathbb{C}_X) \longrightarrow Ext_{\mathcal{D}_X,c}^{2n}(\mathcal{O}_X, \mathcal{O}_X)$$
$$\wr\wr$$
$$H_c^{2n}(X, \mathbb{C}_X) \longrightarrow \mathbb{C},$$

which is nothing but the well known Poincaré duality for X.

The other extreme of a holonomic $\mathcal{D}_X$-Module $\mathcal{O}_X$ is the sheaf $\mathcal{D}_X$. Namely, as mentioned in Chapter III, $\mathcal{D}_X$ is the $\mathcal{D}_X$-Module corresponding to the system of no equations. Its characteristic variety has dimension $2n$, and Serre's duality is obtained by putting $\mathcal{M} = \mathcal{D}_X$ in Mebkhout's theorem. Then, notice that $\mathcal{H}om_{\mathcal{D}_X}(\mathcal{D}_X, -)$ is an exact functor. Hence, the higher cohomology $\mathcal{E}xt_{\mathcal{D}_X}^q(\mathcal{D}_X, \mathcal{O}_X)$ vanishes for $q \neq 0$, and $\mathcal{E}xt_{\mathcal{D}_X}^0(\mathcal{D}_X, \mathcal{O}_X) \cong \mathcal{H}om_{\mathcal{D}_X}(\mathcal{D}_X, \mathcal{O}_X) \cong \mathcal{O}_X$. Hence, the spectral sequence gives us

$$E_2^{p,0} = H^p(X, \mathcal{E}xt_{\mathcal{D}_X}^0(\mathcal{D}_X, \mathcal{O}_X)) \cong H^p(X, \mathcal{O}_X).$$

On the other hand, we have $\mathcal{E}xt_{\mathcal{D}_X}^q(\mathcal{O}_X, \mathcal{D}_X) = 0$ for $q \neq n$, and $\mathcal{E}xt_{\mathcal{D}_X}^n(\mathcal{O}_X, \mathcal{D}_X) \cong \Omega_X^n$, as we showed in the last paragraph preceding Remark 4.2.3. Therefore,

$$E_{2,c}^{n-j,n} = H_c^{n-j}(X, \mathcal{E}xt_{\mathcal{D}_X}^n(\mathcal{O}_X, \mathcal{D}_X)) \cong H_c^{n-j}(X, \Omega_X^n).$$

Their abutments are computed through those collapsing spectral sequences as

$$\begin{cases} Ext_{\mathcal{D}_X,c}^{2n-j}(\mathcal{O}_X, \mathcal{D}_X) \cong E_{2,c}^{n-j,n} = H_c^{n-j}(X, \Omega_X^n) \\ Ext_{\mathcal{D}_X}^j(\mathcal{D}_X, \mathcal{O}_X) \cong E_2^{j,0} = H^j(X, \mathcal{O}_X). \end{cases}$$

Then the pairing becomes

$$H_c^{n-j}(X, \Omega_X^n) \times H^j(X, \mathcal{O}_X) \quad \longrightarrow \quad Ext_{\mathcal{D}_X,c}^{2n}(\mathcal{O}_X, \mathcal{O}_X)$$
$$\wr\wr$$
$$H_c^{2n}(X, \mathcal{C}_X) \quad \longrightarrow \quad \mathcal{C}.$$

Thus, we have obtained Serre duality, i.e. mutually strong duality between the FS-space $H^j(X, \mathcal{O}_X)$ and the DFS-space $H_c^{n-j}(X, \Omega_X^n)$. See [8] for further details.

4.3   Algebraic Local Cohomology

Consider now the case in which the complex manifold X is an open neighborhood of the origin $\{z = 0\}$ in $\mathcal{C}$. Then the stalk $\mathcal{O}_{X,0}$ at $z = 0$ is the ring of convergent power series, which we will denote by $\mathcal{C}\{z\}$. The sheaf $\mathcal{D}_X$ of germs of ordinary differential operators may then be written as

$$\mathcal{D}_X = \bigoplus_{m \geq 0} \mathcal{O}_X \left(\frac{d}{dz}\right)^m.$$

Let $\mathcal{O}_X[z^{-1}]$ be the field of Laurent series. Then we have the canonical exact sequence

$$0 \longrightarrow \mathcal{O}_X \longrightarrow \mathcal{O}_X[z^{-1}] \longrightarrow \mathcal{O}_X[z^{-1}]/\mathcal{O}_X \longrightarrow 0.$$

Since both $\mathcal{O}_X$ and $\mathcal{O}_X[z^{-1}]$ are $\mathcal{D}_X$-Modules, we can define a $\mathcal{D}_X$-Module structure on $\mathcal{O}_X[z^{-1}]/\mathcal{O}_X$. Notice that the support of $\mathcal{O}_X[z^{-1}]/\mathcal{O}_X$ is concentrated at $z = 0$. This observation gives a different explanation for the phenomenon described in section 2 of Chapter I.

Define elements $\delta^{(m)}$ of $\mathcal{O}_X[z^{-1}]/\mathcal{O}_X$ as follows:

$$\delta^{(m)} = [(-1)^m m! z^{-(m+1)}], \quad m = 0, 1, 2. \ldots,$$

where $[\varphi]$ indicates the equivalence class of an element φ in $\mathcal{O}_X[z^{-1}]$ under the canonical epimorphism defined above. It is easy to see that $\{\delta^{(m)}\}_{m=0,1,2,\ldots}$ is a set of generators of $\mathcal{O}_X[z^{-1}]/\mathcal{O}_X$ over $\mathbb{C}$.

We also have

$$\begin{cases} \dfrac{d}{dz}\delta^{(m)} = \delta^{(m+1)} \\ z\delta^{(m)} = -m\delta^{(m-1)}. \end{cases}$$

That is, $\delta^{(m)}$ is the m-th derivative of Dirac delta function $\delta = \delta^{(0)}$.

Consider now the case in which X is an open subset of $\mathbb{C}^n$ and Y is an analytic variety defined by an ideal $\mathcal{J}$ of $\mathcal{O}_X$, i.e., Y is the support of $\mathcal{O}_X/\mathcal{J}$. For any $\mathcal{O}_X$-Module $\mathcal{N}$, the set

$$\mathcal{E}xt^0_{\mathcal{O}_X}(\mathcal{O}_X/\mathcal{J}^n, \mathcal{N}) = \mathcal{H}om_{\mathcal{O}_X}(\mathcal{O}_X/\mathcal{J}^n, \mathcal{N})$$

represents those germs u in $\mathcal{N}$ satisfying $\mathcal{J}^n u = 0$.

That is, given the exact sequence

$$0 \longrightarrow \mathcal{J}^n \longrightarrow \mathcal{O}_X \longrightarrow \mathcal{O}_X/\mathcal{J}^n \longrightarrow 0,$$

we have the long exact sequence of $\mathcal{O}_X$-Modules

$$0 \longrightarrow \mathcal{H}om_{\mathcal{O}_X}(\mathcal{O}_X/\mathcal{J}^n, \mathcal{N}) \longrightarrow \mathcal{H}om_{\mathcal{O}_X}(\mathcal{O}_X, \mathcal{N}) \longrightarrow \mathcal{H}om_{\mathcal{O}_X}(\mathcal{J}^n, \mathcal{N}) \longrightarrow$$

$$\longrightarrow \mathcal{E}xt^1_{\mathcal{O}_X}(\mathcal{O}_X/\mathcal{J}^n, \mathcal{N}) \longrightarrow \mathcal{E}xt^1_{\mathcal{O}_X}(\mathcal{O}_X, \mathcal{N}) \longrightarrow \mathcal{E}xt^1_{\mathcal{O}_X}(\mathcal{J}^n, \mathcal{N})$$

(4.3.1) $$\longrightarrow \mathcal{E}xt^2_{\mathcal{O}_X}(\mathcal{O}_X/\mathcal{J}^n, \mathcal{N}) \longrightarrow \cdots$$

Note however that the groups $\mathcal{E}xt^j_{\mathcal{O}_X}(\mathcal{O}_X, \mathcal{N})$ vanish for $j \geq 1$ since $\mathcal{H}om_{\mathcal{O}_X}(\mathcal{O}_X, -)$ is an exact functor.

Definition 4.3.1 *The algebraic local (or relative) j-th cohomology of Y is defined by*

$$\mathrm{ind\,lim}\, \mathcal{E}xt^j_{\mathcal{O}_X}(\mathcal{O}_X/\mathcal{J}^m, \mathcal{N})$$

and is denoted by $\mathcal{H}_{[Y]}^j(\mathcal{N})$. Furthermore, we define

$$\mathcal{H}_{[X|Y]}^j(\mathcal{N}) = indlim\, \mathcal{E}xt_{\mathcal{O}_X}^j(\mathcal{J}^m, \mathcal{N}).$$

Then we obtain the following exact sequence from (4.3.1)

(4.3.2) $0 \longrightarrow \mathcal{H}_{[Y]}^0(\mathcal{N}) \longrightarrow \mathcal{N} \longrightarrow \mathcal{H}_{[X|Y]}^0(\mathcal{N}) \longrightarrow \mathcal{H}_{[Y]}^1(\mathcal{N}) \longrightarrow 0,$

and the isomorphism for each $j = 1, 2, 3, \ldots$

$$\mathcal{H}_{[X|Y]}^j(\mathcal{N}) \xrightarrow{\approx} \mathcal{H}_{[Y]}^{j+1}(\mathcal{N}).$$

In the case in which $\mathcal{N}$ is a $\mathcal{D}_X$-Module, the $\mathcal{O}_X$-flatness of $\mathcal{D}_X$ implies the isomorphism

$$\mathcal{H}om_{\mathcal{D}_X}(\mathcal{D}_X \mathcal{J}^m, \mathcal{N}) \approx \mathcal{H}om_{\mathcal{O}_X}(\mathcal{J}^m, \mathcal{N}).$$

Since we also have the $\mathcal{D}_X$-isomorphism

$$\mathcal{D}_X \mathcal{J}^m \longrightarrow \mathcal{D}_X \mathcal{J}^{m-1}$$

induced by the inclusion $\partial_{z_i} \mathcal{J}^m \subset \mathcal{J}^{m-1}$, we obtain

(4.3.3) $\mathcal{H}om_{\mathcal{D}_X}(\mathcal{D}_X \mathcal{J}^m, \mathcal{N}) \xrightarrow{\text{``}\partial_{z_i}\text{''}} \mathcal{H}om_{\mathcal{D}_X}(\mathcal{D}_X \mathcal{J}^{m-1}, \mathcal{N}).$

Hence, by taking the direct limit over varying m, one can define a $\mathcal{D}_X$-Module structure on $\mathcal{H}_{[X|Y]}^0(\mathcal{N})$. Namely, (4.3.3) induces a homomorphism

$$\mathcal{H}_{[X|Y]}^0(\mathcal{N}) \xrightarrow{\text{``}\partial_{z_i}\text{''}} \mathcal{H}_{[X|Y]}^0(\mathcal{N}).$$

Since $\mathcal{H}_{[Y]}^0(\mathcal{N})$ is the kernel of

$$\mathcal{N} \longrightarrow \mathcal{H}_{[X|Y]}^0(\mathcal{N}),$$

$\mathcal{H}_{[Y]}^0(\mathcal{N})$ is a $\mathcal{D}_X$-Module. Furthermore, since $\mathcal{H}_{[Y]}^j$ and $\mathcal{H}_{[X|Y]}^j$ are, respectively, the j-th derived functors of $\mathcal{H}_{[Y]}^0$ and $\mathcal{H}_{[X|Y]}^0$, and since the cohomology functor and the direct limit commute, $\mathcal{H}_{[Y]}^j(\mathcal{N})$ and $\mathcal{H}_{[X|Y]}^j(\mathcal{N})$ become $\mathcal{D}_X$-Modules.

Remark 4.3.1   Since the germs of $\mathcal{H}om_{\mathcal{O}_X}(\mathcal{O}_X/\mathcal{J}^m, \mathcal{N})$ are annihilated by $\mathcal{J}^m$ as we noted earlier, their supports are in Y. Hence the natural map

$$\mathcal{H}om_{\mathcal{O}_X}(\mathcal{O}_X/\mathcal{J}^m, \mathcal{N}) \longrightarrow \mathcal{H}_Y^0(\mathcal{N})$$

induces a map from

$$indlim \, \mathcal{H}om_{\mathcal{O}_X}(\mathcal{O}_X/\mathcal{J}^m, \mathcal{N}) \longrightarrow \mathcal{H}^0_Y(\mathcal{N}),$$

where $\mathcal{H}^0_Y(\mathcal{N})$ is the trascendental local (or relative) cohomology sheaf. Consequently, we have the induced map from the algebraic local cohomology sheaves to the trascendental local cohomology sheaves

$$\mathcal{H}^j_{[Y]}(\mathcal{N}) \longrightarrow \mathcal{H}^j_Y(\mathcal{N}).$$

Remark 4.3.2 As we indicated, the algebraic cohomology sheaves $\mathcal{H}^j_{[Y]}(\mathcal{N})$ and $\mathcal{H}^j_{[X|Y]}(\mathcal{N})$ are $\mathcal{D}_X$-Modules. Therefore, the long exact sequence in (4.3.1) is a long exact sequence of $\mathcal{D}_X$-Modules. If $\mathcal{N}^\bullet$ is an object in the derived category of $\mathcal{D}_X$-Modules, then we have a triangle

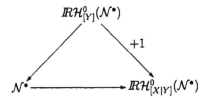

where $+1$ indicates the map from $I\!\!R\mathcal{H}^0_{[X|Y]}(\mathcal{N}^\bullet)$ to $I\!\!R\mathcal{H}^0_{[Y]}(\mathcal{N}^\bullet)[1]$.

4.4 Cohomological Properties of $\mathcal{D}_X$

In this section, we will study several properties of the derived functors of the functor $\mathcal{H}om_{\mathcal{D}_X}(-, \mathcal{D}_X)$, which will be used in Chapter V to give a cohomological characterization of holonomic $\mathcal{D}_X$-Modules. As we observed in Chapter III, the notion of filtration was utilized as a communication device between the category of non-commutative objects and the category of commutative objects. In this section we will further observe the usefulness of the notion of filtration to study the non-commutative ring $\mathcal{D}_X$.

First we will begin with the assertion that the functor $\mathcal{O}_{T^\bullet X} \otimes_{\overline{\mathcal{D}}_X}$ is not only an exact functor, but also a faithful functor from the category of coherent $\overline{\mathcal{D}}_X$-Modules to the category of coherent $\mathcal{O}_{T^\bullet X}$-Modules. In order to prove this, it will be sufficient to show that at each stalk the sheaf $\mathcal{O}_{T^\bullet X}$ is flat over $\overline{\mathcal{D}}_X$. But this follows immediately since $\mathcal{O}_{T^\bullet X}$ is the noetherian ring of convergent power series while $\overline{\mathcal{D}}_X$ is the polynomial ring in the cotangent coordinates $\xi_1, \xi_2, \ldots, \xi_n$. For the ideal $(\xi_1, \ldots, \xi_n)$, these noetherian rings are analytically isomorphic, i.e., their $(\xi_1, \ldots, \xi_n)$-adic completions are isomorphic. Therefore, we obtain the flatness of $\mathcal{O}_{T^*X}$ over the ring $\overline{\mathcal{D}}_X$. See, for example, [19] for the commutative ring argument. For a $\mathcal{D}_X$-Module $\mathcal{M}$ with a good filtration, we have a quasi-free resolution of $\mathcal{M}$

$$(4.4.1) \qquad \mathcal{L}_\bullet : \quad \cdots \xrightarrow{P_1} \mathcal{L}_1 \xrightarrow{P_0} \mathcal{L}_0 \longrightarrow \mathcal{M} \longrightarrow 0.$$

Namely, each $\mathcal{L}_i = \oplus_{k=1}^{r_i} \mathcal{D}_X(l_{i,k})$, and each P_i is a filtration preserving map. Then the contravariant functor $\mathcal{H}om_{\mathcal{D}_X}(-, \mathcal{D}_X)$ induces the following complex from the complex $\mathcal{L}_\bullet$ of (4.4.1):

$$(4.4.2) \qquad {}'\mathcal{L}^\bullet : \quad {}'\mathcal{L}^0 \xrightarrow{{}'P^0} {}'\mathcal{L}_1 \xrightarrow{{}'P^1} {}'\mathcal{L}_2 \xrightarrow{{}'P^2} \cdots .$$

Then the derived functor $\mathcal{E}xt_{\mathcal{D}_X}^h(\mathcal{M}, \mathcal{D}_X)$ of $\mathcal{H}om_{\mathcal{D}_X}(\mathcal{M}, \mathcal{D}_X)$ can be computed as the cohomology $\mathcal{H}^h(\mathcal{H}om_{\mathcal{D}_X}(\mathcal{L}_\bullet, \mathcal{D}_X)) = \mathcal{H}^h({}'\mathcal{L}^\bullet)$. With the induced filtration on ${}'\mathcal{L}^\bullet$ defined by

$$({}'\mathcal{L}^i)^{(k)} = \{t \in \mathcal{H}om_{\mathcal{D}_X}(\mathcal{L}_i, \mathcal{D}_X) = {}'\mathcal{L}^i : t(\mathcal{L}_i^{(l)}) \subset \mathcal{D}_X^{(l+k)}, \quad \forall l \},$$

we obtain a complex of $\overline{\mathcal{D}}_X$-Modules

$$(4.4.3) \qquad \overline{{}'\mathcal{L}^0} \xrightarrow{{}'\overline{P}_0} \overline{{}'\mathcal{L}^1} \xrightarrow{{}'\overline{P}_1} \overline{{}'\mathcal{L}^2} \longrightarrow \cdots .$$

From the canonical commutativity of the diagram

we observe also that complex (4.4.3) is obtained as $\mathcal{H}om_{\overline{\mathcal{D}}_X}(\overline{\mathcal{L}}_\bullet, \overline{\mathcal{D}}_X)$, i.e., from

$$(4.4.4) \qquad \overline{\mathcal{L}}_\bullet : \quad \cdots \longrightarrow \overline{\mathcal{L}}_2 \xrightarrow{\overline{P}_1} \overline{\mathcal{L}}_1 \xrightarrow{\overline{P}_0} \overline{\mathcal{L}}_0.$$

We can also get (4.4.3) by applying the functor $\mathcal{H}om_{\overline{\mathcal{D}}_X}(-, \overline{\mathcal{D}}_X)$. Since $\overline{\mathcal{L}}_\bullet$ is a free resolution of $\overline{\mathcal{M}}$, $\mathcal{E}xt_{\overline{\mathcal{D}}_X}^h(\overline{\mathcal{M}}, \overline{\mathcal{D}}_X)$ can be computed as the h-th cohomology of complex $\mathcal{H}om_{\overline{\mathcal{D}}_X}(\overline{\mathcal{L}}_\bullet, \overline{\mathcal{D}}_X) = {}'\overline{\mathcal{L}^\bullet}$ of $\overline{\mathcal{D}}_X$-Modules. The functor $\mathcal{O}_{T^*X} \otimes_{\overline{\mathcal{D}}_X}$ is a faithful and exact functor from the category of coherent $\overline{\mathcal{D}}_X$-Modules to the category of coherent $\mathcal{O}_{T^*X}$-Modules.

Therefore, we obtain the isomorphism

$$(4.4.5) \quad \mathcal{O}_{T^*X} \otimes_{\overline{\mathcal{D}}_X} \mathcal{E}xt_{\overline{\mathcal{D}}_X}^h(\overline{\mathcal{M}}, \overline{\mathcal{D}}_X) = \mathcal{E}xt_{T^*X}^h(\mathcal{O}_{T^*X} \otimes_{\overline{\mathcal{D}}_X} \overline{\mathcal{M}}, \mathcal{O}_{T^*X}),$$

and we note that $\mathcal{O}_{T^*X} \otimes_{\overline{\mathcal{D}}_X} \overline{\mathcal{M}}$ is coherent as an $\mathcal{O}_{T^*X}$-Module. Then we have

$$\text{codim supp}(\mathcal{E}xt^h_{\mathcal{O}_{T^*X}}(\mathcal{O}_{T^*X} \otimes_{\overline{\mathcal{D}_X}} \mathcal{M}, \mathcal{O}_{T^*X})) \geq h$$

from the theory of commutative algebras. Hence, in order to prove that

(4.4.6)   codim $V(\mathcal{E}xt^h_{\mathcal{D}_X}(\mathcal{M}, \mathcal{D}_X)) \geq h,$

it is sufficient to show

$$V(\mathcal{E}xt^h_{\mathcal{D}_X}(\mathcal{M}, \mathcal{D}_X)) \subset \text{supp } (\mathcal{E}xt^h_{\mathcal{O}_{T^*X}}(\mathcal{O}_{T^*X} \otimes_{\overline{\mathcal{D}_X}} \mathcal{M}, \mathcal{O}_{T^*X})).$$

As was pointed out above, those higher extension sheaves $\mathcal{E}xt^h_{\mathcal{D}_X}(\mathcal{M}, \mathcal{D}_X)$ and $\mathcal{E}xt^h_{\mathcal{O}_{T^*X}}(\mathcal{O}_{T^*X} \otimes_{\overline{\mathcal{D}_X}} \mathcal{M}, \mathcal{O}_{T^*X})$ are computed as cohomologies of complexes (4.4.2) and (4.4.3), respectively, through (4.4.5). We have the following lemma.

Lemma 4.4.1 *Let* $\mathcal{M}^\bullet$ *be a complex of filtered* $\mathcal{D}_X$-*Modules with good filtrations, and let* $\mathcal{O}_{T^*X} \otimes_{\overline{\mathcal{D}_X}} \mathcal{M}^\bullet$ *be the induced complex. Then we have*

$$V(\mathcal{H}^h(\mathcal{M}^\bullet)) \subset \text{supp } (\mathcal{H}^h(\mathcal{O}_{T^*X} \otimes_{\overline{\mathcal{D}_X}} \mathcal{M}^\bullet)).$$

Proof. One can define a filtration on $ker(\mathcal{M}^h \longrightarrow \mathcal{M}^{h+1})$ by inducing it from the filtration on $\mathcal{M}^h$. Then the induced map

$$\mathcal{O}_{T^*X} \otimes_{\overline{\mathcal{D}_X}} \overline{ker(\mathcal{M}^h \longrightarrow \mathcal{M}^{h+1})} \longrightarrow \mathcal{O}_{T^*X} \otimes_{\overline{\mathcal{D}_X}} \overline{\mathcal{M}^{h+1}}$$

is the zero map. Outside the support of $(\mathcal{H}^h(\mathcal{O}_{T^*X} \otimes_{\overline{\mathcal{D}_X}} \mathcal{M}^\bullet))$, the sequence

$$\cdots \longrightarrow \mathcal{O}_{T^*X} \otimes_{\overline{\mathcal{D}_X}} \overline{\mathcal{M}^{h-1}} \longrightarrow \mathcal{O}_{T^*X} \otimes_{\overline{\mathcal{D}_X}} \overline{\mathcal{M}^h} \longrightarrow \mathcal{O}_{T^*X} \otimes_{\overline{\mathcal{D}_X}} \overline{\mathcal{M}^{h+1}} \longrightarrow \cdots$$

is exact at the degree h. Since we have an inclusion

$$\mathcal{O}_{T^*X} \otimes_{\overline{\mathcal{D}_X}} \overline{ker(\mathcal{M}^h \longrightarrow \mathcal{M}^{h+1})} \hookrightarrow \mathcal{O}_{T^*} \otimes_{\overline{\mathcal{D}_X}} \overline{\mathcal{M}^h},$$

the map

$$\mathcal{O}_{T^*X} \otimes_{\overline{\mathcal{D}_X}} \overline{\mathcal{M}^{h-1}} \longrightarrow \mathcal{O}_{T^*X} \otimes_{\overline{\mathcal{D}_X}} \overline{ker(\mathcal{M}^h \longrightarrow \mathcal{M}^{h+1})}$$

is an epimorphism and we have the split described in the diagram below:

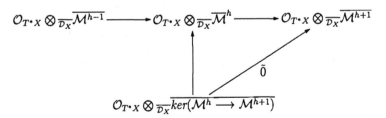

Next consider $\mathcal{H}^h(\mathcal{M}^\bullet)$ as the filtered object induced by the epimorphism

$$\mathcal{M}^h \supset ker(\mathcal{M}^h \longrightarrow \mathcal{M}^{h+1}) \longrightarrow \mathcal{H}^h(\mathcal{M}^\bullet) \longrightarrow 0.$$

Now we have two epimorphisms

$$\mathcal{O}_T^\bullet X \otimes_{\overline{\mathcal{D}_X}} \overline{\mathcal{M}^{h-1}} \to \mathcal{O}_{T^\bullet X} \otimes_{\overline{\mathcal{D}_X}} \overline{ker(\mathcal{M}^h \to \mathcal{M}^{h+1})} \to \mathcal{O}_{T^\bullet X} \otimes_{\overline{\mathcal{D}_X}} \overline{\mathcal{H}^h(\mathcal{M}^\bullet)}$$

whose composition is the zero map at each stalk outside the support of $\mathcal{H}^h(\mathcal{O}_{T^\bullet X} \otimes_{\overline{\mathcal{D}_X}} \overline{\mathcal{M}^\bullet})$. Therefore $\mathcal{O}_{T^\bullet X} \otimes_{\overline{\mathcal{D}_X}} \overline{\mathcal{H}(\mathcal{M}^\bullet)} = 0$ over $T^*X \backslash (\text{supp } (\mathcal{H}^h(\mathcal{O}_{T^\bullet X} \otimes_{\overline{\mathcal{D}_X}} \overline{\mathcal{M}^\bullet})))$. Namely, we obtain

$$V(\mathcal{H}^h(\mathcal{M}^\bullet)) = \text{ supp } (\mathcal{O}_{T^\bullet X} \otimes_{\overline{\mathcal{D}_X}} \overline{\mathcal{H}^h(\mathcal{M}^\bullet)}) \subset \text{ supp } (\mathcal{H}^h(\mathcal{O}_{T^\bullet X} \otimes_{\overline{\mathcal{D}_X}} \overline{\mathcal{M}^\bullet})).$$

□

We can now use this lemma to prove a vanishing of higher cohomology theorem for $\mathcal{E}xt_{\mathcal{D}_X}^h(\mathcal{M}, \mathcal{D}_X)$ as follows.

Theorem 4.4.1 *If $h < codim\ V(\mathcal{M})$, then*

$$\mathcal{E}xt_{\mathcal{D}_X}^h(\mathcal{M}, \mathcal{D}_X) = 0.$$

Proof.   From the isomorphism in (4.4.5), we have

$$\mathcal{E}xt_{\mathcal{O}_{T^\bullet X}}^h(\mathcal{O}_{T^\bullet X} \otimes_{\overline{\mathcal{D}_X}} \overline{\mathcal{M}}, \mathcal{O}_{T^\bullet X}) = 0$$

for all $h < codim\ \text{supp } (\mathcal{O}_{T^\bullet X} \otimes_{\overline{\mathcal{D}_X}} \overline{\mathcal{M}})$. See, e.g., [68] for its proof. Since $\mathcal{O}_{T^\bullet X} \otimes_{\overline{\mathcal{D}_X}}$ is faithful and since the functor assigning the $\overline{\mathcal{D}_X}$-Module $\overline{\mathcal{M}}$ to the $\mathcal{D}_X$-Module $\mathcal{M}$ with the good filtration is faithful, we have the vanishing of $\mathcal{E}xt_{\mathcal{D}_X}^h(\mathcal{M}, \mathcal{D}_X)$ for $h < codim\ V(\mathcal{M})$.   □

We are now ready to prove the final theorem of this Chapter:

Theorem 4.4.2 *For $h > n = \dim X$, we have*

$$\mathcal{E}xt_{\mathcal{D}_X}^h(\mathcal{M}, \mathcal{D}_X) = 0.$$

Proof.   Since the characteristic variety $V(\mathcal{M})$ is involutive (see e.g. Chapter VI), for $V(\mathcal{M}) \not\subset X$ we have $\dim V(\mathcal{M}) \geq n$, see [18]. Then, from (4.4.6) we have $\dim V(\mathcal{E}xt_{\mathcal{D}_X}^h(\mathcal{M}, \mathcal{D}_X)) \leq 2n - (h+1)$. In particular, for $h > n$ we obtain $\dim V(\mathcal{E}xt_{\mathcal{D}_X}^h(\mathcal{M}, \mathcal{D}_X)) \leq n-2$. Consequently, we have that $V(\mathcal{E}xt_{\mathcal{D}_X}^h(\mathcal{M}, \mathcal{D}_X))$ is a subvariety of X. However, from Theorem 5.2.3 in Chapter V, there exists an integer l for which it is

$$\mathcal{E}xt^h_{\mathcal{D}_X}(\mathcal{M}, \mathcal{D}_X) \cong \mathcal{O}^l_X$$

as a $\mathcal{D}_X$-Module. Let now $\mathcal{M} = \cup \mathcal{M}^{(k)}$ be a good filtration. Choose, in the complement of a nowhere dense subset of X, a point x such that $(\mathcal{M}^{(k)}/\mathcal{M}^{(k-1)})_x$ is a free $\mathcal{O}_{X,x}$-Module, and let U be a neighborhood of x over which $\mathcal{M}^{(k)}/\mathcal{M}^{(k-1)}$ is a locally free $\mathcal{O}_X$-Module. Then, for a sufficiently large k we obtain a locally free resolution of $\mathcal{M}$ over U, i.e., the first Spencer sequence

$$0 \longrightarrow \mathcal{D}_X \otimes_{\mathcal{O}_X} \wedge^n \Theta \otimes_{\mathcal{O}_X} \mathcal{M}^{(k-n)} \longrightarrow \ldots \longrightarrow \mathcal{D}_X \otimes_{\mathcal{O}_X} \mathcal{M}^{(k)} \longrightarrow \mathcal{M} \longrightarrow 0$$

is exact. Therefore, we have

$$\mathcal{E}xt^h_{\mathcal{D}_X}(\mathcal{M}, \mathcal{D}_X) = \mathcal{H}^h(\mathcal{H}om_{\mathcal{D}_X}(\mathcal{D}_X \otimes_{\mathcal{O}_X} \wedge^\bullet \Theta \otimes_{\mathcal{O}_X} \mathcal{M}^{(k-\bullet)}, \mathcal{D}_X) = 0 \text{ for } h > n,$$

which concludes the proof. $\qquad\qquad\qquad\qquad\qquad\qquad\qquad\qquad\qquad\qquad\qquad$ $\square$

Chapter 5

Holonomic $\mathcal{D}$-modules

5.1 Introduction

With the preparation in Chapters III and IV, we focus on basic themes in this chapter. The Cauchy-Kowalewsky theorem in the language of $\mathcal{D}$-modules is given in Section 2. In Section 3, the direct image of a $\mathcal{D}$-module is defined. The most significant theorem in the fundamental theory of holonomic $\mathcal{D}$-modules, due to Kashiwara, is discussed in Section 4.

5.2 Inverse Image and Cauchy Problem

Let $f : Y \to X$ be a holomorphic map of complex manifolds. For the sheaf $\mathcal{D}_X$ of germs of holomorphic linear partial differential operators of finite order, we will define a left $\mathcal{D}_Y$-Module, which is also a right $f^{-1}\mathcal{D}_X$-Module as follows.

Definition 5.2.1 $\mathcal{D}_{Y \to X} = \mathcal{O}_Y \otimes_{f^{-1}\mathcal{O}_X} f^{-1}\mathcal{D}_X$.

Example 5.2.1

(i) For a closed embedding $f : Y \hookrightarrow X$, we have

$$\mathcal{D}_{Y \hookrightarrow X} = \mathcal{O}_Y \otimes_{\mathcal{O}_{X|Y}} \mathcal{D}_{X|Y}.$$

If Y is given by equations $z_1 = z_2 = \ldots = z_r = 0$, then we can express $\mathcal{D}_{Y \hookrightarrow X}$ as

$$\mathcal{D}_{Y \hookrightarrow X} = \mathcal{D}_X / z_1 \mathcal{D}_X + \cdots + z_r \mathcal{D}_X.$$

Hence an arbitrary operator $P(y, \partial)$ in $\mathcal{D}_{Y \hookrightarrow X}$ may be written as

$$P(y, \partial) = \sum_\alpha a_\alpha(z_{r+1}, \ldots, z_n)\partial_1^{\alpha_1}\partial_2^{\alpha_2} \ldots \partial_n^{\alpha_n},$$

where $y \in Y$ and $\partial_j = \partial/\partial z_j$, $j = 1, 2, \ldots, n$. Then for a function $h \in \mathcal{O}_X$ we get a function $P(y, \partial)h$ in $\mathcal{O}_Y$:

$$P(y, \partial)h = \sum_\alpha a_\alpha(z_{r+1}, \ldots, z_n)\partial^\alpha h|_Y.$$

(ii) For the projection $Y = X \times Z \to X$, defined by $(z_1, \ldots, z_n, w_1, \ldots, w_l) \to (z_1, \ldots, z_n)$, the $\mathcal{D}_Y$-Module $\mathcal{D}_{Y \to X}$ is the sheaf of differential operators on X whose coefficients are extended to $\mathcal{O}_Y$. Namely, we get

$$\mathcal{D}_{Y \to X} = \mathcal{D}_Y / \sum_{i=1}^l \mathcal{D}_Y \cdot \frac{\partial}{\partial w_i}.$$

An element $P(z, w, \partial_z)$ in $\mathcal{D}_{Y \to X}$ may be written as follows:

$$P(z, w, \partial_z) = \sum_\alpha a_\alpha(z_1, \ldots, z_n, w_1, \ldots, w_l)\partial_1^{\alpha_1} \ldots \partial_n^{\alpha_n},$$

where $(z, w) \in Y = X \times Z$ and $\partial_j = \partial/\partial z_j$, $j = 1, 2, \ldots, n$. For a function h in $\mathcal{O}_X$, we get a function in $\mathcal{O}_Y$ as follows:

$$P(z, w, \partial_z)h = \sum_\alpha a_\alpha(z, w)\partial^\alpha h.$$

Note that for holomorphic maps $Z \xrightarrow{g} Y \xrightarrow{f} X$, we have a map

$$\mathcal{D}_{Z \to Y} \times g^{-1}\mathcal{D}_{Y \to X} \to \mathcal{D}_{Z \to X}.$$

The above map is defined by

$$(b(z)\partial_y^\beta, d(y)\partial_x^\delta) \to b(z)\partial_y^\alpha d(y)\partial_x^\delta,$$

where $y = g(z)$ and $x = f(y)$ for $z \in Z$. That is, one can compose a differential operator from X to Y along f with a differential operator from Y to Z along g to get a differential operator from X to Z along $f \circ g$.

Definition 5.2.2 *Let $f : Y \to X$ be a holomorphic map. For a $\mathcal{D}_X$-Module $\mathcal{M}$, define the $\mathcal{D}_Y$-Module $f^*\mathcal{M}$ using the right $f^{-1}\mathcal{D}_X$-Module $\mathcal{D}_{Y \to X}$ by*

(5.2.1) $$f^*\mathcal{M} = \mathcal{D}_{Y \to X} \otimes_{f^{-1}\mathcal{D}_X} f^{-1}\mathcal{M}.$$

By the definition of $\mathcal{D}_{Y \to X}$, the $\mathcal{D}_Y$-Module $f^*\mathcal{M}$ is also given by

$$f^*\mathcal{M} = \mathcal{O}_Y \otimes_{f^{-1}\mathcal{O}_X} f^{-1}\mathcal{M}.$$

Notice also that we have

$$f^*\mathcal{O}_X = \mathcal{O}_Y \text{ and } f^*\mathcal{D}_X = \mathcal{D}_{Y \to X}.$$

Let $\mathcal{M}$ be a $\mathcal{D}_X$-Module defined by

$$\mathcal{M} = \mathcal{D}_X u_1 + \cdots + \mathcal{D}_X u_m$$

such that

(5.2.2) $$\sum_{j=1}^{m} P_{ij} u_j = 0, \; i = 1, 2, \ldots l.$$

Namely, we have an isomorphism

$$\mathcal{M} \overset{\approx}{\longleftarrow} \mathcal{D}_X^m / \mathcal{D}_X^l \cdot P,$$

$P = [P_{ij}]_{\substack{1 \leq i \leq l \\ 1 \leq j \leq m}}$, as we mentioned in 3.2 of Chapter III. When f is the projection from $Y = X \times Z \to X$, defined by $(z_1, \ldots, z_n, w_1, \ldots, w_k) \to (z_1, \ldots, z_n)$, the $\mathcal{D}_Y$-Module $f^*\mathcal{M}$ may be regarded as not only $\mathcal{D}_Y^m / \mathcal{D}_Y^l P$ but also $\partial_{w_i} u_j' = 0$, $i = 1 \ldots, k$, where $u_j' = f^* u_j$, $j = 1, \ldots, m$.

In the case where $Y \hookrightarrow X$ such that $X = \mathbb{C}^2$ and $Y = \{(x, y) \in \mathbb{C}^2 | x = 0\}$, let $\mathcal{M} = \mathcal{D}_X u$ satisfying $Pu = 0$. Then $f^*\mathcal{M}$ is a differential equation over Y, i.e., $\mathcal{D}_Y$-Module. For $\partial_x^\alpha u$, we have $f^*(\partial_x^\alpha u) = f^*(\partial_x^\alpha u(x,y)) = \partial_x^\alpha u(x,y)|_Y = \partial_x^\alpha u(0,y)$. On the other hand, for an element $\partial_y^\beta u$, we have $f^*(\partial_y^\beta u) = f^*(\partial_y^\beta u(x,y)) = \partial_y^\beta (f^* u(x,y)) = \partial_y^\beta u(0,y)$. The relation among those elements over Y is exactly the $\mathcal{D}_Y$-Module $f^*\mathcal{M}$. We will give explicit examples.

Example 5.2.2 If $\mathcal{M} \cong \mathcal{D}_X$, namely there are no equations, then $f^*\mathcal{M} \cong f^*\mathcal{D}_X \cong \mathcal{D}_{Y \to X} = \mathcal{D}_X / x \mathcal{D}_X = \sum \mathcal{O}_Y \partial_y^\beta \partial_x^\alpha = \sum \mathcal{D}_Y \partial_x^\alpha$. See example 5.2.1, (i).

If $\mathcal{M}$ is given by $\partial_y u = 0$, i.e., $\mathcal{M} = \mathcal{D}_X / \mathcal{D}_X \partial_y$, then we have

$$f^*\mathcal{M} = \mathcal{D}_X / \mathcal{D}_X \partial_y + x \mathcal{D}_X = \sum \mathcal{O}_Y \partial_x^\alpha.$$

Notice that $Y = \{(x,y) \in \mathbb{C}^2 | x = 0\}$ is characteristic to $\mathcal{M}$ and that the $\mathcal{D}_Y$-Module $f^*\mathcal{M}$ is not coherent.

For $\mathcal{M} = \mathcal{D}_X / \mathcal{D}_X \partial_x$, we have $f^*\mathcal{M} = \sum \mathcal{O}_Y \partial_y^\beta = \mathcal{D}_Y$. That is to say the generator u for $\mathcal{M}$ satisfying the relation $\partial_x u = 0$ over X need not satisfy any further relations over Y.

Before we state the Cauchy problem, in the case of embedding, we will give the $\mathcal{D}$-Module version of the classical Cauchy-Kowalewsky theorem. Let

$Y = \{z_1 = 0\}$ in X, where $(z_1, \ldots, z_n)$ are the coordinates of X. Suppose Y is non-characteristic to a cyclic $\mathcal{D}_X$-Module $\mathcal{M}$ represented by a partial differential operator $P \in \mathcal{D}_X$ of order m. Then the principal part may be written as a Weierstrass type:

$$(5.2.3) \qquad \partial_1^m + P_1(z, \partial')\partial_1^{m-1} + \cdots + P_m(z, \partial'),$$

where $z' = (z_2, \ldots, z_n)$, $\partial' = (\partial_2, \partial_3, \ldots, \partial_n)$ and $P_i(z, \partial')$ is of order i. With relation (5.2.3), $\mathcal{D}_X/\mathcal{D}_X P = \mathcal{M}$ is generated by ∂^α, $\alpha_1 < m$, over $\mathcal{O}_X$. Furthermore, the generators ∂^α, $\alpha_1 < m$; are linearly independent over $\mathcal{O}_X$. Hence we have

$$\mathcal{M} = \oplus_{\alpha_1 < m} \mathcal{O}_X \partial^\alpha, \text{ a free } \mathcal{O}_X\text{-Module.}$$

We have, for $f : Y \hookrightarrow X$,

$$f^*\mathcal{M} = \mathcal{D}_{Y \to X} \otimes_{f^{-1}\mathcal{D}_X} f^{-1}\mathcal{M} = \mathcal{O}_Y \otimes_{f^{-1}\mathcal{O}_X} f^{-1}\mathcal{M}$$
$$= \bigoplus_{\alpha_1 < m} \mathcal{O}_Y \partial^\alpha = \bigoplus_{\alpha_1 < m} \mathcal{D}_Y \partial_1^{\alpha_1}.$$

That is, the $\mathcal{D}_Y$-Module $f^*\mathcal{M}$ is isomorphic to $\mathcal{D}_Y^m$. Then we obtain $\mathcal{H}om_{\mathcal{D}_Y}(f^*\mathcal{M}, \mathcal{O}_Y) \approx \mathcal{O}_Y^m$. Let $\varphi \in \mathcal{H}om_{\mathcal{D}_X}(\mathcal{M}, \mathcal{O}_X)$ and let $h \in \varphi(u) \in \mathcal{O}_X$, where $Pu = 0$. Then $f^{-1}h = h|_{z_1=0} = h(0, z') \in \mathcal{O}_Y$. The assignment

$$(5.2.4) \qquad h|_{z_1=0} \to (h|_{z_1=0}, \partial_1 h|_{z_1=0}, \ldots, \partial_1^{m-1} h|_{z_1=0}) \in \mathcal{O}_Y^m$$

is nothing but the map

$$(5.2.5) \qquad f^{-1}\mathcal{H}om_{\mathcal{D}_X}(\mathcal{M}, \mathcal{O}_X) \to \mathcal{H}om_{\mathcal{D}_Y}(f^*\mathcal{M}, f^*\mathcal{O}_X) \approx \mathcal{O}_Y^m.$$

The classical Cauchy-Kowalewsky theorem states that assignment (5.2.4) is onto and one-to-one, if $Y = (z_1 = 0)$ is non-characteristic to P. Namely, we obtain an isomorphism in (5.2.5). Furthermore, the map $\mathcal{O}_X \overset{P}{\to} \mathcal{O}_X$ is an epimorphism, i.e., for any holomorphic function f there exists $u \in \mathcal{O}_X$ such that $Pu = f$. Therefore $\mathcal{E}xt_{\mathcal{D}_X}^1(\mathcal{M}, \mathcal{O}_X) = 0$. Next, let Y be a hyperface of codimension 1 on X, and let $\mathcal{M}$ be a coherent $\mathcal{D}_X$-Module such that Y is non-characteristic to $\mathcal{M}$, i.e., the singular support of $\mathcal{M}$ does not intersect with the conormal projective bundle to T, $V(\mathcal{M}) \cap P_Y^*(X) = \emptyset$. For a set of generators $u_1, u_2, \ldots, u_m$ for $\mathcal{M}$, one can find P_j in $\mathcal{D}_X$ satisfying $P_j u_j = 0$, $j = 1, \ldots, m$. Namely, we have the following free resolution of $\mathcal{M}$:

$$\mathcal{D}_X^m \cdot \begin{bmatrix} P_1 & 0 & \cdots & 0 \\ 0 & P_2 & \cdots & 0 \\ \vdots & & \cdots & 0 \\ 0 & \cdots & 0 & P_m \end{bmatrix} \mathcal{D}_X^m \overset{u}{\to} \mathcal{M} \to 0.$$

In terms of relations, the above exact sequence means

$$\left\{ \begin{array}{llll} P_1 u_1 & & & = 0 \\ & P_2 u_2 & & = 0 \\ & & \cdots & \\ & & P_m u_m & = 0. \end{array} \right.$$

(5.2.6) $$0 \longleftarrow \mathcal{M} \longleftarrow \oplus_{j=1}^m \mathcal{D}_X/\mathcal{D}_X P_j \longleftarrow \mathcal{N} \longleftarrow 0,$$

where $V(\oplus_{j=1}^m \mathcal{D}_X/\mathcal{D}_X P_j) \cap P_Y^* X = \emptyset$. Notice also that $\oplus_{j=1}^m \mathcal{D}_X/\mathcal{D}_X P_j$ is a free $\mathcal{O}_X$-Module as we observed in the case of classical Cauchy-Kowalewsky theorem. In particular, the higher $\mathcal{T}or_h^{\mathcal{O}_X}(\mathcal{O}_Y, \oplus_{j=1}^m \mathcal{D}_X/\mathcal{D}_x P_j)$ vanishes for all $h \geq 1$. Hence, from short exact sequence (5.2.6) we obtain, through the functor $f^* = \mathcal{O}_Y \otimes_{\mathcal{O}_X} -$,

(5.2.7) $$0 \longleftarrow f^* \mathcal{M} \longleftarrow f^* \left(\oplus_{j=1}^m \mathcal{D}_X/\mathcal{D}_X P_j \right) \longleftarrow f^* \mathcal{N} \longleftarrow 0$$

From exact sequences (5.2.6) and (5.2.7), the following commutative diagram is induced

$$
\begin{array}{ccc}
0 \longrightarrow f^{-1}\mathcal{H}om_{\mathcal{D}_X}(\mathcal{M}, \mathcal{O}_X) \longrightarrow f^{-1}\mathcal{H}om_{\mathcal{D}_X}(\oplus_{j=1}^m \mathcal{D}_X/\mathcal{D}_X P_j, \mathcal{O}_Y) \longrightarrow \\
\downarrow \gamma^0 \qquad\qquad\qquad \wr\wr\, \prime\, \gamma^0 \\
0 \longrightarrow \mathcal{H}om_{\mathcal{D}_X}(f^* \mathcal{M}, \mathcal{O}_Y) \longrightarrow \mathcal{H}om_{\mathcal{D}_Y}(f^*(\oplus_{j=1}^m \mathcal{D}_X/\mathcal{D}_X P_j), \mathcal{O}_Y) \longrightarrow
\end{array}
$$

$$
\begin{array}{c}
\longrightarrow f^{-1}\mathcal{H}om_{\mathcal{D}_X}(\mathcal{N}, \mathcal{O}_X) \longrightarrow \cdots \\
\downarrow \prime\prime \gamma^0 \\
\longrightarrow \mathcal{H}om_{\mathcal{D}_Y}(f^* \mathcal{N}, \mathcal{O}_Y) \longrightarrow \cdots
\end{array}
$$

$$
\begin{array}{cc}
\longrightarrow f^{-1}\mathcal{E}xt_{\mathcal{D}_X}^{i-1}(\oplus_{j=1}^m \mathcal{D}_X/\mathcal{D}_X P_j, \mathcal{O}_X) \longrightarrow f^{-1}\mathcal{E}xt_{\mathcal{D}_X}^{i-1}(\mathcal{N}, \mathcal{O}_X) \longrightarrow \\
\downarrow \prime\gamma^{i-1} \qquad\qquad\qquad \downarrow \prime\prime\gamma^{i-1} \\
\longrightarrow \mathcal{E}xt_{\mathcal{D}_Y}^{i-1}(f^* \oplus_{j=1}^m \mathcal{D}_X/\mathcal{D}_X P_j, \mathcal{O}_Y) \longrightarrow \mathcal{E}xt_{\mathcal{D}_Y}^{i-1}(f^* \mathcal{N}, \mathcal{O}_Y) \longrightarrow
\end{array}
$$

$$
\begin{array}{cc}
\longrightarrow f^{-1}\mathcal{E}xt_{\mathcal{D}_X}^{i}(\mathcal{M}, \mathcal{O}_X) \longrightarrow f^{-1}\mathcal{E}xt_{\mathcal{D}_X}^{i}(\oplus_{j=1}^m \mathcal{D}_X/\mathcal{D}_X P_j, \mathcal{O}_X) \longrightarrow \cdots \\
\downarrow \gamma^{i} \qquad\qquad\qquad \downarrow \prime\gamma^{i} \\
\longrightarrow \mathcal{E}xt_{\mathcal{D}_Y}^{i}(f^* \mathcal{M}, \mathcal{O}_Y) \longrightarrow \mathcal{E}xt_{\mathcal{D}_Y}^{i}(f^* \oplus_{j=1}^m \mathcal{D}_X/\mathcal{D}_X P_j, \mathcal{O}_Y) \longrightarrow \cdots
\end{array}
$$

Then all the $'\gamma^i$ are isomorphisms from the classical Cauchy-Kowalewsky theorem.   From a well known lemma in homological algebra, $\ker \gamma^0 \to \ker \,'\gamma^0 \to \ker \,''\gamma^0$ is exact. Furthermore, since γ^1 is monomorphic, there exists a connecting homomorphism δ^0 so that

$$\ker \gamma^0 \to \ker \,'\gamma^0 \to \ker \,''\gamma^0 \xrightarrow{\delta^0} \operatorname{coker} \gamma^0 \to \operatorname{coker} \,'\gamma^0 \to \operatorname{coker} \,''\gamma^0$$

may be exact. Consequently, we obtain

$$\operatorname{coker} \gamma^0 = 0$$

since $\ker \,''\gamma^0 = 0$ as well. That is, γ^0 is an isomorphism. Repeating this process, we have

$$\gamma^i, \; i = 1, 2, \ldots$$

are isomorphisms.

Next, we will consider the general case. Let Y be of codimension r in X, and let Y' be a hypersurface containing Y defined by $z_1 = 0$. Since Y' is locally non-characteristic for $\mathcal{M}$, we have

(5.2.8) $f^{-1}\mathcal{H}om_{\mathcal{D}_X}(\mathcal{M}, \mathcal{O}_X) \xrightarrow{\approx} \mathcal{H}om_{\mathcal{D}_{Y'}}(f'^*\mathcal{M}, \mathcal{O}_{Y'})$

as in the above, where f' is an inclusion from Y' to X. The equation codim $(Y, Y') = \operatorname{codim} Y - 1$ provides the proof by induction on the codimension. Since the characteristic variety of $f'^{-1}\mathcal{M}$ is smaller than the projection of the characteristic variety of $\mathcal{M}$ on T^*Y', Y is non-characteristic for $f'^{-1}\mathcal{M}$. Hence for $f : Y \hookrightarrow Y'$ we get

(5.2.9) $f^{-1}\mathcal{H}om_{\mathcal{D}_{Y'}}(f'^{-1}\mathcal{M}, \mathcal{O}_{Y'}) \xrightarrow{\approx} \mathcal{H}om_{\mathcal{D}_Y}(f^*(f'^*\mathcal{M}), \mathcal{O}_Y).$

Note that $f^*(f'^*\mathcal{M}) = (f' \circ f)^*\mathcal{M}$ holds, since, formally speaking, we have

$$\mathcal{D}_{Y \hookrightarrow X} \otimes_{f^{-1}\mathcal{D}_{Y'}} f^{-1}(\mathcal{D}_{Y' \hookrightarrow X} \otimes_{f'^{-1}\mathcal{D}_X} f^{-1}\mathcal{M}) =$$
$$= \mathcal{O}_Y \otimes_{f^{-1}\mathcal{O}_{Y'}} f^{-1}\mathcal{D}_{Y'} \otimes_{f^{-1}\mathcal{D}_{Y'}} f^{-1}\mathcal{O}_{Y'} \otimes_{(f' \circ f)^{-1}\mathcal{O}_X} (f' \circ f)^{-1}\mathcal{M} =$$
$$= \mathcal{O}_Y \otimes_{(f' \circ f)^{-1}\mathcal{O}_X} (f' \circ f)^{-1}\mathcal{M} =$$
$$= (f' \circ f)^*\mathcal{M}.$$

Consequently, we have from (5.2.8) and (5.2.9)

$$f^{-1}f'^{-1}\mathcal{H}om_{\mathcal{D}_X}(\mathcal{M}, \mathcal{O}_X) = (f' \circ f)^{-1}\mathcal{H}om_{\mathcal{D}_X}(\mathcal{M}, \mathcal{O}_X) \xrightarrow{\approx}$$
$$f^{-1}\mathcal{H}om_{\mathcal{D}_{Y'}}(f'^*\mathcal{M}, \mathcal{O}_{Y'}) \xrightarrow{\approx} \mathcal{H}om_{\mathcal{D}_Y}((f' \circ f)^*\mathcal{M}, \mathcal{O}_Y).$$

For the higher cohomology groups for the general case, we repeat the inductive proof for the hypersurface case. We have obtained the following theorem which generalizes the classical Cauchy-Kowalewsky theorem.

Theorem 5.2.1 *For a $\mathcal{D}_X$-Module $\mathcal{M}$, let $Y \overset{f}{\hookrightarrow} X$ be of an arbitrary codimension that is non-characteristic for $\mathcal{M}$. Then the induced maps*

$$f^{-1}\mathcal{E}xt^h_{\mathcal{D}_X}(\mathcal{M}, \mathcal{O}_X) \to \mathcal{E}xt^h_{\mathcal{D}_Y}(f^*\mathcal{M}, \mathcal{O}_Y), \quad h = 0, 1, 2 \ldots$$

are isomorphisms.

Remark 5.2.1 Let $Y = (z_1 = 0) \overset{f}{\hookrightarrow} X$. Then, as we saw, $\mathcal{D}_{Y \hookrightarrow X} = \mathcal{D}_X/z_1 \mathcal{D}_X$ holds. That is, we obtain the exact sequence:

$$\mathcal{D}_X \overset{z_1;}{\to} \mathcal{D}_X \to \mathcal{D}_{Y \hookrightarrow X} \to 0,$$

furthermore inducing the exactness of

$$f^{-1}\mathcal{D}_X \to f^{-1}\mathcal{D}_X \to f^{-1}\mathcal{D}_{Y \hookrightarrow X} \to 0.$$

Note that $f^{-1}\mathcal{D}_{Y \hookrightarrow X} = f^{-1}(\mathcal{O}_Y \otimes_{f^{-1}\mathcal{O}_X} f^{-1}\mathcal{D}_X) = \mathcal{D}_{Y \hookrightarrow X}$. Tensoring the above with $\otimes_{f^{-1}\mathcal{D}_X} f^{-1}\mathcal{M}$, we get

$$f^{-1}\mathcal{M} \overset{z_1;}{\to} f^{-1}\mathcal{M} \to \mathcal{D}_{Y \hookrightarrow X} \otimes_{f^{-1}\mathcal{D}_X} f^{-1}\mathcal{M} = f^*\mathcal{M} \to 0.$$

Namely, $f^*\mathcal{M} = \mathcal{M}|_Y = f^{-1}\mathcal{M}/z_1 f^{-1}\mathcal{M}$. Let $\psi \in \mathcal{H}om_{\mathcal{D}_X}(\mathcal{M}, \mathcal{O}_X)$, i.e., for $u \in \mathcal{M}$, ψ assigns $\psi(u) = \psi_u \in \mathcal{O}_X$, and let $1_{Y \hookrightarrow X}$ be the unit in $\mathcal{D}_{Y \hookrightarrow X}$, i.e., $1_{Y \hookrightarrow X} \otimes \psi \in \mathcal{H}om_{\mathcal{D}_Y}(\mathcal{D}_{Y \hookrightarrow X} \otimes_{f^{-1}\mathcal{D}_X} f^{-1}\mathcal{M}, \mathcal{D}_{Y \hookrightarrow X} \otimes_{f^{-1}\mathcal{D}_X} f^{-1}\mathcal{O}_X) \cong \mathcal{H}om_{\mathcal{D}_Y}(\mathcal{M}/z_1\mathcal{M}, ($

$$
\begin{array}{ccc}
f^*\mathcal{M} = \mathcal{D}_{Y \hookrightarrow X} \otimes \mathcal{M} & \overset{\approx}{\longleftarrow} & \mathcal{M}/z_1\mathcal{M} \\
\downarrow {\scriptstyle 1 \otimes \psi} & & \downarrow \\
\mathcal{O}_Y = \mathcal{D}_{Y \hookrightarrow X} \otimes \mathcal{O}_X & \longleftarrow & \mathcal{O}_X/z_1\mathcal{O}_X.
\end{array}
$$

If $\mathcal{M}$ is given by $\mathcal{M} = \mathcal{D}_X/\mathcal{D}_X P$, where $P = \partial_1^m + P_1(z, \partial')\partial_1^{m-1} + \cdots + P_m(z, \partial') \in \mathcal{D}_X^{(m)}$ as before, then $f^*\mathcal{M}$ is generated by $(1 \otimes u, \ldots, 1 \otimes \partial_1^{m-1}u)$ as a $\mathcal{D}_Y$-Module. That is, $f^*\mathcal{M} \approx \mathcal{D}_Y^m$ as we observed earlier. Then the corresponding $\psi^* \in \mathcal{H}om_{\mathcal{D}_Y}(f^*\mathcal{M}, \mathcal{O}_Y)$ may be defined as follows: for generators $1 \otimes \partial_1^j u$ of $f^*\mathcal{M}$

$$\psi^*(1 \otimes \partial_1^j u) = \partial_1^j \psi(1 \otimes u)|_Y \in \mathcal{O}_Y, \quad j = 1, \ldots, m - 1.$$

Theorem 5.2.2 *Let Y be a subvariety of X, $Y \overset{f}{\hookrightarrow} X$, and let $\mathcal{M}^{\bullet}$ be a bounded complex of $\mathcal{D}_X$-Modules, i.e., $\mathcal{H}^j(\mathcal{M}^{\bullet}) \neq 0$ for only finitely many j. If Y is non-characteristic for all non-trivial $\mathcal{H}^j(\mathcal{M})$, then the complex $\mathcal{O}_Y \otimes^{\mathbb{L}} f^{-1}\mathcal{M}^{\bullet}$ has coherent complex cohomologies, and furthermore, we have an isomorphism*

$$f^{-1}\mathbb{R}\mathcal{H}om_{\mathcal{D}_X}(\mathcal{M}^{\bullet}, \mathcal{O}_X) \overset{\approx}{\to} \mathbb{R}\mathcal{H}om_{\mathcal{D}_Y}(\mathcal{O}_Y \otimes^{\mathbb{L}}_{f^{-1}\mathcal{D}_X} f^{-1}\mathcal{M}^{\bullet}, \mathcal{O}_Y).$$

Remark 5.2.2   For the complex $\mathcal{M}^{\bullet}$ concentrated at degree zero in Remark 5.2.1, the complex $\mathcal{O}_Y \otimes^{\mathbb{L}}_{f^{-1}\mathcal{O}_X} f^{-1}\mathcal{M} = \mathcal{D}_{Y \hookrightarrow X} \otimes^{\mathbb{L}}_{f^{-1}\mathcal{D}_X} f^{-1}\mathcal{M}$ may be computed by the projective resolution of $\mathcal{D}_{Y \hookrightarrow X}$ in the above remark as follows.

The cohomologies of $\mathcal{O}_Y \otimes^{\mathbb{L}}_{f^{-1}\mathcal{O}_X} f^{-1}\mathcal{M} = \mathcal{D}_{Y \hookrightarrow X} \otimes^{\mathbb{L}}_{f^{-1}\mathcal{D}_X} f^{-1}\mathcal{M}$ are isomorphic to the cohomologies of the complex

$$0 \to f^{-1}\mathcal{M} \overset{z_1;}{\to} f^{-1}\mathcal{M} \to 0.$$

In particular, the 0-th cohomology $\mathcal{H}^0(\mathcal{D}_{Y \hookrightarrow X} \otimes^{\mathbb{L}}_{f^{-1}\mathcal{D}_X} f^{-1}\mathcal{M})$ is isomorphic to $f^{-1}\mathcal{M}/z_1 f^{-1}\mathcal{M}$. As we saw in Remark 5.2.1, for $\mathcal{M} = \mathcal{D}_X/\mathcal{D}_X P$, $P \in \mathcal{D}^{(m)}$, the $\mathcal{D}_Y$-Module $f^*\mathcal{M} = f^{-1}\mathcal{M}/z_1 f^{-1}\mathcal{M}$ is given as $\mathcal{D}_Y^m$. Next, the 1-st cohomology $\mathcal{H}^1(\mathcal{D}_{Y \hookrightarrow X} \otimes^{\mathbb{L}}_{f^{-1}\mathcal{D}_X} f^{-1}\mathcal{M}) = \mathcal{T}or_1^{f^{-1}\mathcal{D}_X}(\mathcal{D}_{Y \hookrightarrow X}, f^{-1}\mathcal{M})$ is isomorphic to the sheaf $Ker(f^{-1}\mathcal{M} \overset{z_1;}{\to} f^{-1}\mathcal{M})$.   However,   in   general,   the $\mathcal{D}_Y$-Module $\mathcal{T}or_1^{f^{-1}\mathcal{D}_X}(\mathcal{D}_{Y \hookrightarrow X}, f^{-1}\mathcal{M})$ is isomorphic to $\mathcal{T}or_1^{f^{-1}\mathcal{O}_X}(\mathcal{O}_Y, f^{-1}\mathcal{M})$. This is because $\mathcal{D}_X$ is $\mathcal{O}_X$-flat. Hence $\mathcal{T}or_j^{f^{-1}\mathcal{O}_X}(\mathcal{O}_Y, f^{-1}\mathcal{M}) = \mathcal{H}_j(\mathcal{O}_Y \otimes_{f^{-1}\mathcal{O}_X} f^{-1}\mathcal{D}_X^{\bullet}) = \mathcal{H}_j(\mathcal{D}_{Y \hookrightarrow X} \otimes_{f^{-1}\mathcal{D}_X} f^{-1}\mathcal{D}_X^{\bullet}) = \mathcal{T}or_j^{f^{-1}\mathcal{D}_X}(\mathcal{D}_{Y \hookrightarrow X}, f^{-1}\mathcal{M})$, where $\mathcal{D}_X^{\bullet}$ is a free resolution of $\mathcal{M}$. Then, as we observed earlier, $\mathcal{M} = \mathcal{D}_X/\mathcal{D}_X P$ is a free $\mathcal{O}_X$-Module. Therefore, the higher $\mathcal{T}or_j^{f^{-1}\mathcal{D}_X}(\mathcal{D}_{Y \hookrightarrow X}, f^{-1}\mathcal{M}) \cong \mathcal{T}or_j^{f^{-1}\mathcal{O}_X}(\mathcal{O}_Y, f^{-1}\mathcal{M})$ vanishes, for $j = 1, 2, \ldots$.

Proof of the Theorem 5.2.2.   Regard $f^{-1}\mathbb{R}\mathcal{H}om_{\mathcal{D}_X}(\mathcal{M}^{\bullet}, \mathcal{O}_X)$ as the abutment of

$$f^{-1'}E_2^{p,q} = f^{-1}\mathcal{E}xt_{\mathcal{D}_X}^p(\mathcal{H}^q(\mathcal{M}^{\bullet}), \mathcal{O}_X),$$

and also regard $\mathbb{R}\mathcal{H}om_{\mathcal{D}_X}(\mathcal{O}_Y \otimes^{\mathbb{L}}_{f^{-1}\mathcal{O}_X} f^{-1}\mathcal{M}^{\bullet}, \mathcal{O}_Y)$ as the abutment of

$$'E_2^{p,q} = \mathcal{E}xt_{\mathcal{D}_X}^p('E_2^{p',q'}) = \mathcal{E}xt_{\mathcal{D}_Y}^p(\mathcal{T}or_{p'}^{f^{-1}\mathcal{O}_X}(\mathcal{O}_Y, \mathcal{H}^{q'}(f^{-1}\mathcal{M}^{\bullet})), \mathcal{O}_Y),$$

where $q = p' + q'$ and $'E_2^{p',q'} \Rightarrow \mathcal{H}^q(\mathcal{O}_Y \otimes^{\mathbb{L}} f^{-1}\mathcal{M}^{\bullet})$. Therefore, by induction on the codimension of Y and by Remark 5.2.1, the proof is reduced to the case of Theorem 5.2.1.   □

As an application of Theorem 5.2.1, we have the following theorem.

Theorem 5.2.3 *For a $\mathcal{D}_X$-Module $\mathcal{M}$, if the characteristic variety $V(\mathcal{M})$ is contained in X, then $\mathcal{M}$ is isomorphic to $\mathcal{O}_X^l$ for some $l \geq 0$.*

Proof. Let $\mathcal{M} = \mathcal{D}_X u_1 + \mathcal{D}_X u_2 + \cdots + \mathcal{D}_X u_m$. Then the $\mathcal{O}_X$-coherency of $\mathcal{M}$ follows from the $\mathcal{O}_X$-coherency of $\mathcal{D}_X u_j$, $j = 1, 2, \ldots, m$. That is, it is enough to prove the cyclic $\mathcal{D}_X$-Module case. Let $\mathcal{D}_X u_j \cong \mathcal{D}_X/\mathcal{J}$, where $\mathcal{J}$ is $\ker \cdot u_j$ of the epimorphism $\mathcal{D}_X \overset{\cdot u_j}{\to} \mathcal{D}_X u_j \to 0$. Since $V(\mathcal{D} u_j) = V(\sqrt{\mathcal{J}}) \subset (\xi_1 = \cdots = \xi_n = 0)$, there exists an integer N so that we may have

$$\xi_i^N = \bar{P}_{ij}, \quad j = 1, 2, \ldots, m$$

for some P_{ij} in $\mathcal{J}$, $i = 1, 2, \ldots, n$. Hence, the good filtration on $\mathcal{M}$ becomes stationary, i.e.,

$$\mathcal{M}^{(k)} = \mathcal{M} \quad \text{for} \quad k \gg 0.$$

Since $\mathcal{D}^{(k-N)}$ is $\mathcal{O}_X$-coherent, $\mathcal{M}^{(k)}$ is $\mathcal{O}_X$-coherent.

Next we will prove $\mathcal{M} \cong \mathcal{O}_X^l$ for some l. Let $Y = (0)$. Then we have for $f : Y \hookrightarrow X$

$$f^*\mathcal{M} \cong \mathcal{D}_{(0)\hookrightarrow X} \otimes_{f^{-1}\mathcal{D}_X} f^{-1}\mathcal{M} \cong \mathcal{O}_{(0)} \otimes_{\mathcal{O}_{X,0}} \mathcal{M}_0 \cong \mathbb{C}_{(0)} \otimes_{\mathcal{O}_{X,0}} \mathcal{M}_0,$$

where $\mathcal{M}_0$ and $\mathcal{O}_{X,0}$ are stalks at $Y = (0)$. We have

$$f^{-1}\mathcal{H}om_{\mathcal{D}_X}(\mathcal{M}, \mathcal{O}_X) \cong \mathcal{H}om_{\mathcal{D}_Y}(f^*\mathcal{M}, \mathcal{O}_Y) \cong \mathcal{H}om_{\mathbb{C}_{(0)}}(\mathbb{C}_{(0)} \otimes_{\mathcal{O}_{X,0}} \mathcal{M}_0, \mathbb{C}_0) \cong \mathbb{C}_{(0)}^l,$$

where l is the number of generators of $\mathcal{M}_0$ over $\mathcal{O}_{X,0}$. Hence there exists a non-zero element φ in $\mathcal{H}om_{\mathcal{D}_X}(\mathcal{M}, \mathcal{O}_X)_0$ inducing the following epimorphism

$$(5.2.10) \qquad \mathcal{O}_Y \otimes_{f^{-1}\mathcal{O}_X} f^{-1}\mathcal{M} \overset{\varphi'}{\to} \mathcal{O}_Y \otimes_{f^{-1}\mathcal{O}_X} f^{-1}\mathcal{O}_X^l \to 0.$$

From the exact sequence

$$\mathcal{M} \overset{\varphi''}{\to} \mathcal{O}_X^l \to \mathcal{O}_X^l/\text{Im}\varphi'' \to 0$$

and (5.2.10), the fibre $\mathcal{O}_Y \otimes_{f^{-1}\mathcal{O}_X} f^{-1}(\mathcal{O}_X^l/\text{Im}\varphi'') = 0$ holds. Then Nakayama-Azumaya's lemma implies $\mathcal{O}_X^l/\text{Im}\varphi'' = 0$. Namely, $\mathcal{M} \overset{\varphi''}{\to} \mathcal{O}_X^l \to 0$ is exact. Let $\mathcal{N}$ be $\ker \varphi''$. We have the exact sequence

$$0 \to \mathcal{N} \to \mathcal{M} \to \mathcal{O}_X^l \to 0.$$

Since $\mathcal{M}$ and $\mathcal{O}_X^l$ are $\mathcal{O}_X$-coherent, $\mathcal{N}$ is also a coherent $\mathcal{O}_X$-Module. From the exact sequence

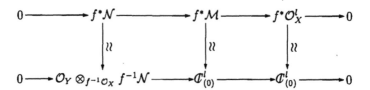

We therefore obtain $\mathcal{N} = 0$ and the proof is concluded.   □

5.3   Direct Image

Let $f : Y \to X$ be a holomorphic map. We are going to define the direct image of a left $\mathcal{D}_Y$-Module $\mathcal{M}$ via a bimodule $\mathcal{D}_{X \leftarrow Y}$. Let also Ω_X^n and Ω_Y^m be the sheaves of holomorphic forms of highest degrees on X and Y, respectively.

First we note that for the right $\mathcal{D}_X$-Module Ω_X^n we have the left $\mathcal{D}_X$-Module $(\Omega_X^n)^{-1} = \mathcal{H}om_{\mathcal{O}_X}(\Omega_X^n, \mathcal{O}_X)$. Therefore, for the $\mathcal{O}_X$-Module $(\Omega_X^n)^{-1}$, $\mathcal{D}_X \otimes_{\mathcal{O}_X} (\Omega_X^n)^{-1}$ is a left $\mathcal{D}_X$-Module, since $\mathcal{D}_X$ is a right $\mathcal{D}_X$-Module. Consequently, $f^{-1}(\mathcal{D}_X \otimes_{\mathcal{O}_X} (\Omega_X^n)^{-1})$ becomes a left $f^{-1}\mathcal{D}_X$-Module. Then we define the left $f^{-1}\mathcal{D}_X$-Module $\mathcal{D}_{X \leftarrow Y}$ by

$$\mathcal{D}_{X \leftarrow Y} = f^{-1}(\mathcal{D}_X \otimes_{\mathcal{O}_X} (\Omega_X^n)^{-1}) \otimes_{f^{-1}\mathcal{O}_X} \Omega_Y^m.$$

Furthermore,

$$f^{-1}(\mathcal{D}_X \otimes_{\mathcal{O}_X} (\Omega_X^n)^{-1}) \otimes_{f^{-1}\mathcal{O}_X} \Omega_Y^m =$$
$$= \mathcal{O}_Y \otimes_{f^{-1}\mathcal{D}_X} f^{-1}\mathcal{D}_X \otimes_{f^{-1}\mathcal{D}_X} f^{-1}(\mathcal{D}_X) \otimes_{f^{-1}\mathcal{O}_X} f^{-1}((\Omega_X^n)^{-1}) \otimes_{\mathcal{O}_Y} \Omega_Y^m$$
$$= \mathcal{D}_{Y \to X} \otimes_{f^{-1}\mathcal{O}_X} f^{-1}((\Omega_X^n)^{-1}) \otimes_{\mathcal{O}_Y} \Omega_Y^m.$$

obtaining the relation with $\mathcal{D}_{Y \to X}$. Recall that $\mathcal{D}_{Y \to X}$ is not only a left $\mathcal{D}_Y$-Module, but also a right $f^{-1}\mathcal{D}_X$-Module, as seen in Section 5.2. On the other hand, we have that

$$\mathcal{D}_{Y \to X} \otimes_{f^{-1}\mathcal{D}_X} f^{-1}(\mathcal{D}_X \otimes_{\mathcal{O}_X} (\Omega_X^n)^{-1}) \otimes_{\mathcal{O}_Y} \Omega_Y^m$$

is $f^*(\mathcal{D}_X \otimes_{\mathcal{O}_X} (\Omega_X^n)^{-1}) \otimes_{\mathcal{O}_Y} \Omega_Y^m$. Then, since the left multiplication defines a left $\mathcal{D}_X$-Module structure on $\mathcal{D}_X \otimes_{\mathcal{O}_X} (\Omega_X^n)^{-1}$, the inverse image $f^*(\mathcal{D}_X \otimes_{\mathcal{O}_X} (\Omega_X^n)^{-1})$ becomes a left $\mathcal{D}_Y$-Module. Summarizing what we have observed in the above is the following. The sheaf

$$\mathcal{D}_{X \leftarrow Y} = f^{-1}(\mathcal{D}_X \otimes_{\mathcal{O}_X} (\Omega_X^n)^{-1}) \otimes_{f^{-1}\mathcal{O}_X} \Omega_Y^m$$
$$f^*(\mathcal{D}_X \otimes_{\mathcal{O}_X} (\Omega_X^n)^{-1}) \otimes_{\mathcal{O}_Y} \Omega_Y^m$$

is not only a left $f^{-1}\mathcal{D}_X$-Module, but also a right $\mathcal{D}_Y$-Module. Furthermore, $\mathcal{D}_{X \leftarrow Y}$ is connected with $\mathcal{D}_{Y \to X}$ as

$$\mathcal{D}_{X \leftarrow Y} = \mathcal{D}_{Y \to X} \otimes_{f^{-1}\mathcal{O}_X} f^{-1}((\Omega_X^n)^{-1}) \otimes_{\mathcal{O}_Y} \Omega_Y^m$$
$$= \mathcal{H}om_{f^{-1}\mathcal{O}_X}(f^{-1}\Omega_X^n, \Omega_Y^m \otimes_{\mathcal{O}_Y} \mathcal{D}_{Y \to X}).$$

Definition 5.3.1 *For a left $\mathcal{D}_Y$-Module $\mathcal{M}$, the direct image of $\mathcal{M}$ is defined by*

$$\mathbb{R}f_*(\mathcal{D}_{X \leftarrow Y} \otimes_{\mathcal{D}_Y}^{\mathbb{L}} \mathcal{M}),$$

denoted as $\int_f \mathcal{M}$. We also write its j-th cohomology as

$$\int_f^j \mathcal{M} = \mathbb{R}^j f_*(\mathcal{D}_{X \leftarrow Y} \otimes_{\mathcal{D}_Y}^{\mathbb{L}} \mathcal{M}).$$

Example 5.3.1 Let $Y \overset{\imath}{\hookrightarrow} X$ be a closed embedding such that Y is defined by $Y = (z_1 = \ldots = z_d = 0)$. Then Ω_X^n and Ω_Y^{n-d} are respectively isomorphic to $\mathcal{O}_X$ and $\mathcal{O}_Y$. Then the right $\mathcal{D}_Y$-Module and the left $\imath^{-1}(\mathcal{D}_X)$-Module $\mathcal{D}_{X \leftarrow Y}$ can be written as

$$\imath^{-1}(\mathcal{D}_X) \otimes_{\imath^{-1}\mathcal{O}_X} \mathcal{O}_Y = \mathcal{D}_X \otimes_{\mathcal{O}_X} \mathcal{O}_X/(z_1, z_2, \ldots, z_d),$$

which is isomorphic to $\mathcal{D}_X/(\mathcal{D}_X z_1 + \cdots + \mathcal{D}_X z_d)$.

Recall that the left $\mathcal{D}_Y$-Module and right $\imath^{-1}(\mathcal{D}_X)$-Module $\mathcal{D}_{Y \to X}$ in Example 5.2.1 is isomorphic to $\mathcal{D}_{Y \to X} \cong \mathcal{D}_X/(z_1 \mathcal{D}_X + \cdots + z_d \mathcal{D}_X)$. The description of $\mathcal{D}_{X \leftarrow Y}$ in the above is just as expected since the leftness and the rightness of the module structures of $\mathcal{D}_{X \leftarrow Y}$ and $\mathcal{D}_{Y \to X}$ are exactly reversed.

Remark 5.3.1 For a holomorphic map f from Y to X and for a right $\mathcal{D}_Y$-Module $\mathcal{N}$, the direct image of $\mathcal{N}$ for f is defined by

$$\int_f \mathcal{N} = \mathbb{R}f_*(\mathcal{N} \otimes_{\mathcal{D}_Y} \mathcal{D}_{Y \to X})$$

when, $\int_f \mathcal{N}$ is a right $\mathcal{D}_X$-Module.

Example 5.3.2 Let us consider the case where the above holomorphic map is the projection π from $Y = X \times Z$ onto X defined as $\pi(z, w) = z$. Note that the general case is given by the composition of embedding case as in Example 5.3.1 and the projection case. Namely, any holomorphic map $f : Y \to X$ can be factored as

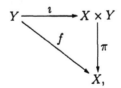

where $\imath(w) = (f(w), w)$ and $\pi(f(w), w) = f(w)$. Then we have

$$\mathcal{D}_{X \leftarrow Y} = \imath^{-1}(\mathcal{D}_{X \leftarrow X \times Y}) \otimes_{\imath^{-1} \mathcal{D}_{X \times Y}} \mathcal{D}_{X \times Y \leftarrow Y}.$$

Let $\mathcal{M}$ be a left $\mathcal{D}_Y$-Module. We will compute $\int_\pi \mathcal{M}$ for π as follows. The right $\mathcal{D}_Y$-Module $\mathcal{D}_{X \leftarrow Y}$ has a locally free resolution, see [19]:

(5.3.1)

$$0 \longrightarrow \Omega^0_{Y/X} \otimes_{\mathcal{O}_Y} \mathcal{D}_Y \longrightarrow \Omega^1_{Y/X} \otimes_{\mathcal{O}_Y} \mathcal{D}_Y \longrightarrow \cdots \longrightarrow \Omega^l_{Y/X} \otimes_{\mathcal{O}_Y} \mathcal{D}_Y \longrightarrow 0$$

$$\downarrow{\varepsilon}$$

$$\mathcal{D}_{X \leftarrow Y},$$

where $l = \dim(Y)$, i.e., the relative dimension of the projection $\pi : X \times Y \to X$, and $\Omega^p_{Y/X}$ denotes the relative p-form. This is because $\mathcal{D}_{X \leftarrow Y} \overset{\text{def}}{=} f^*(\mathcal{D}_X \otimes_{\mathcal{O}_X} (\Omega^n_X)^{-1}) \otimes_{\mathcal{O}_Y} \Omega^m_Y$ is isomorphic to $\Omega^l_{Y/X} \otimes_{\mathcal{O}_Y} f^*\mathcal{D}_X$, where $f^*\mathcal{D}_X = \mathcal{D}_{Y \to X}$ which is $\mathcal{O}_Y \otimes_{f^{-1}\mathcal{O}_X} f^{-1}\mathcal{D}_X$ by definition. Then the epimorphism from $\mathcal{D}_Y \to \mathcal{D}_{Y \to X}$ induces the epimorphism $\varepsilon : \Omega^l_{Y/X} \otimes_{\mathcal{O}_Y} \mathcal{D}_Y \to \mathcal{D}_{X \leftarrow Y}$ giving a locally free resolution of the right $\mathcal{D}_Y$-Module $\mathcal{D}_{X \leftarrow Y}$. The only non-vanishing l-th cohomology $\mathcal{H}^l(\Omega^*_{Y/X} \otimes_{\mathcal{O}_Y} \mathcal{D}_Y)$ is given by $\mathcal{D}_Y/(\partial_1 \mathcal{D}_Y + \cdots + \partial_l \mathcal{D}_Y)$ as in Section 4.2. Hence, $\mathcal{D}_{X \leftarrow Y}$ is isomorphic to $\mathcal{D}_Y/(\partial_1 \mathcal{D}_Y + \cdots + \partial_l \mathcal{D}_Y)$. Compare again with $\mathcal{D}_{Y \to X} = \mathcal{D}_Y/(\mathcal{D}_Y \partial_1 + \cdots + \mathcal{D}_Y \partial_l)$, in Section 5.2, i.e., the reversion of leftness and rightness is observed. Note also that the epimorphism $\mathcal{D}_Y \to f^*\mathcal{D}_X = \mathcal{D}_{Y \to X}$ induces a locally free resolution $\overset{.}{\wedge}\Theta_{Y/X} \otimes_{\mathcal{O}_Y} \mathcal{D}_Y$ of $\mathcal{D}_{Y \to X}$. Therefore, one can compute $\mathbb{R}\mathcal{H}om_{\mathcal{D}_Y}(\mathcal{D}_{Y \to X}, \mathcal{D}_Y)$ as follows

$$\mathbb{R}\mathcal{H}om_{\mathcal{D}_Y}(\mathcal{D}_{Y \to X}, \mathcal{D}_Y)$$
$$\cong \mathcal{H}om_{\mathcal{D}_Y}(\overset{.}{\wedge}\Theta_{Y/X} \otimes_{\mathcal{O}_Y} \mathcal{D}_Y, \mathcal{D}_Y)$$
$$\cong \mathcal{H}om_{\mathcal{O}_Y}(\overset{.}{\wedge}\Theta_{Y/X}, \mathcal{D}_Y)$$
$$\cong \Omega^*_{Y/X} \otimes_{\mathcal{O}_Y} \mathcal{D}_Y$$

which is the locally free resolution (5.3.1) for $\mathcal{D}_{X \leftarrow Y}$.

Next we will express the direct image $\int_f \mathcal{M}$ of a left $\mathcal{D}_Y$-Module $\mathcal{M}$ as a hyperderived functor of f_* using the resolution of $\mathcal{D}_{X \leftarrow Y}$ in (5.3.1). Since $\mathcal{D}_{X \leftarrow Y} \otimes^{\mathbb{L}}_{\mathcal{D}_Y} \mathcal{M}$ can be computed by $\Omega^*_{Y/X} \otimes_{\mathcal{O}_Y} \mathcal{D}_Y[l] \otimes_{\mathcal{D}_Y} \mathcal{M}$, we obtain

$$\int_f^j \mathcal{M} = \mathbb{R}^{j+l} f_*(\Omega^*_{Y/X} \otimes_{\mathcal{O}_Y} \mathcal{M}).$$

Notice that for $\mathcal{M} = \mathcal{O}_Y$ we obtain

$$\int_f^j \mathcal{M} = \int_f^j \mathcal{O}_Y = I\!R^{j+l} f_*(\Omega^\bullet_{Y/X}).$$

That is $\int_f^j \mathcal{O}_Y = I\!R^{j+l} f_*(\Omega^\bullet_{Y/X})$ is nothing but the usual relative version of de Rham cohomology sheaf, on which the Gauss-Mannin connection is defined, often denoted as $\mathcal{H}^{j+l}_{DR}(Y/X)$. More generally, if a $\mathcal{D}_Y$-Module $\mathcal{M}$ is locally free of finite rank as an $\mathcal{O}_Y$-Module, then the hyperderived functor of f_* evaluated at $\Omega^\bullet_{Y/X} \otimes_{\mathcal{O}_Y} \mathcal{M}$, i.e. the direct image $\int_f \mathcal{M}$, is the relative de Rham cohomology sheaf of $\mathcal{O}_X$-Modules with the Gauss-Mannin connection. See [129] for the construction. The observation of the direct image as the hyperderived functor gives the following Proposition.

Proposition 5.3.1 *For an exact sequence of $\mathcal{D}_Y$-Modules*

$$0 \to \mathcal{M}' \to \mathcal{M} \to \mathcal{M}'' \to 0,$$

a long exact sequence of $\mathcal{D}_X$-Modules is induced as follows

$$\dots \to \int_f^j \mathcal{M}' \to \int_f^j \mathcal{M} \to \int_f^j \mathcal{M}'' \to \int_f^{j+1} \mathcal{M}' \to \dots.$$

Remark 5.3.2 1) Let $f : Y \to X$ be a holomorphic map of relative dimension l. As an application of spectral sequences (4.2.5) and (4.2.6) in Remark 4.2.5 in Chapter IV, let

$$'\mathcal{N}^j = \begin{cases} \mathcal{D}_{X \leftarrow Y} & \text{for} j = 0 \\ 0 & \text{for} j \neq 0 \end{cases}$$

for a complex $'\mathcal{N}^\bullet$ of right $\mathcal{D}_Y$-Modules. Then (4.2.5) becomes

$$E^1_{0,l-h} = \mathcal{T}or^{\mathcal{D}_Y}_{l-h}(\mathcal{D}_{X \leftarrow Y}, \mathcal{M})$$

abutting to $(h - l)$-th cohomology of the complex $\mathcal{D}_{X \leftarrow Y} \otimes^{I\!L}_{\mathcal{D}_Y} \mathcal{M}$. From the locally free resolution (5.3.1):

$$\Omega^\bullet_{Y/X} \otimes_{\mathcal{O}_Y} \mathcal{D}_Y[l] \to \mathcal{D}_{X \leftarrow Y},$$

the abutment is isomorphic to $\mathcal{H}^h(\Omega^\bullet_{Y/X} \otimes_{\mathcal{O}_Y} \mathcal{M})$. The collapsing spectral sequence implies the following isomorphism:

$$\mathcal{T}or^{\mathcal{D}_Y}_{l-h}(\mathcal{D}_{X \leftarrow Y}, \mathcal{M}) \approx \mathcal{H}^h(\Omega^\bullet_{Y/X} \otimes_{\mathcal{O}_Y} \mathcal{M}).$$

2) We have the direct image $\int_f^\bullet \mathcal{M}$ as

$$I\!Rf_*(\mathcal{D}_{X \leftarrow Y} \otimes^{I\!L}_{\mathcal{D}_Y} \mathcal{M}).$$

We denote the h-th cohomology as $\int_f^h \mathcal{M} = \mathbb{R}^h f_*(\mathcal{D}_{X\leftarrow Y} \otimes^{\mathbb{L}}_{\mathcal{O}_Y} \mathcal{M})$, where the right hand side is the hyperderived functor of f_*. Then (4.2.5) becomes

$$E_1^{p,q} = \mathbb{R}^q f_*(\Omega^p_{Y/X} \otimes_{\mathcal{O}_Y} \mathcal{M}),$$

inducing

$$E_2^{p,q} = \mathcal{H}^p(\mathbb{R}^q f_*(\Omega^\bullet_{Y/X} \otimes_{\mathcal{O}_Y} \mathcal{M}),$$

and (4.2.6) becomes

$$E_2^{p,q} = \mathbb{R}^p f_*(\mathcal{H}^q(\Omega^\bullet_{Y/X} \otimes_{\mathcal{O}_Y} \mathcal{M}))$$
$$= \mathbb{R}^p f_*(\mathcal{T}or^{\mathcal{D}_Y}_{l-q}(\mathcal{D}_{X\leftarrow Y}, \mathcal{M})),$$

with abutment $\int_f^h \mathcal{M} = \mathbb{R}^{l+h} f_*(\Omega^\bullet_{Y/X} \otimes_{\mathcal{O}_Y} \mathcal{M})$.

5.4   Holonomic $\mathcal{D}$-Modules

For any $\mathcal{D}_X$-Module $\mathcal{M}$ we have $\dim V(\mathcal{M}) \geq n$, (see [206]). One may ask which $\mathcal{D}_X$-Modules are most strongly determined by the relations among generators so that solutions may not have any free variables. Such systems of partial differential equations are traditionally called maximally overdetermined systems.

Definition 5.4.1 *A $\mathcal{D}_X$-Module $\mathcal{M}$ is said to be holonomic if $\dim V(\mathcal{M}) = n$ or $\mathcal{M} = 0$ holds.*

Let

$$0 \to \mathcal{M}' \to \mathcal{M} \to \mathcal{M}'' \to 0$$

be a short exact sequence of $\mathcal{D}_X$-Modules. Then we have $V(\mathcal{M}) = V(\mathcal{M}') \cup V(\mathcal{M}'')$. Suppose $\dim V(\mathcal{M}) \leq n$. Then we have $\dim V(\mathcal{M}') \cup V(\mathcal{M}'') \leq n$. Consequently, $\dim V(\mathcal{M}') \leq n$ and $\dim V(\mathcal{M}'') \leq n$, must hold. Conversely, if $\dim V(\mathcal{M}') \leq n$ and $\dim V(\mathcal{M}'') \leq n$, then the dimension of $V(\mathcal{M}') \cup V(\mathcal{M}'')$ cannot strictly be greater than n. Namely, we have the following

Proposition 5.4.1 *For a short exact sequence*

$$0 \to \mathcal{M}' \to \mathcal{M} \to \mathcal{M}'' \to 0$$

of $\mathcal{D}_X$-Modules, $\mathcal{M}$ is holonomic if and only if $\mathcal{M}'$ and $\mathcal{M}''$ are holonomic.

Remark 5.4.1 Let $X = \mathbb{C}$, and let $\mathcal{M} = \mathcal{D}_{\mathbb{C}} u$ such that

$$\mathcal{D}_{\mathbb{C}} \xrightarrow{\bullet P} \mathcal{D}_{\mathbb{C}} \xrightarrow{\bullet u} \mathcal{M} \to 0$$

is exact. Namely, we are considering an ordinary differential equation $Pu = 0$ of order k, for some $k \geq 0$. Then the dimension of the zero set of the principal symbol

$$\sigma(P) = g(z)\xi^k, \quad g(z) \in \mathcal{O}_{\mathbb{C}}$$

is one, i.e., $V(\mathcal{M}) = 1 = \dim\mathbb{C}$. That is, the $\mathcal{D}_{\mathbb{C}}$-Module $\mathcal{M}$ is holonomic. The classical Cauchy's theorem states that the sheaf of solutions $\mathcal{H}om_{\mathcal{D}_{\mathbb{C}}}(\mathcal{M}, \mathcal{O})$ is locally free of finite rank as a $\mathbb{C}_X$-vector space outside the singularities of the operator P.

As we defined in Section 4.2, an $\mathcal{O}_X$-Module $\mathcal{M}$ with an integrable connection is a $\mathcal{D}_X$-Module that is a free $\mathcal{O}_X$-Module satisfying the axioms in Section 4.1. Namely, the structure of an integrable connection on $\mathcal{M} = \oplus_{i=1}^m \mathcal{O}_X u_i$ is defined by an $\mathcal{O}_X$-Algebra map:

$$\Theta \to \mathcal{E}nd_{\mathbb{C}_X}(\oplus_{i=1}^m \mathcal{O}_x u_i, \oplus_{i=1}^m \mathcal{O}_x u_i).$$

Therefore, for an integrable connection $\mathcal{M} = \oplus_{i=1}^m \mathcal{O}_X u_i$ and each ∂_j in Θ, we have a representation by n equations:

(5.4.1) $$\partial_j u_i = P_{j1} u_1 + P_{j2} u_2 + \cdots + P_{jm} u_m, \quad j = 1, \ldots, n.$$

System (5.4.1) gives rise to a free resolution of the left $\mathcal{D}_X$-Module $\mathcal{M}$, namely, the integrable connection $\mathcal{M}$. By Frobenius theorem, there exists a base $(v_1, v_2, \ldots, v_m)$ for the free $\mathcal{O}_X$-Module $\mathcal{M}$ so that system (5.4.1) is transformed into a Pfaff system of m equations:

(5.4.2) $$\partial_j v_i = 0, \quad j = 1, \ldots, n.$$

Since the characteristic variety $V(\mathcal{O}_X) = T_X^* X \cong X$, by Example 3.2.1, and any integrable connection is isomorphic to a finite direct sum of the de Rham system $\mathcal{O}_X$, as observed in the above, integrable connections are holonomic.

The goal of this section is to prove Kashiwara's theorem on the constructibility of $\mathcal{E}xt_{\mathcal{D}_X}^h(\mathcal{M}, \mathcal{O}_X)$ for a holonomic $\mathcal{D}_X$-Module $\mathcal{M}$.

Let us consider a $\mathcal{D}_X$-Module $\mathcal{M}$ that is locally free of finite rank as an $\mathcal{O}_X$-Module. That is, $\mathcal{M}$ is a vector bundle with a holomorphic integral connection over X. That is, locally we have an isomorphism $\mathcal{M} \xleftarrow{\approx} \oplus_{i=1}^m \mathcal{O}_X u_j$, where $\{u_1, u_2, \ldots, u_m\}$ is a set of generators of $\mathcal{M}$ over $\mathcal{O}_X$. Then we can define a good filtration on $\mathcal{M}$ as follows:

$$\begin{cases} \mathcal{M}^{(k)} = 0 & \text{for } k < 0 \\ \mathcal{M}^{(k)} = \mathcal{M} = \oplus_{j=1}^m \mathcal{O}_X u_j = \mathcal{D}_X^{(0)} u_1 \oplus \cdots \oplus \mathcal{D}_X^{(0)} u_m & \text{for } k \geq 0. \end{cases}$$

The induced $\overline{\mathcal{D}_X}$-Module $\overline{\mathcal{M}}$ is annihilated by $\xi_i = \sigma_1(D_i)$, $D_i = \dfrac{\partial}{\partial z_i}$. That is, $\xi_1, \ldots, \xi_n$ belong to the annihilator ideal $\mathcal{J}(\mathcal{M})$. Hence, the zero set of $\sqrt{\mathcal{J}(\mathcal{M})}$, the characteristic variety $V(\mathcal{M})$, is the zero section X of $\pi : T^*X \to X$.

Conversely, if a holonomic $\mathcal{D}_X$-Module $\mathcal{M}$ satisfies $V(\mathcal{M}) = X$, i.e. the zero section of T^*X, then we have $\mathcal{M} \cong \mathcal{O}_X^l$ for some l, (compare with Theorem 5.2.3).

For an integrable connection $\mathcal{M}$ we have, by the Cauchy existence and uniqueness theorem, $\mathcal{E}xt^0_{\mathcal{D}_X}(\mathcal{M}, \mathcal{O}_X) = \mathcal{H}om_{\mathcal{D}_X}(\mathcal{M}, \mathcal{O}_X)$ is a local system over X, establishing an equivalence between integrable connections and local systems. We also have, by Poincaré Lemma, $\mathcal{E}xt^h_{\mathcal{D}_X}(\mathcal{M}, \mathcal{O}_X) = 0$, $h = 1, 2, \ldots$, (see Section 4.2). Recall that a **local system** is a $\mathbb{C}_X$-Module that is locally free of finite rank. Let us state the above as a type of Frobenius existence theorem as follows.

Theorem 5.4.1 *The de Rham functor $\mathcal{H}om_{\mathcal{D}_X}(\mathcal{O}_X, -)$ and the solution functor $\mathcal{H}om_{\mathcal{D}_X}(-, \mathcal{O}_X)$ induce an equivalence between the category of left $\mathcal{D}_X$-Modules that are locally free of finite rank as $\mathcal{O}_X$-Modules and the category of local systems over X as $\mathbb{C}_X$-Modules. The converse functor can be given by $\mathcal{O}_X \otimes_{\mathbb{C}_X} -$.*

Remark 5.4.2 For a holonomic $\mathcal{D}_X$-Module $\mathcal{M}$, we have the duality

$$\mathbb{R}\mathcal{H}om_{\mathcal{D}_X}(\mathcal{O}_X, \mathcal{M}) \xrightarrow{\approx} \mathbb{R}\mathcal{H}om_{\mathbb{C}_X}(\mathbb{R}\mathcal{H}om_{\mathcal{D}_X}(\mathcal{M}, \mathcal{O}_X), \mathbb{C}_X)$$

in the derived category of $\mathbb{C}$-constructible sheaves.

Note that a $\mathbb{C}$-Module $\mathcal{F}$ over X is said to be $\mathbb{C}$-constructible if there exists an increasing sequence of finitely many closed analytic subspaces

$$\emptyset = X_0 \subset X_1 \subset \cdots \subset X_i \subset \cdots \subset X$$

such that $\mathcal{F}|_{X_i - X_{i-1}}$ is isomorphic to $\mathbb{C}_X^{r_i}$ for some $r_i \in \mathbb{N}$, i.e., $\mathcal{F}|_{X_i - X_{i-1}}$ is a local system over the locally closed $X_i - X_{i-1}$.

One can generalize the duality in Remark 5.4.2 to a bounded complex $\mathcal{M}^\bullet$ of $\mathcal{D}_X$-Modules whose cohomologies are holonomic. That is, we have

$$(5.4.3) \quad \mathbb{R}\mathcal{H}om_{\mathcal{D}_X}(\mathcal{O}_X, \mathcal{M}^\bullet) \xrightarrow{\approx} \mathbb{R}\mathcal{H}om_{\mathbb{C}_X}(\mathbb{R}\mathcal{H}om_{\mathcal{D}_X}(\mathcal{M}^\bullet, \mathcal{O}_X), \mathbb{C}_X).$$

Proof. Let $\mathcal{J}^\bullet$ be a complex of injective $\mathcal{D}_X$-Modules such that $\mathcal{M}^\bullet \to \mathcal{J}^\bullet$ is a quasi-isomorphism, and let $'\mathcal{J}^\bullet$ be an injective resolution of the de Rham $\mathcal{D}_X$-Module $\mathcal{O}_X$. Consider the following diagram

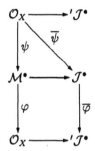

As we saw in Section 4.2, the Poincaré Lemma tells us $\mathbb{C}_X$ is quasi-isomorphic to $\mathbb{R}\mathcal{H}om_{\mathcal{D}_X}(\mathcal{O}_X, \mathcal{O}_X)$. Hence we can define a map Φ of complexes:

$$\mathcal{H}om_{\mathcal{D}_X}(\mathcal{O}_X, \mathcal{J}^\bullet) \xrightarrow{\Phi} \mathcal{H}om_{\mathbb{C}_X}(\mathcal{H}om_{\mathcal{D}_X}(\mathcal{J}^\bullet, {}'\mathcal{J}^\bullet), \mathcal{H}om_{\mathcal{D}_X}(\mathcal{O}_X, {}'\mathcal{J}^\bullet))$$

namely, for $\bar{\psi} \in \mathcal{H}om_{\mathcal{D}_X}(\mathcal{O}_X, \mathcal{J}^\bullet)$, assign a $\mathbb{C}_X$-linear map

$$\Phi_{\bar{\psi}}(\bar{\varphi}) = \bar{\varphi} \circ \bar{\psi} \quad \text{for} \quad \bar{\varphi} \in \mathcal{H}om_{\mathcal{D}_X}(\mathcal{J}^\bullet, {}'\mathcal{J}^\bullet).$$

Next, we must show the map Φ

$$\mathbb{R}\mathcal{H}om_{\mathcal{D}_X}(\mathcal{O}_X, \mathcal{M}^\bullet) \xrightarrow{\approx} \mathbb{R}\mathcal{H}om_{\mathbb{C}_X}(\mathbb{R}\mathcal{H}om_{\mathcal{D}_X}(\mathcal{M}^\bullet, \mathcal{O}_X), \mathbb{C}_X)$$

is an isomorphism. Since $\mathcal{H}^h(\mathcal{M}^\bullet)$ is holonomic, as in the proof of Theorem 5.2.2, it is sufficient to prove the case of a single holonomic $\mathcal{D}_X$-Module $\mathcal{M}$. For an arbitrary $x \in X$, we will compute the stalk at x:

$$\mathbb{R}\mathcal{H}om_{\mathbb{C}}(\mathbb{R}\mathcal{H}om_{\mathbb{C}}(\mathbb{R}\mathcal{H}om_{\mathcal{D}_X}(\mathcal{M}, \mathcal{O}_X), \mathbb{C}_X), \mathbb{C}_x)_x,$$

where $\mathbb{C}_x$ denotes the sheaf concentrated at $\{x\}$ whose stalk at x is $\mathbb{C}$. Rewrite the above stalk as follows:

$$= \mathbb{R}\mathcal{H}om_{\mathbb{C}}(\mathbb{R}\mathcal{H}om_{\mathbb{C}}(\mathbb{C}_x, \mathbb{C}), \mathbb{R}\mathcal{H}om_{\mathcal{D}_X}(\mathcal{M}, \mathcal{O}_X))_x$$
$$= \mathbb{R}\mathcal{H}om_{\mathbb{C}}(\mathbb{C}_x[-2n], \mathbb{R}\mathcal{H}om_{\mathcal{D}_X}(\mathcal{M}, \mathcal{O}_X))_x$$
$$= \mathbb{R}\Gamma_{\{x\}}(\mathbb{R}\mathcal{H}om_{\mathcal{D}_X}(\mathcal{M}, \mathcal{O}_X))_x[2n]$$
$$= \mathbb{R}\mathcal{H}om_{\mathcal{D}_X}(\mathcal{M}, \mathbb{R}\Gamma_{\{x\}}(\mathcal{O}_X))[2n]$$
$$= \mathbb{R}\mathcal{H}om_{\mathbb{C}}(\mathbb{R}\mathcal{H}om_{\mathcal{D}_X}(\mathcal{O}_X, \mathcal{M})_x, \mathbb{C}_x)_x$$

Therefore, we have a quasi-isomorphism

$$\mathbb{R}\mathcal{H}om_{\mathcal{D}_X}(\mathcal{O}_X, \mathcal{M})_x \to \mathbb{R}\mathcal{H}om_{\mathbb{C}}(\mathbb{R}\mathcal{H}om_{\mathcal{D}_X}(\mathcal{M}, \mathcal{O}_X), \mathbb{C}_X)_x.$$

We are ready to prove the most significant result for a holonomic $\mathcal{D}$-Module, namely, Kashiwara's constructibility theorem.

Theorem 5.4.2 *For a holonomic $\mathcal{D}$-Module $\mathcal{M}$, all the higher cohomology sheaves of solutions, i.e., $\mathcal{E}xt^h_{\mathcal{D}_X}(\mathcal{M}, \mathcal{O}_X)$ are $\mathbb{C}$-constructible. That is, there exists a whitney stratification of $X = \sqcup_\alpha X_\alpha$, independently of each h, so that*

$$(5.4.4) \qquad\qquad V(\mathcal{M}) \subset \sqcup_\alpha T^*_{X_\alpha} X$$

and $\mathcal{E}xt^h_{\mathcal{D}_X}(\mathcal{M}, \mathcal{O}_X)|_{X_\alpha}$ is a locally constant sheaf of finite rank over $\mathbb{C}$.

Remark 5.4.3 A partition $\{X_\alpha\}$ of a complex manifold X is said to be a **stratification** of $X = \sqcup_\alpha X_\alpha$ when $\{X_\alpha\}$ is a locally finite partition, each X_α is a locally closed submanifold of X, and $X_\alpha \subset \overline{X_{\alpha'}} - X_{\alpha'}$ holds for $X_\alpha \cap X_{\alpha'} \neq \emptyset$. When the last condition is satisfied, we often write $X_\alpha \prec X_{\alpha'}$.
Furthermore, a stratification $X = \sqcup X_\alpha$ is said to be a Whitney stratification if

1. $\sqcup_\alpha T^*_{X_\alpha} X$ is a closed set of T^*X, and

2. for $X_\alpha \prec X_{\alpha'}$, i.e., X_α is contained in $\overline{X_{\alpha'}}$ for $X_\alpha \cap \overline{X_{\alpha'}} \neq \emptyset$, let x be any point in X_α and consider sequences $\{x_n\} \subset X_\alpha$ and $\{x'_n\} \subset X_{\alpha'}$ satisfying $\lim_{n\to\infty} x_n = \lim_{x\to\infty} x'_n = x$ in X_α. If the sequence of tangent spaces $\{T_{x'_n} X_{\alpha'}\}$ converges to $\tau \subset T_x X$ and if $\mathbb{C}(x_n - x'_n)$ converges to a line l in $T_x X$, we have $l \subset \tau$.

Whitney proved that any stratification $X = \sqcup_\alpha X_\alpha$ has a finer stratification $X = \sqcup_\beta X'_\beta$ of X (such that X'_β is contained in some X_α, i.e., for each α there is an index set B_α such that $X_\alpha = \sqcup_{\beta \in B_\alpha} X'_\beta$) satisfying the above conditions (1) and (2).

Proof. First we will establish the existence of a Whitney stratification satisfying (5.4.4). Let $\Lambda = V(\mathcal{M})$. Note that the holonomicity of $\mathcal{M}$, i.e., $V(\mathcal{M})$ is of dimension n, implies that $T_p(V(\mathcal{M}))$ is Lagrangian at a non-singular point p. Since $V(\mathcal{M})$ is an involutive analytic set in T^*X, the converse is also true. As before let $\pi : T^*X \to X$ be the projection. Then $X - \pi(\Lambda) = X - \pi(V(\mathcal{M}))$ is an open dense set in X. This is because the dimension of $\pi(V(\mathcal{M}))$ is less than $(n-1)$. Let X'_0 be the set of non-singular points of $\pi(\Lambda)$. Then we have $T^*_{X'_0} X \subset \Lambda$, and furthermore $\overline{\Lambda - T^*_{X'_0} X}$ is also a Lagrangian analytic set in T^*X. Since $\pi(\Lambda - T^*_{X'_0} X)$ can not equal X'_0, the dimension of $\pi(\overline{\Lambda - T^*_{X'_0} X})$ is strictly less than the dimension of X'_0. Let $X_1 = \pi(\overline{\Lambda - T^*_{X'_0} X})$, and let $\Lambda_1 = \overline{\Lambda - T^*_{X'_0} X}$. Then, denoting the set of non-singular points of $\pi(\Lambda_1)$ by X'_1, we can define inductively as follows:

$$\Lambda_2 = \overline{\Lambda_1 - T^*_{X'_1} X}$$

$X'_2 = $ the non $-$ singular locus of $\pi(\Lambda_2)$

$\cdots\cdots\cdots$

$$\Lambda_{\alpha+1} = \overline{\Lambda_\alpha - T^*_{X'_\alpha} X}$$

$X'_{\alpha+1} = $ the non $-$ singular locus of $\pi(\Lambda_{\alpha+1})$

$\cdots\cdots\cdots$

Then we obtain a stratification by non-singular manifolds so that

$$X = \sqcup_\alpha X'_\alpha$$

and

$$\Lambda \subset \cup_\alpha T^*_{X'_\alpha} X.$$

Then there exists a finer stratification $\{X'_\beta\}$ than $\{X'_\alpha\}$, such that

$$X = \sqcup_\beta X'_\beta$$

satisfying Whitney conditions (1) and (2) in Remark 5.4.3. Before we prove the finite dimensionality of the locally constant sheaf $\mathcal{E}xt^h_{\mathcal{D}}(\mathcal{M}, \mathcal{O})|_{X_\alpha}$ for a Whitney stratification $\{X_\alpha\}$, we need to know under what conditions the restriction map

$$Ext^h_{\mathcal{D}}(\Omega', \mathcal{M}, \mathcal{O}_X) \to Ext^h_{\mathcal{D}_X}(\Omega, \mathcal{M}, \mathcal{O}_X)$$

becomes an isomorphism, where Ω and Ω' are open sets satisfying $\Omega \subset \Omega'$.

Lemma 5.4.1 *For an arbitrary point x_0 in X_α, there exists a neighborhood U of x_0 such that for a small enough ε the boundary of a ball $B(x', \varepsilon) = \{x \in X : |x - x'| < \varepsilon\}$, $x' \in X_\alpha \cap U$, is non-characteristic for $\mathcal{M}$.*

Proof. If the statement of the above lemma is not true, there are sequences $\{x_n\}$ in X_α and $\{y_n\}$ in X such that $x_n \to x_0$ and $y_n \to x_0$ as $n \to \infty$, $x_n \neq y_n$ and such that $d\varphi_{x_n}(y_n) \in V(\mathcal{M})$, where $\varphi_x(y) = |x - y|$. Then we may assume $y_n \in X_\beta$ for some X_β satisfying $X_\beta \succ X_\alpha$ (take some subsequence $\{y_n\}$ if necessary). Consequently, we can find a sequence in $\mathbb{C}$ so that $\mathbb{C}(y_n - x_n)$ converges to l in $T_{x_0} X$ and $T_{y_n} X_\beta$ approaches τ in $T_{x_0} X$. Therefore, $d\varphi_{x_n}(y_n)$ converges to the dual vector l^* of l. By the assumption, $d\varphi_{x_n}(y_n) = 0$ on $T_{y_n} X_\beta$. Then $l^* = 0$ on τ, which contradicts to $l \subset \tau$.

Next, we will prove the restriction map ρ

(5.4.5) $\qquad Ext^h_{\mathcal{D}_X}(B(x_0, \varepsilon'), \mathcal{M}, \mathcal{O}_X) \to Ext^h_{\mathcal{D}_X}(B(x_0, \varepsilon), \mathcal{M}, \mathcal{O}_X)$

is an isomorphism for $\varepsilon' \geq \varepsilon$, $x_0 \in X_\alpha$. Notice that this isomorphism implies the following .

$$\mathcal{E}xt^h_{\mathcal{D}_X}(\mathcal{M}, \mathcal{O}_X) = \varinjlim_{\varepsilon \to 0} Ext^h_{\mathcal{D}_X}(B(x_0, \varepsilon), \mathcal{M}, \mathcal{O}_X)$$
$$\approx Ext^h_{\mathcal{D}_X}(B(x_0, \varepsilon), \mathcal{M}, \mathcal{O}_X)$$

for a small enough ε. From a well known lemma in Functional Analysis,

$$Ext^h_{\mathcal{D}_X}(B(x_0, \varepsilon), \mathcal{M}, \mathcal{O}_X)$$

is finite dimensional. See Lemma 5.4.5. As a consequence, at the stalk at x_0,

$$\dim_{\mathbb{C}} \mathcal{E}xt^h_{\mathcal{D}_X}(\mathcal{M}, \mathcal{O}_X)_{x_0}$$

is finite. Therefore, for x_0' satisfying $|x_0 - x_0'| < \varepsilon$ we obtain

$$\mathcal{E}xt^h_{\mathcal{D}_X}(\mathcal{M}, \mathcal{O}_X)_{x_0'} \cong Ext^h_{\mathcal{D}_X}(B(x_0, \varepsilon), \mathcal{M}, \mathcal{O}_X) \cong \mathcal{E}xt^h_{\mathcal{D}_X}(\mathcal{M}, \mathcal{O}_X)_{x_0},$$

proving the finite dimensionality of the locally constant sheaf $\mathcal{E}xt^h_{\mathcal{D}_X}(\mathcal{M}, \mathcal{O}_X)|_{X_\alpha}$. We must prove that the map (5.4.5) is an isomorphism. Since we have a restriction map ρ'' from $Ext^h_{\mathcal{D}_X}(B(x_0, \varepsilon''), \mathcal{M}, \mathcal{O}_X)$ to $Ext^h_{\mathcal{D}_X}(B(x_0, \varepsilon), \mathcal{M}, \mathcal{O}_X)$ for $\varepsilon'' > \varepsilon' > \varepsilon$ satisfying $\rho'' = \rho \circ \rho'$, where ρ' is the restriction map from $Ext^h_{\mathcal{D}_X}(B(x_0, \varepsilon''), \mathcal{M}, \mathcal{O}_X)$ to $Ext^h_{\mathcal{D}_X}(B(x_0, \varepsilon'), \mathcal{M}, \mathcal{O}_X)$, $\{Ext^h_{\mathcal{D}_X}(B(x_0, \varepsilon), \mathcal{M}, \mathcal{O}_X)\}$ forms an inverse system. Consider a flabby resolution of $\mathcal{O}_X$ by the sheaf of forms of $(0, *)$-type with coefficients in hyperfunctions:

$$0 \to \mathcal{O}_X \to \mathcal{B}_X^{(0,1)} \xrightarrow{\bar{\partial}} \mathcal{B}_X^{(0,2)} \to \ldots \to \mathcal{B}^{(0,n)} \to 0.$$

Then, $Ext^h_{\mathcal{D}_X}(B(x_0, \varepsilon), \mathcal{M}, \mathcal{O}_X)$ is the h-th cohomology of the complex

$$Hom_{\mathcal{D}_X}(B(x_0, \varepsilon), \mathcal{M}, \mathcal{B}_X^{(0,\bullet)}).$$

By the definition of flabbiness, the restriction map for $\varepsilon' > \varepsilon$

$$Hom_{\mathcal{D}_X}(B(x_0, \varepsilon'), \mathcal{M}, \mathcal{B}_X^{(0,i)}) \to Hom_{\mathcal{D}_X}(B(x_0, \varepsilon), \mathcal{M}, \mathcal{B}_X^{(0,i)})$$

is an epimorphism. That is, the inverse system $\{Hom_{\mathcal{D}_X}(B(x_0, \varepsilon), \mathcal{M}, \mathcal{B}_X^{(0,i)})\}$ is satisfying the following condition, called Mittag-Leffler condition: for any $\varepsilon > 0$ there exists ε_0 so that the decreasing sequence satisfies

$$\mathrm{Im}(Hom_{\mathcal{D}_X}(B(x_0, \varepsilon'), \mathcal{M}, \mathcal{B}_X^{(0,i)})) \xrightarrow{\rho'} Hom_{\mathcal{D}_X}(B(x_0, \varepsilon), \mathcal{M}, \mathcal{B}_X^{(0,i)})$$

$$= \mathrm{Im}(Hom_{\mathcal{D}_X}(B(x_0, \varepsilon_0), \mathcal{M}, \mathcal{B}_X^{(0,i)})) \xrightarrow{\rho_0} Hom_{\mathcal{D}_X}(B(x_0, \varepsilon), \mathcal{M}, \mathcal{B}_X^{(0,i)})$$

for $\varepsilon' > \varepsilon_0$. Furthermore $Ext_{\mathcal{D}_X}^{h-1}(B(x_0,\varepsilon),\mathcal{M},\mathcal{O}_X)$ is finite dimensional. Hence, $Ext_{\mathcal{D}_X}^{h-1}(B(x_0,\varepsilon),\mathcal{M},\mathcal{O}_X)$ satisfies Mittag-Leffler condition. Then the epimorphism

$$H^h(Hom_{\mathcal{D}_X}(B(x_0,\varepsilon),\mathcal{M},\mathcal{B}_X^{(0,\bullet)})) \overset{\text{def}}{=} Ext_{\mathcal{D}_X}^h(B(x_0,\varepsilon),\mathcal{M},\mathcal{O}_X) \to$$

$$\varprojlim_{\varepsilon'\to\varepsilon} Ext_{\mathcal{D}_X}^h(B(x_0,\varepsilon'),\mathcal{M},\mathcal{O}_X) \overset{\text{def}}{=} \varprojlim_{\varepsilon'\to\varepsilon} H^h(Hom_{\mathcal{D}_X}(B(x_0,\varepsilon'),\mathcal{M},\mathcal{B}_X^{(0,\bullet)}))$$

becomes an isomorphism. Namely, we need a general lemma on the commutativity of the inverse limit with the cohomology.

Lemma 5.4.2 Let $\{V_c\}$, $c \in \mathbb{R}$, be an inverse system satisfying Mittag-Leffler condition, i.e., for an arbitrary $c \in \mathbb{R}$ there exists $c_0 \in \mathbb{R}$ so that for $c' > c_0$, $Im(V_{c'} \to V_c) = Im(V_{c_0} \to V_c)$ holds.
Then
$$H^h(\varprojlim_c V_c^\bullet) \to \varprojlim_c H^h(V_c^\bullet)$$

is always an epimorphism. Moreover, when $\{H^{h-1}(V_c^\bullet)\}$, $c \in \mathbb{R}$, satisfies Mittag-Leffler condition, the above map is a monomorphism.

The next lemma implies that the restriction map in (5.4.5) is actually an isomorphism.

Lemma 5.4.3 Let $\{V_c\}_{c\in\mathbb{R}}$ be an inverse system of finite dimensional vector spaces. If

$$\varinjlim_{c'>c} V_{c'} \to V_c \quad \text{and} \quad \varprojlim_{c'<c} V_{c'} \leftarrow V_c$$

are monomorphisms (epimorphisms), then $V_{c'} \to V_c$ is also a monomorphism (an epimorphism).

We omit the proofs of these two lemmas due to Kashiwara.
Remark 5.4.4 The finite dimensionality of $Ext_{\mathcal{D}_X}^h(B(x_0,\varepsilon),\mathcal{M},\mathcal{O}_X)$ follows from the following lemma on a compact operator between Fréchet spaces.

Lemma 5.4.4 Let $V^\bullet$ and $W^\bullet$ be bounded complexes of Fréchet spaces with continuous linear differentials, and let $u^\bullet$ be a map from $V^\bullet$ to $W^\bullet$ such that $u^h : V^h \to W^h$ is a compact operator for each h. When $u^\bullet$ is a quasi-isomorphism from $V^\bullet$ to $W^\bullet$, for each h, induced isomorphic cohomology groups $H^h(V^\bullet)$ and $H^h(W^\bullet)$ are vector spaces over $\mathbb{C}$ of finite dimension.

See [8] for its proof. From this lemma, we will be able to prove the finite
dimensionality of

$$Ext^h_{\mathcal{D}_X}(B(x_0, \varepsilon), \mathcal{M}, \mathcal{O}_X)$$

as follows. Consider a free resolution of $\mathcal{M}$ in a small neighborhood $B(x_0, \varepsilon')$ of
x_0

$$\ldots \to \mathcal{D}^l_X \to \mathcal{D}^m_X \to \mathcal{M} \to 0$$

as in Chapter III. Then left exact contravariant functor $Hom_{\mathcal{D}_X}(B(x_0, \varepsilon'), -, \mathcal{O}_X)$
induces a complex of Fréchet spaces with continuous linear differential:

$$\mathcal{O}^m_X(B(x_0, \varepsilon')) \to \mathcal{O}^l(B(x_0, \varepsilon')) \to \ldots.$$

The restriction map $\mathcal{O}_X(B(x_0, \varepsilon')) \to \mathcal{O}_X(B(x_0, \varepsilon))$ is a compact map for $\varepsilon < \varepsilon'$.
Then

$$H^h(\mathcal{O}^m_X(B(x_0, \varepsilon')) \to \ldots) = Ext^h_{\mathcal{D}_X}(B(x_0, \varepsilon'), \mathcal{M}, \mathcal{O}_X)$$

$$\downarrow$$

$$H^h(\mathcal{O}^m_X(B(x_0, \varepsilon)) \to \ldots) = Ext^h_{\mathcal{D}_X}(B(x_0, \varepsilon), \mathcal{M}, \mathcal{O}_X)$$

is an isomorphism between finite dimensional complex vector spaces.

Remark 5.4.5 For an elliptic $\mathcal{D}_M$-Module $\mathcal{M}$ on a real analytic manifold M
with its complexification X, we have the following finite dimensionality theorem
of Kawai.

Recall that a coherent $\mathcal{D}_M$-Module $\mathcal{M}$ is elliptic if $V(\mathcal{M}) \cap S^*_M X = \emptyset$, where
$\mathcal{D}_M = \mathcal{D}_X|_M$ and $S^*_M X = \sqrt{-1}S^*M$, the cosphere bundle of M with respect to
X. Kawai, T. proved the following theorem. See Kawai, T. "Theorems on the
finite dimensionality of cohomology groups", I, II, III, IV and V, Proc. Japan
Acad. N° 48, 70-72, 287-289 (1972), N° 49, 243-246, 655-658 and 782-784 (1973),
respectively.

Theorem 5.4.3 *Let Ω be a relatively compact open subset of M so that the
boundary $\partial\Omega$ is a real analytic hypersurface in M. When the restriction of an
elliptic $\mathcal{D}_M$-Module $\mathcal{M}$ to the boundary $\partial\Omega$ is elliptic on $\partial\Omega$,*

$$Ext^h_{\mathcal{D}_M}(\Omega, \mathcal{M}, \mathcal{A}_M)$$

*is finite dimensional over $\mathbb{C}$, where $\mathcal{A}_M$ is the sheaf of real analytic functions on
M.*

Notice that Kashiwara's constructibility theorem is a vast and ideal generalization of Kawai's.

Let $\mathcal{M}$ be a holonomic $\mathcal{D}_X$-Module. Namely, we have $\dim V(\mathcal{M}) = \dim X = n$. From theorem 4.4.1, for a holonomic $\mathcal{D}_X$-Module $\mathcal{M}$, we have

$$\mathcal{E}xt^h_{\mathcal{D}_X}(\mathcal{M}, \mathcal{D}_X) = 0 \quad \text{for} \quad h < n = \text{codim } V(\mathcal{M}).$$

On the other hand, theorem 4.4.2 implies, for $h > n = \dim V(\mathcal{M})$,

$$\mathcal{E}xt^h_{\mathcal{D}_X}(\mathcal{M}, \mathcal{D}_X) = 0.$$

For this non-vanishing n-th cohomology sheaf, we have

$$\text{codim } V(\mathcal{E}xt^n_{\mathcal{D}_X}(\mathcal{M}, \mathcal{D}_X)) \geq n$$

from (4.4.6). From the involutivity of $V(\mathcal{M})$, we get

$$\text{codim} V(\mathcal{E}xt^n_{\mathcal{D}_X}(\mathcal{M}, \mathcal{D}_X)) = n.$$

Namely, $\mathcal{E}xt^n_{\mathcal{D}_X}(\mathcal{M}, \mathcal{D}_X)$ is a right holonomic $\mathcal{D}_X$-Module, which is denoted as $\mathcal{M}^*$.

Let

$$0 \to {}'\mathcal{M} \to \mathcal{M} \to {}''\mathcal{M} \to 0$$

be a short exact sequence of left holonomic $\mathcal{D}_X$-Modules. The left exact contravariant functor $\mathcal{H}om_{\mathcal{D}_X}(-, \mathcal{D}_X)$ induces a long exact sequence of cohomologies. Then the purely codimensionality of $\mathcal{E}xt^h_{\mathcal{D}_X}(\mathcal{M}, \mathcal{D}_X) = 0$, $h \neq n$ implies the following exact sequence

$$0 \to {}''\mathcal{M}^* \to \mathcal{M}^* \to {}'\mathcal{M}^* \to 0.$$

That is, $\mathcal{E}xt^n_{\mathcal{D}_X}(-, \mathcal{D}_X)$ is an exact functor from the category of left holonomic $\mathcal{D}_X$-Modules to the category of right holonomic $\mathcal{D}_X$-Modules.

Similarly, $\mathcal{M}^{**} = (\mathcal{M}^*)^*$ is a left holonomic $\mathcal{D}_X$-Module. One can show actually $\mathcal{M}^{**} \approx \mathcal{M}$. See the following Remark 5.4.6 for its proof. In general the characteristic variety of $\mathcal{E}xt^n_{\mathcal{D}_X}(\mathcal{M}, \mathcal{D}_X)$ is smaller, i.e. contained in, than that of $\mathcal{M}$. Therefore, we have codim $V(\mathcal{M}) \leq$ codim $V(\mathcal{E}xt^n_{\mathcal{D}_X}(\mathcal{M}, \mathcal{D}_X))$, (or from (4.4.6) for $h = n$). Similarly,

$$V(\mathcal{E}xt^n_{\mathcal{D}_X}(\mathcal{M}, \mathcal{D}_X)) \supset V(\mathcal{E}xt^n_{\mathcal{D}_X}(\mathcal{E}xt^n_{\mathcal{D}_X}(\mathcal{M}, \mathcal{D}_X), \mathcal{D}_X)\mathcal{M} \overset{\text{def}}{=} V(\mathcal{M}^{**}) = V(\mathcal{M})$$

implies

$$n \leq \text{codim } V(\mathcal{E}xt^n_{\mathcal{D}_X}(\mathcal{M}, \mathcal{D}_X)) \leq \text{codim } V(\mathcal{M}) \leq n.$$

Consequently, codim $V(\mathcal{M}) = n$ holds. That is, the cohomological statement

$$\text{``}Ext^h_{\mathcal{D}_X}(\mathcal{M}, \mathcal{D}_X) = 0, \quad h \neq n\text{''}$$

characterize the holonomicity of $\mathcal{M}$.

Remark 5.4.6 Consider a projective resolution of a $\mathcal{D}_X$-Module $\mathcal{M}$:

$$\ldots \to P^{-2} \to P^{-1} \to P^0 \to \mathcal{M} \to 0.$$

Then the contravariant functor $\mathcal{H}om_{\mathcal{D}_X}(-, \mathcal{D}_X)$ induces a complex as follows.

$$0 \to \mathcal{H}om_{\mathcal{D}_X}(\mathcal{M}, \mathcal{D}_X) \to \mathcal{H}om_{\mathcal{D}_X}(P^0, \mathcal{D}_X) \to \mathcal{H}om_{\mathcal{D}_X}(P^{-1}, \mathcal{D}_X) \to \ldots$$

Denote $\mathcal{H}om_{\mathcal{D}_X}(P^{-h}, \mathcal{D}_X)$ as $\bar{P}^h$. Then we have a Cartan-Eilenberg resolution of $\bar{P}^\bullet$ as follows.

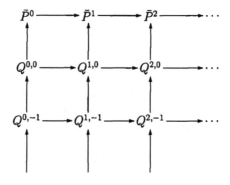

The above fourth quadrant double complex is taken by the contravariant functor $\mathcal{H}om_{\mathcal{D}_X}(-, \mathcal{D}_X)$ to the second quadrant double complex as follows.

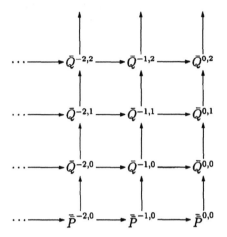

where $\bar{\bar{P}}^{-h,0} = \mathcal{H}om_{\mathcal{D}_X}(\bar{P}^h, \mathcal{D}_X)$.

Then spectral sequences associated with the above second quadrant double complex is induced as follows. (See introduction of Chapter III).

$$E_0^{-p,q} = \bar{Q}^{-p,q}, \quad {}'E_0^{p,-q} = \bar{Q}^{-q,p}$$

$$E_1^{-p,q} =_{(0,1)} \mathcal{H}^q(\bar{Q}^{-p,*}), \quad {}'E_1^{p,-q} =_{(0,1)} \mathcal{H}^{-q}(\bar{Q}^{*,p})$$

$$E_2^{-p,q} =_{(1,0)} \mathcal{H}^{-p}(_{(0,1)}\mathcal{H}^q(\bar{Q}^{**})), \quad {}'E_2^{p,-q} =_{(0,1)} \mathcal{H}^p(_{(1,0)}\mathcal{H}^{-q}(\bar{Q}^{**}))$$

abutting to the total cohomology $\mathcal{H}^n(\bar{Q}^*)$. Notice that $E_1^{-p,q} = 0$ for $q \neq 0$ and $E_1^{-p,0} =_{(0,1)} \mathcal{H}^0(\bar{Q}^{-p,*})$. Since a projective object is reflexive, we obtain

$$
\begin{array}{ccc}
\cdots \longrightarrow E_1^{-p,0} \longrightarrow & E_1^{-p+1,0} \longrightarrow \cdots \\
\| & \| \\
\cdots \longrightarrow \mathcal{H}om_{\mathcal{D}_X}(\bar{P}^{p,0}, \mathcal{D}_X) \longrightarrow \mathcal{H}om_{\mathcal{D}_X}(\bar{P}^{p-1,0}, \mathcal{D}_X) \longrightarrow \cdots \\
\| & \| \\
\cdots \longrightarrow \bar{\bar{P}}^{-p-1,0} \longrightarrow \bar{\bar{P}}^{-p,0} \longrightarrow \bar{\bar{P}}^{-p+1,0} \longrightarrow \cdots \\
\|\wr & \|\wr & \|\wr \\
\cdots \longrightarrow P^{-p-1} \longrightarrow P^{-p} \longrightarrow P^{-p+1} \longrightarrow \cdots
\end{array}
$$

Therefore, we have $E_2^{-p,0} = 0$ for $p \neq 0$ and $E_2^{0,0} \approx \mathcal{M}$. That is, $E_\infty^{0,0} \approx E^0 = H^0(\bar{Q}^\bullet)$, where $\bar{Q}^n = \oplus_{p+q=n} \bar{Q}^{p,q}$. On the other hand, the 0-th term second spectral sequence

$$\cdots \; \to \; 'E_0^{p,-q} \; \to \; 'E_0^{p,-q-1} \; \to \; \cdots$$

$$\| \qquad\qquad \|$$

$$\cdots \; \to \; \bar{Q}^{-q,p} \; \to \; \bar{Q}^{-q-1,p} \; \to \; \cdots$$

gives

$$
\begin{array}{ccc}
\vdots & & \vdots \\
\uparrow & & \uparrow \\
'E_1^{2,-q} & =_{(1,0)} & H^{-q}(\bar{Q}^{\bullet,2}) \\
\uparrow & & \uparrow \\
'E_1^{1,-q} & =_{(1,0)} & H^{-q}(\bar{Q}^{\bullet,1}) \qquad \uparrow \\
\uparrow & & \uparrow \\
'E_1^{0,-q} & =_{(1,0)} & H^{-q}(\bar{Q}^{\bullet,0}) \\
\uparrow & & \uparrow \\
0 & & 0
\end{array}
$$

Therefore, we get

$$'E_2^{p,-q} = \mathcal{E}xt_{\mathcal{D}_X}^p(\mathcal{E}xt_{\mathcal{D}_X}^q(\mathcal{M}, \mathcal{D}_X), \mathcal{D}_X)$$

abutting the $(p - q)$-th derived functor of the identity, i.e. $E^0 = \mathcal{M}$ with $p - q = 0$.

From the sequence

$$\cdots \; \to \; 'E_2^{p-2,-q+1} \; \overset{'d_2^{p-2,-q+1}}{\longrightarrow} \; 'E_2^{p,-q} \; \overset{'d_2^{p,-q}}{\longrightarrow} \; 'E_2^{p+2,-q-1}$$

$$\| \qquad\qquad\qquad\qquad\qquad \|$$

$$0 \qquad\qquad\qquad\qquad\qquad\qquad 0$$

we obtain $'E_\infty^{p,-p} = F^p(\mathcal{M})/F^{p+1}(\mathcal{M})$, where $\mathcal{M} = \oplus_{p=0}^\infty E_\infty^{p,-p}$. This is because the total degrees of $'d_2^{p,-q}$ and $'d_2^{p-2,-q+1}$ increase by $+1$. For a holonomic $\mathcal{D}_X$-Module $\mathcal{M}$, we have

$$'E_2^{n,-n} = \mathcal{E}xt_{\mathcal{D}_X}^n(\mathcal{E}xt_{\mathcal{D}_X}^n(\mathcal{M}, \mathcal{D}_X), \mathcal{D}_X) \approx E_\infty^{n,-n} \approx \mathcal{M}.$$

Proposition 5.4.2 *For a holonomic $\mathcal{D}_X$-Module $\mathcal{M}$ we have an isomorphism*

$$\mathcal{E}xt^h_{\mathcal{D}_X}(\mathcal{M}, \mathcal{O}_X) \xrightarrow{\approx} \mathcal{E}xt^h_{\mathcal{D}_X}(\mathcal{O}_X, \mathcal{M}^* \otimes_{\mathcal{O}_X} (\Omega^n)^{-1}),$$

where $- \otimes_{\mathcal{O}_X} (\Omega^n)^{-1}$ *is a quasi-inverse of* $- \otimes \Omega^n$, *i.e.*

$$\mathcal{M}^* \otimes_{\mathcal{O}_X} (\Omega^n)^{-1} = \mathcal{H}om_{\mathcal{O}_X}(\Omega^n_X, \mathcal{M}^*)$$

(see Chapter IV, Introduction).

Remark 5.4.7 Namely, the above isomorphism says that the higher solution sheaves are nothing but the de Rham cohomology sheaves of the dual system. See Remark 5.4.8.

Proof. Since the de Rham Module $\mathcal{O}_X$ is of finite presentation as a $\mathcal{D}_X$-Module, we have the spectral sequence

$$E_2^{p,q} = \mathcal{T}or^{\mathcal{D}_X}_{-p}(\mathcal{E}xt^q_{\mathcal{D}_X}(\mathcal{M}, \mathcal{D}_X), \mathcal{O}_X))$$

abutting to $\mathcal{E}xt^h_{\mathcal{D}_X}(\mathcal{M}, \mathcal{O}_X)$. See Chapter IV, (4.2.7). Since $\mathcal{M}$ is holonomic, $E_2^{p,q} = 0$ for $q \neq n$. Hence we have an isomorphism

$$E_2^{h-n,n} = \mathcal{T}or^{\mathcal{D}_X}_{n-h}(\mathcal{E}xt^n_{\mathcal{D}_X}(\mathcal{M}, \mathcal{D}_X), \mathcal{O}_X) \xrightarrow{\approx} \mathcal{E}xt^h_{\mathcal{D}_X}(\mathcal{M}, \mathcal{O}_X).$$

Then we have

$$\begin{aligned}
E_2^{h-n,n} &= \mathcal{T}or^{\mathcal{D}_X}_{n-h}(\Omega^n \otimes_{\mathcal{O}_X} \mathcal{M}^* \otimes_{\mathcal{O}_X} (\Omega^n)^{-1}, \mathcal{O}_X) \\
&\approx \mathcal{T}or^{\mathcal{D}_X}_{n-h}(\Omega^n, \mathcal{M}^* \otimes_{\mathcal{O}_X} (\Omega^n)^{-1}) \\
&\approx \mathcal{H}_{-h}((\Omega^\bullet \otimes_{\mathcal{O}_X} \mathcal{D}_X) \otimes_{\mathcal{D}_X} \mathcal{M}^* \otimes_{\mathcal{O}_X} (\Omega^n)^{-1}),
\end{aligned}$$

where $\Omega^\bullet \otimes_{\mathcal{O}_X} \mathcal{D}_X[n]$ is the projective resolution of Ω^n in (4.2.2). From the above, we have further an isomorphism

$$\begin{aligned}
&\approx \mathcal{H}^h(\Omega^\bullet \otimes_{\mathcal{O}_X} \mathcal{M}^* \otimes_{\mathcal{O}_X} (\Omega^n)^{-1}) \\
&= \mathcal{E}xt^h_{\mathcal{D}_X}(\mathcal{O}_X, \mathcal{M}^* \otimes_{\mathcal{O}_X} (\Omega^n)^{-1}).
\end{aligned}$$

$\square$

Remark 5.4.8 Notice that by replacing $\mathcal{M}$ by $\mathcal{M}^* \otimes_{\mathcal{O}_X} (\Omega^n)^{-1}$ in the previous Proposition 5.4.2, we also have

$$\mathcal{E}xt^h_{\mathcal{D}_X}(\mathcal{M}^* \otimes_{\mathcal{O}_X} (\Omega^n)^{-1}, \mathcal{O}_X) \xrightarrow{\approx} \mathcal{E}xt^h_{\mathcal{D}_X}(\mathcal{O}_X, \mathcal{M}).$$

The generalization to the derived category version of the above Proposition becomes as follows.

Let $\mathcal{M}^\bullet$ be a bounded complex of left $\mathcal{D}$-Modules whose cohomologies are of finite presentation. Then $\mathbb{R}\mathcal{H}om_{\mathcal{D}_X}(\mathcal{O}_X, \mathcal{M}^\bullet)$ is isomorphic to

$I\!R\mathcal{H}om_{\mathcal{D}_X}((\mathcal{M}^\bullet)^* \otimes_{\mathcal{O}_X} (\Omega^n)^{-1}[n], \mathcal{O}_X)$,   where $(\mathcal{M}^\bullet)^*$ is   defined   by $I\!R\mathcal{H}om_{\mathcal{D}_X}(\mathcal{M}^\bullet, \mathcal{D}_X)$. The proof is an exercise in Derived Category:

$$I\!R\mathcal{H}om_{\mathcal{D}_X}((\mathcal{M}^\bullet)^* \otimes_{\mathcal{O}_X} (\Omega^n)^{-1}[n], \mathcal{O}_X)$$
$$\cong I\!R\mathcal{H}om_{\mathcal{D}_X}(I\!R\mathcal{H}om_{\mathcal{D}_X}(\mathcal{M}^\bullet, \mathcal{D}_X), \Omega^n)[-n]$$
$$\cong \Omega^n \otimes_{\mathcal{D}_X}^{I\!L} \mathcal{M}^\bullet[-n]$$
$$\cong I\!R\mathcal{H}om_{\mathcal{D}_X}(\mathcal{O}_X, \mathcal{M}^\bullet).$$

Remark 5.4.9 Let $\mathcal{M}$ be a left $\mathcal{D}_X$-Module that is a locally free $\mathcal{O}_X$-Module, i.e., there locally exists a set of generators $\{u_i\}_{1 \le i \le m}$ such that $\mathcal{M} = \mathcal{O}_X u_1 \oplus \cdots \oplus \mathcal{O}_X u_m$. Then the $\mathcal{D}_X$-Module structure is determined by the assignment of the generators for each $\partial_i = \dfrac{\partial}{\partial z_i}$. Since $\mathcal{M}$ is locally free as an $\mathcal{O}_X$-Module, there exists holomorphic functions $g_{j_1}, g_{j_2}, \ldots, g_{j_m}$ so that we may have

$$\partial_i u_j = g_{j_1} u_1 + g_{j_2} u_2 + \cdots + g_{j_m} u_m, \quad j \le m.$$

Namely,

(5.4.6)
$$\begin{bmatrix} \partial_i & 0 & \ldots & 0 \\ 0 & \partial_i & \ldots & 0 \\ 0 & \ldots & \ldots & 0 \\ 0 & \ldots & 0 & \partial_i \end{bmatrix} \begin{bmatrix} u_1 \\ \vdots \\ \vdots \\ u_m \end{bmatrix} = \begin{bmatrix} g_{11} & \ldots & g_{1m} \\ g_{21} & \ldots & g_{2m} \\ & \ldots & \\ g_{m1} & \ldots & g_{mm} \end{bmatrix} \begin{bmatrix} u_1 \\ \vdots \\ \vdots \\ u_m \end{bmatrix}.$$

(See Chapter III, §2). On the other hand, the associated complex with a connection

$$\mathcal{M}^\bullet \otimes_{\mathcal{O}_X} (\Omega^n)^{-1} \to (\mathcal{M}^\bullet \otimes_{\mathcal{O}_X} (\Omega^n)^{-1}) \otimes_{\mathcal{O}_X} \Omega^1 \to \ldots \to (\mathcal{M}^\bullet \otimes_{\mathcal{O}_X} (\Omega^n)^{-1}) \otimes_{\mathcal{O}_X} \Omega^n$$

is integrable when, for a section $g_1 u_1 + \cdots + g_m u_m$ of $\mathcal{M}$

(5.4.7)
$$\begin{bmatrix} \partial_i & 0 & \ldots & 0 \\ 0 & \partial_i & \ldots & 0 \\ 0 & \ldots & \ldots & 0 \\ 0 & \ldots & 0 & \partial_i \end{bmatrix} \begin{bmatrix} g_1 \\ \vdots \\ \vdots \\ g_m \end{bmatrix} = - \begin{bmatrix} g_{11} & \ldots & g_{1m} \\ g_{21} & \ldots & g_{2m} \\ & \ldots & \\ g_{m1} & \ldots & g_{mm} \end{bmatrix}^t \begin{bmatrix} g_1 \\ \vdots \\ \vdots \\ g_m \end{bmatrix}$$

holds, i.e., the 0-th cohomology of the above complex. That is, the above equations 5.4.6 and 5.4.7 correspond to the 0-th cohomologies in Proposition 5.4.2.

5.5 Historical Notes

The next significant step from Kashiwara's constructibility theorem is the Riemann-Hilbert correspondence theorem. First recall the concept of an equivalence of categories.

Let F be a covariant functor from a category $\mathcal{C}$ to another category $\mathcal{C}'$. Then F is said to be faithful when the map induced by F:

$$\mathcal{H}om_{\mathcal{C}}(X,Y) \to \mathcal{H}om_{\mathcal{C}'}(F(X),F(Y))$$

is injective for arbitrary objects X and Y in $\mathcal{C}$. Furthermore, F is full when the above map is surjective. For a fully faithful, i.e., full and faithful, functor F is said to be an equivalence between $\mathcal{C}$ and $\mathcal{C}'$ when for any X' in $\mathcal{C}'$ there exists X in $\mathcal{C}$ satisfying

$$F(X) \xrightarrow{\approx} X'.$$

Since we have

$$I\!R\mathcal{H}om_{\mathcal{D}_X}(\mathcal{O}_X, \mathcal{M}^\bullet) \xrightarrow{\approx} I\!R\mathcal{H}om_{\mathcal{C}_X}(\mathcal{H}om_{\mathcal{D}_X}(\mathcal{M}^\bullet, \mathcal{O}_X), \mathcal{C}_X)$$

for a bounded holonomic complex of $\mathcal{D}_X$-Modules, (i.e., $\mathcal{H}^h(\mathcal{M}^\bullet)$ being holonomic) Kashiwara's constructibility theorem can be stated through the de Rham functor as follows: $I\!R\mathcal{H}om_{\mathcal{D}_X}(\mathcal{O}_X, \mathcal{M}^\bullet)$ is a bounded complex of $\mathcal{C}_X$-Modules whose cohomologies are constructible.

The above de Rham functor restricted to a full subcategory of the derived category of bounded holonomic complex of $\mathcal{D}_X$-Modules induces an equivalence to the derived category of bounded constructible complex of $\mathcal{C}_X$-Modules. The desired subcategory is the derived category of regular holonomic complex of $\mathcal{D}_X$-Modules. One of the equivalent definitions of a regular holonomic complex of $\mathcal{D}_X$-Modules is the following:

$$I\!R\mathcal{H}om_{\mathcal{D}_X}(\mathcal{M}^\bullet, \mathcal{O}_X)_x \xrightarrow{\approx} I\!R\mathcal{H}om_{\mathcal{D}_{X,x}}(\mathcal{M}_x^\bullet, \hat{\mathcal{O}}_{X,x}),$$

where $\hat{\mathcal{O}}_{X,x}$ is the completion at x, i.e., $\hat{\mathcal{O}}_{X,x} = \varprojlim_i \mathcal{O}_X/\mathfrak{m}_x^i$.

The above isomorphism may be interpreted as formal solutions being actually holomorphic (converging) ones. Regular holonomic $\mathcal{D}_X$-Modules are studied "correctly" via the microlocal view in the magnificent 166-page paper by Kashiwara and Kawai [120]. For example, the above definition of the regularity of a single holonomic $\mathcal{D}_X$-Module $\mathcal{M}$ is equivalent to the following: there exists locally a good filtration $\{\mathcal{M}^{(k)}\}$ on $\mathcal{M}$ such that the annihilator ideal $\mathcal{J}(\mathcal{M})$ of the $\bar{\mathcal{D}}_X$-Module $\bar{\mathcal{M}}$, i.e., $\mathcal{J}(\mathcal{M}) = \{\bar{P} \in \bar{\mathcal{D}}_X : \bar{P}\bar{\mathcal{M}} = \bar{0}\}$, is reduced in $\bar{\mathcal{D}}_X$. (See Chapter III). That is, if $\sigma_k(P)$, $P \in \mathcal{D}_X^{(k)}$, vanishes on the characteristic variety $V(\mathcal{M}) = V(\sqrt{\mathcal{J}(\mathcal{M})})$, then $\sigma_k(P)$ annihilates $\bar{\mathcal{M}}$, i.e., for any $l \in \mathbb{Z}$, $P\mathcal{M}^{(l)} \subset \mathcal{M}^{(l+k-1)}$ holds. Riemann-Hilbert-Kashiwara Correspondence Theorem asserts that the de Rham functor $I\!R\mathcal{H}om_{\mathcal{D}_X}(\mathcal{O}_X, \mathcal{M}^\bullet)$, (or the solution functor $I\!R\mathcal{H}om_{\mathcal{D}_X}(\mathcal{M}^\bullet, \mathcal{O}_X) = I\!R\mathcal{H}om_{\mathcal{C}_X}(I\!R\mathcal{H}om_{\mathcal{D}_X}(\mathcal{O}_X, \mathcal{M}^\bullet), \mathcal{C}_X))$ provides an equivalence from the derived category of bounded complex of $\mathcal{D}_X$-Modules $\mathcal{M}^\bullet$ whose cohomologies are holonomic $\mathcal{D}_X$-Modules, to the derived category of bounded complex of $\mathbb{C}_X$-Modules whose cohomologies are $\mathbb{C}_X$-constructible. See Kashiwara [116].

Kashiwara constructs the inverse contravariant functor to the solution functor whose compositions are identities for those derived categories.

Chapter 6

Systems of Microdifferential Equations

6.1 Introduction

The purpose of this final Chapter is to present to the reader a series of applications of the theory developed so far to the study of systems of microdifferential equations. The single most remarkable result of the early theory of microfunctions, is arguably the celebrated Fundamental Theorem of Sato, which (though classical by now) was probably the first result which clarified the need for the apparently cumbersome machinery of microfunctions theory.

As we have already seen in Chapter 2, the notion of parametrix was developed in the theory of partial differential equations with the purpose of "inverting" elliptic differential operators and, analogously, pseudo-differential and microdifferential operators have been introduced and studied with the goal of generalizing parametrices and providing inverses for a larger class of operators. Sato's Fundamental Theorem builds up on this theory to show the conditions under which such an inversion is always possible. Before we give the statement of Sato's result, let us try to put it in perspective.

In the theory of linear differential equations with constant coefficients, the most general and powerful result of these last fifty years has certainly been the so called Fundamental Principle of Ehrenpreis-Palamodov [52], [178] which has shown that when studying the structure of the (e.g. distributional) solutions u of a linear constant coefficients differential equation

$$P(D)u = 0$$

($P = P(z_1, \ldots, z_n)$ a polynomial in n complex variables, $D = (-i\frac{\partial}{\partial x_1}, \ldots, -i\frac{\partial}{\partial x_n})$, $u \in \mathcal{D}'(\mathbb{R}^n)$), then the central object to be reckoned with is the so called "characteristic variety"

$$V_P = \{z \in \mathbb{C}^n : P(z) = 0\}$$

and, indeed, if multiplicities are taken into account, such a variety completely characterizes and describes the space of solutions

$$\{u \in \mathcal{D}'(\mathbb{R}^n) : P(D)u = 0\}.$$

This result has of course its origin in Euler's Fundamental Principle (on exponential sums representation for linear constant coefficients ordinary differential equations), and, more recently, in L. Schwartz's work on mean periodic functions, but before its surprising announcement in 1961 by Ehrenpreis, few people (if any) really believed in the possibility of such a characterization. The problem for variable coefficients operators has always, of course, been much more delicate, in view of the lack of elementary solution in the form of a family of exponential solutions. The first general result in the direction of a complete characterization of the sheaf of solutions of variable coefficients equations is, indeed, Sato's Fundamental Principle, which, on the other hand, does indeed require the notion of microfunction, in order to be fully stated. It might be added that from our point of view, and after the preparation in Chapter 2, its proof is rather natural, as it will essentially rely on the explicit construction of an inverse, much in the same way in which one would construct the inverse of a polynomial. From a historical point of view, however, this construction was neither trivial nor evident.

There are two different formulations for this important result. One of them we have already discussed in the framework of $\mathcal{E}$-modules, when (in Chapter 3) we have shown that microdifferential operators of finite order can be microlocally inverted wherever their symbol does not vanish. More interesting, the inverse is, itself, a microdifferential operator. The other form, which is the one we will describe in this Chapter, only shows the invertibility in the sheaf of microlocal operators.

As a consequence of this important result, one sees that the structure of solutions of microdifferential equations is essentially trivial outside the characteristic variety, and one is therefore lead to the consideration of what this structure may be on the characteristic variety. One of the early goals of the theory developed by Sato was to show that systems of microdifferential equations could essentially be classified in a very simple way, according to some properties of their characteristic variety. This fact may be considered as a powerful generalization of the fundamental result from classical mechanics which states that every partial differential equation of the first order can be transformed into a standard form by means of the Jacobi canonical transformations. In order to be able to prove a result of this nature for higher order linear differential equations, we will need to generalize the class of transformations which we allow for consideration. This new class is constituted by the so called quantized contact transformations, which in some form were first introduced by Maslov [163], and later on made more accessible by Egorov [45].

The plan of this Chapter is as follows. In section 2 we state and prove Sato's Fundamental Theorem, for the case of linear differential operators of finite order. In section 3 we discuss some special cases of differential equations, namely the wave equation and hyperbolic equations, which provide a first hint of what is the microlocal aspect of the study of differential equations. Quantized contact transformations are dealt with in section 4, while the last section provides the final classification theorems for systems of microdifferential equations. A short historical appendix concludes this Chapter.

6.2 The Invertibility of Microlocal Operators

This section is devoted to the detailed proof of the so called weak form of Sato's Fundamental Theorem, which states the invertibility for any finite order differential operator in the ring of microlocal operators. To properly state this result we first review some preliminary definitions (see also Chapter III): given

$$P(x, D) = \sum_{|\alpha| \leq m} a_\alpha(x) D^\alpha$$

a linear differential operator of order m, its **principal symbol** $\sigma(P)(x, \xi)$ is the holomorphic function defined (on $T^* I\!\!R^n$) by

$$\sigma(P)(x, \xi) := \sum_{|\alpha| = m} a_\alpha(x) \xi^\alpha.$$

Note that the principal symbol of such a differential operator is invariant under a coordinate transformation; this property is clearly not satisfied by the lower order terms of $P(x, D)$. The set of points where the principal symbol of an operator vanishes is known as the characteristic variety of the operator, i.e.

$$V(P) = \{(x, \xi) \in T^* I\!\!R^n : \sigma(P)(x, \xi) = 0\}.$$

The reader is invited to compare the above definition with Definition 3.3.1 and the examples that follow. We can then state Sato's theorem:

Theorem 6.2.1 *A linear differential operator of finite order $P(x, D)$ is left and right invertible in the ring $\mathcal{L}$ of microlocal operators over $T^* I\!\!R^n \backslash V(P)$ or, in terms of cospherical tangent bundle over*

$$\{(x, \xi_\infty) \in S^* I\!\!R^n : \sigma(P)(x, \xi) \neq 0\}.$$

The proof of Theorem 6.2.1 is not particularly surprising as it is essentially the microlocalization of F. John's construction of a fundamental solution for elliptic differential equations. This construction, on the other hand, is based on the classical superposition principle.

We will need two simple analytical lemmas, as well as a lemma on composition and conjugation of microlocal operators:

Lemma 6.2.1 *Let* $u = u(x_1, x')$ *be a real analytic function of*

$$x = (x_1, x') = (x_1, x_2, \ldots, x_n)$$

and let $H(t)$ *be the usual Heaviside function. Then*

(6.2.1)
$$\frac{\partial^j}{\partial x_1^j} (u(x) \cdot H(x_1)) =$$

$$= \left(\frac{\partial^j}{\partial x^j} u(x) \right) \cdot H(x_1) + \sum_{k=1}^{j} \frac{\partial^{j-k}}{\partial x_1^{j-k}} u(0, x') \cdot \delta^{(k-1)}(x_1).$$

Proof.   For $j = 0$, the lemma simply claims that

$$u(x) \cdot H(x_1) = u(x) \cdot H(x_1)$$

and is therefore immediately true. For $j = 1$, the lemma claims that

$$\frac{\partial}{\partial x_1} (u(x) \cdot H(x_1)) = \frac{\partial}{\partial x_1} u(x) \cdot H(x_1) + u(0, x') \cdot \delta(x_1)$$

which, once again, is immediately true by the fact that $H'(x_1) = \delta(x_1)$. We now proceed by induction, assuming that (6.2.1) holds for a given value of j. Then

$$\frac{\partial^{j+1}}{\partial x_1^{j+1}} (u(x) \cdot H(x_1)) = \left(\frac{\partial^{j+1}}{\partial x_1^{j+1}} u(x) \right) \cdot H(x_1) + \left(\frac{\partial^j}{\partial x_1^j} u(x) \right) \cdot \delta(x_1) +$$

$$+ \sum_{k=1}^{j} \frac{\partial^{j-k}}{\partial x_1^{j-k}} u(0, x') \delta(x_1) = \left(\frac{\partial^{j+1}}{\partial x_1^{j+1}} u(x) \right) \cdot H(x) + \sum_{k=1}^{j+1} \frac{\partial^{j+1-k}}{\partial x_1^{j+1-k}} u(0, x') \delta^{(k-1)}(x_1),$$

where the last equality is a consequence of the inductive hypothesis. The lemma is therefore proved.   □

Lemma 6.2.2 *Let* $\sigma(P)(x, d\psi(x)) \neq 0$. *Any real analytic function* $u(x)$ *which satisfies*

$$\begin{cases} P(x, D)u(x) = 1 \\ u \equiv 0 \mod (\psi(x))^m \end{cases}$$

also satisfies

$$P(x, D) (u(x) \cdot H(\psi(x))) = H(\psi(x)).$$

Proof. Since $d\psi \neq 0$, we can use a suitable coordinate transformation to obtain, without loss of generality, $\psi = x_1$. We now rewrite $P(x, D)$ as a differential polynomial in $\frac{\partial}{\partial x_1}$, by setting $D' = (\frac{\partial}{\partial x_2}, \ldots, \frac{\partial}{\partial x_n})$,

$$A_j(x, D') = \sum_{\alpha'} a_{\alpha'}(x) D'^{\alpha'}$$

and, finally

$$P(x, D) = \sum_{j=0}^{m} A_j(x, D') \frac{\partial^j}{\partial x_1^j}.$$

Now, since $u \equiv 0 \mod x_1^m$ (hypothesis) and since $j \cdot k < m$, we immediately obtain

$$\frac{\partial^{j-k}}{\partial x_1^{j-k}} u(0, x') = 0;$$

from this last equality we deduce

$$P(x, D) \left(u(x) H(x_1) \right) = \sum_{j=0}^{m} A_j(x, D') \left(\frac{\partial^j}{\partial x_1^j} u(x) \right) H(x_1) = P(x, D) u(x) \cdot H(x_1).$$

Since, by hypothesis, $P(x, D) u(x) = 1$, the lemma is completely proved. □

We recall here (see Chapter II) that every microlocal operator K is essentially defined by integration against a kernel of the form

$$k(x, y) dy$$

for $k(x, x')$ a suitably defined microfunction. In this case, the kernel

$$k(y, x) dy$$

also defines a microlocal operator which is said to be the adjoint (or conjugate) operator of K and is denoted by K^*.

Incidentally, we note that if K is well defined in a neighborhood of $(x_0, i\xi_0\infty)$, then K^* is a microlocal operator well defined in a neighborhood of $(x_0, -i\xi_0\infty)$. From this definition we immediately see that (when all these objects are well defined)

$$(K_1 K_2)^* = K_2^* K_1^*$$

and

$$(K^*)^* = K.$$

For the proof of Sato's theorem we need to compute explicitly the adjoint of a linear partial differential operator.

Lemma 6.2.3 *Let*

$$P(x, D) = \sum_{|\alpha| \leq m} a_\alpha(x) D^\alpha$$

be a linear partial differential operator with real analytic coefficients. Then its adjoint operator is the linear differential operator defined by

$$P^*(x, D)u(x) = \sum_{|\alpha| \leq m} (-1)^{|\alpha|} D^\alpha \left(a_\alpha(x) \cdot u(x) \right).$$

Proof. We will proceed step by step. First we compute the adjoint of the operator induced by multiplication by $a_\alpha(x)$. Since, as a microlocal operator, this multiplication is represented by the microfunction

$$a_\alpha(x)\delta(y - x),$$

we see that $(a_\alpha(x)\cdot)^*$ is represented, by definition, by the kernel

$$a_\alpha(y)\delta(x - y),$$

i.e.

$$(a_\alpha(x)\cdot)^* = (a_\alpha(x)\cdot).$$

We now proceed to compute the adjoint of the differential operator

$$\frac{\partial}{\partial x_j},$$

whose kernel microfunction is

$$\left(\frac{\partial}{\partial x_j} \right) \delta(y - x).$$

Therefore, the kernel which corresponds to $\left(\frac{\partial}{\partial x_j} \right)$ is given, in view of the definition, by

$$\left(\frac{\partial}{\partial y_j} \right) \delta(x - y) = \left(\frac{\partial}{\partial y_j} \right) \delta(x_1 - y_1) \cdot \ldots \cdot \delta(x_n - y_n) =$$

$$= -\left(\frac{\partial}{\partial y_j} \right) \delta(y_1 - x_1) \cdot \ldots \cdot \delta(y_n - x_n),$$

so that we conclude that

$$(D^\alpha)^* = \left(\frac{\partial^{\alpha_1}}{\partial x_1^{\alpha_1}} \cdot \ldots \cdot \frac{\partial^{\alpha^n}}{\partial x_n^{\alpha_n}}\right)^* = (-1)^{|\alpha|} D^\alpha.$$

Finally, by the linearity of conjugation, and again by the properties of the conjugate of composite operators, we obtain that

$$P^*(x, D)u(x) = \sum_{|\alpha| \le m} (a_\alpha(x)D^\alpha)^* u(x) = \sum_{|\alpha| \le m} (D^\alpha)^*(a_\alpha(x)u(x))^* =$$

$$= \sum_{|\alpha| \le m} (-1)^{|\alpha|} D^\alpha(a_\alpha(x)u(x)),$$

which concludes the proof of the Lemma. □

We finally have all the tools to prove Sato's Fundamental Theorem, which we restate here for the sake of readability:

Theorem 6.2.2 *A linear differential operator of finite order $P(x, D)$ is left and right invertible in the ring $\mathcal{L}$ of microlocal operators over $T^* I\!\!R^n \backslash V(P)$ or, in terms of cospherical tangent bundle over*

$$\{(x, \xi_\infty) \in S^* I\!\!R^n : \sigma(P)(x, \xi) \ne 0\}.$$

Proof. The nature of this result is obviously local, and we therefore consider it in a neighborhood of the origin $x = 0$. Let ξ_0 be a point such that

$$\sigma(P)(0, i\xi_0\infty) \ne 0.$$

By the well known Cauchy-Kowalewsky Theorem, see e.g. [123], there exists a real analytic function $u(x, \xi, p)$, defined in a neighborhood of $(0, \xi_0, 0) \in T^* I\!\!R^n \times I\!\!R$, such that

$$P(x, D)u(x, \xi, p) = 1$$

and

$$u \equiv 0 \mod (x \cdot \xi - p)^m;$$

note that we have used here in a crucial way the hypothesis of the non vanishing of $\sigma(P)$. The solution u which we have constructed is a unitary solution (or also a Leray solution). Set now

$$v(x, \xi, p) := u(x, \xi, p) \cdot H(x \cdot \xi - p).$$

By Lemma 6.2.2, whose hypotheses are immediately satisfied, we have that

$$P(x, D)v(x, \xi, p) = H(x \cdot \xi - p).$$

If we take the n-th derivative with respect to p (which is a real variable) we obtain

$$P(x, \xi, p)\left(\frac{\partial^n}{\partial p^n}\right)v(x, \xi, p) = (-1)^n \delta^{(n-1)}(x \cdot \xi - p),$$

i.e.

$$P(x, \xi, p)w(x, \xi, p) = (-1)^n \delta^{(n-1)}(x \cdot \xi - p)$$

where we have set

$$w(x, \xi, p) = \frac{\partial^n}{\partial p^n}(v(x, \xi, p)),$$

and, finally,

$$(6.2.2) \qquad P(x, D)w(x, \xi, y \cdot \xi) = (-1)^n \delta^{(n-1)}((x - y) \cdot \xi),$$

once we have chosen p to be $y \cdot \xi$. In order to obtain a fundamental solution for $P(x, D)$, we now integrate (6.2.2) with respect to ξ, in a neighborhood U of ξ_0. By the plane wave decomposition of the delta microfunction (see Chapter II), we have that, in a neighborhood of $(0, 0; \xi_0(x - y)\infty)$, it is

$$P(x, D)\int_U w(x, \xi, y \cdot \xi)\omega(\xi) = (-1)^n \int_U \delta^{(n-1)}((x - y) \cdot \xi)\omega(\xi)$$

$$= \left(\frac{1}{-2\pi i}\right)^{n-1}\delta(x - y).$$

We now therefore define

$$E(x, y) := \left(\frac{1}{-2\pi i}\right)^{n-1}\int_U w(x, \xi, y \cdot \xi)\omega(\xi),$$

so that, in a neighborhood of $(0, 0; \xi_0(x - y)\infty)$, one has

$$P(x, D)E(x, y) = \delta(x - y).$$

Now (by the definition of the sheaf $\mathcal{L}$) $E(x, y)dy$ will define a microlocal operator in $\mathcal{L}$ if we can show that, in a sufficiently small neighborhood, its singularity spectrum is contained in $T^*(\mathbb{R}^n \times \mathbb{R}^n)$. As a matter of fact, it is indeed immediate to verify that $E(x, y)dy$ would then be, in $\mathcal{L}$, a right inverse of P, since, quite simply,

$$P(x, D)E(x, y)dy = \left(\int P(x, D)\delta(x - y)E(z, y)dz\right)dy =$$

$$= P(x, D)E(x, y)dy = \delta(x - y)dy.$$

As for the singularity spectrum of $E(x, y)dy$, we recall that u was taken as a real analytic function defined in a neighborhood of $(x, \xi, p) = (0, \xi_0, 0)$; as a consequence, $v(x, \xi, p)$ is a hyperfunction defined in that same neighborhood, whose singularity spectrum is contained in the set

$$\{x \cdot \xi = p, \pm(x\xi - p)\infty\}.$$

Thus, by the results in Chapter II on indefinite integrals of hyperfunctions, we obtain the desired estimate on the singularity spectrum of $E(x, y)dy$ which therefore defines a microlocal operator. To conclude the proof, we will show that $E = E(x, y)dy$ is also a left inverse for the operator $P = P(x, D)$. In other words we will show that in addition to $PE = 1$ we also have $EP = 1$. To begin with, we note that if P^* is the adjoint operator of P, then (by Lemma 6.2.3)

$$\sigma(P^*)(x, -i\xi) = \sigma(P)(x, \xi) \neq 0.$$

Therefore, by what we have just proved, there exists a microlocal operator E' such that (in a suitable open set)

$$P^* E' = 1.$$

The adjoint $(E')^*$ of E' is still a microlocal operator and now

$$(E')^* P = (E')^*(P^*)^* = (P^* E')^* = 1,$$

which shows that $(E')^*$ is a left inverse of P. Finally,

$$(E')^* = (E')^* PE = (E'^* P)E = E,$$

which concludes our proof.   □

A simple, but quite fundamental, corollary can be immediately established, thus giving a form of Weyl's lemma (see e.g. [123]) for partial differential equations on hyperfunctions.

Theorem 6.2.3 *Let u and f be hyperfunctions which satisfy*

(6.2.3) $$P(x, D)u(x) = f(x).$$

Then the following inclusion holds:

(6.2.4) $$S.S.(u) \subseteq \{(x, \xi\infty) \in S^* \mathbb{R}^n : \sigma(P)(x, i\xi\infty) = 0\} \cup S.S.(f).$$

In particular, if $P(x, D)$ is elliptic at a point $x_0 \in \mathbb{R}^n$, and if f is real analytic, then also u is real analytic.

Proof. If we consider equation (6.2.3) in the space of hyperfunctions, and we lift it to the level of microfunctions (i.e. we read every hyperfunction as the microfunction it defines in the quotient space) we obtain

$$P(x, D)\mathrm{sp}(u(x)) = \mathrm{sp}(f(x)).$$

From Theorem 6.2.1 we know that at those points (x, ξ) for which $\sigma(P)(x, i\xi\infty) \neq 0$, then there exists a microlocal left inverse E such that $EP = 1$. Therefore

$$\mathrm{sp}(u) = (EP)\mathrm{sp}(u) = E\mathrm{sp}(f),$$

which proves the first assertion, i.e. (6.2.4). As for the second, it follows immediately from (6.2.4).   □

Another more or less immediate consequence, with which we close this section, is the following:

Theorem 6.2.4 *Let $P(x, D)$ be an elliptic operator at a point x_0. Then*

$$P(x, D) : \mathcal{B}_{x_0} \longrightarrow \mathcal{B}_{x_0}$$

is surjective.

Proof. By ellipticity, Sato's Fundamental Theorem immediately implies that $P(x, D)$ has a right inverse at each point $(x_0, \xi\infty)$. This inverse, in particular, is unique since it is also a left inverse. In turn, this implies that there exists a right inverse operator for P, in a full neighborhood of $\pi^{-1}(x_0)$. From Sato's Fundamental Exact Sequence we know the exactness of

$$\mathcal{B} \xrightarrow{\mathrm{sp}} \pi_*\mathcal{C} \longrightarrow 0$$

so that there exists $h \in \mathcal{B}_{x_0}$ such that

$$E\mathrm{sp}(f) = \mathrm{sp}(h),$$

i.e.

$$\mathrm{sp}(Ph - f) = 0$$

in a neighborhood of $\pi^{-1}(x_0)$. By Theorem 6.2.2,

$$g := Ph - f$$

is a real analytic function while, by Cauchy-Kowalewsky, there exists a real analytic function v, again defined in a neighborhood of x_0, such that

$$Pv = g.$$

Our theorem is now proved by taking $u = h - v$, so that, in a neighborhood of x_0,

$$Pu = f.$$

$\square$

6.3   A First Approach to Bicharacteristic Strips

In the previous section we have constructed the inverse of a linear differential operator as an element of $\mathcal{L}$ under the assumption that $\sigma(P)(x, i\xi\infty) \neq 0$. The proof, as we have seen, was based on the construction of a fundamental solution for P. A specific case in which this can be done explicitly is when P is the wave operator, according to the following definition:

Definition 6.3.1 *The wave operator on $\mathbb{R} \times \mathbb{R}^n$ is defined by*

$$Q = \frac{\partial}{\partial t^2} - \left(\frac{\partial}{\partial x_1^2} + \cdots + \frac{\partial}{\partial x_n^2} \right) = \frac{\partial^2}{\partial t^2} - \Delta.$$

We are interested in exploring the Cauchy problem for such an operator, namely the problem of finding a hyperfunction $u(x)$ such that

(6.3.1)
$$\begin{cases} Qu = 0 \\ u(0, x) = \varphi_0(x) \\ \dfrac{\partial u}{\partial t}(0, x) = \varphi_1(x). \end{cases}$$

The reader is of course familiar with the Cauchy-Kowalewsky Theorem which guarantees the existence of a real analytic solution if real analytic initial data are given. The situation for hyperfunctions is quite different and, in general, we do not even know whether a hyperfunction solution to (6.3.1) exists. When it does though, such a solution is unique in view of Holmgren's Theorem. Let us consider the following hyperfunction on $\mathbb{R} \times \mathbb{R}^n$:

$$u_+ = \left(x_1^2 + \cdots + x_n^2 - (t + i0)^2 \right)^{\frac{1-n}{2}}$$

and

$$u_- = \left(x_1^2 + \cdots + x_n^2 - (t - i0)^2 \right)^{\frac{1-n}{2}}.$$

It suffices a simple direct computation to show that

$$Qu_+ = Qu_- = 0.$$

Since one can see that

$$S.S.(u_\pm) \subseteq \left\{ \left(x, t; i(\frac{x_1}{t}, \ldots, \frac{x_n}{t}, \mp 1)\infty \right) : t \ne 0, t^2 = x_1^2 + \cdots + x_n^2 \right\} \cup$$

$$\left\{ (0, 0; i(\xi_1, \ldots, \xi_n, \eta)\infty) : \eta^2 \ge \xi_1^2 + \cdots + \xi_n^2, \pm\eta > 0 \right\},$$

then by Theorem 6.2.1 (i.e. essentially by Sato's Fundamental Theorem), we have that

$$(S.S.(u_\pm)) \cap \{t = x = 0\} \subseteq \left\{ (0, 0; i(\xi_1, \ldots, \xi_n, \eta)\infty) : \eta^2 = \xi_1^2 + \cdots \xi_n^2, \pm\eta > 0 \right\}.$$

The theory of traces of hyperfunctions which we have developed in Chapter II shows that $u_\pm(t, x)$, as well as its derivative $\frac{\partial u_t}{\partial t}$, can be restricted to $t = 0$. Now, if $x \ne 0$, one has that

$$u_+(0, x) - u_-(0, x) = 0$$

and that

$$\frac{\partial u_t}{\partial t}_{|t=0} = 0,$$

so that these three hyperfunctions are defined on $\mathbb{R}^n$, and supported at the origin. But now we may recall that hyperfunctions supported at the origin admit a particularly simple representation as series of derivatives of the δ-function (see Chapter I), i.e., to be precise, if $u \in B_{\{0\}}(\mathbb{R}^n)$, then

(6.3.2)
$$u(x) = \sum_\alpha a_\alpha \frac{\partial^{|\alpha|}\delta}{\partial x^\alpha}(x),$$

where the sum is extended to all non-negative multi-indices α, and where, for every $\varepsilon > 0$, there exists $C_\varepsilon > 0$ such that for all α,

$$|a_\alpha| \le C_\varepsilon \frac{\varepsilon^{|\alpha|}}{|\alpha|!}.$$

Notice that

$$\left(\sum_{i=1}^n x_i \frac{\partial}{\partial x_i} \right) \delta^{(\alpha)}(x) = -(u + |\alpha|)\delta^{(\alpha)}(x),$$

and therefore if u as in (6.3.2) satisfies

(6.3.3)
$$\left(\sum_{i=1}^{n} x_i \frac{\partial}{\partial x_i} - \ll\right) u(x) = 0,$$

then

$$0 = \left(\sum_{i=1}^{n} x_i \frac{\partial}{\partial x_i} - \ll\right) u(x) = \sum_{\alpha} a_\alpha \left(\sum_{i=1}^{n} x_i \frac{\partial}{\partial x_i} - \ll\right) \delta^{(\alpha)}(x)$$

(6.3.4)
$$= -\sum_{\alpha}(u + |\alpha| - \ll) a_\alpha \delta^{(\alpha)}(x).$$

As a consequence we obtain the following lemma.

Lemma 6.3.1 *If $u \in \mathcal{B}_{\{0\}}(\mathbb{R}^n)$ is homogeneous of degree λ, then:*

(i) $u(x) = 0$ if $\lambda \neq -n, -n-1, -n-2, \ldots$

(ii) $u(x) = \sum_{|\alpha|=m} a_\alpha \delta^{(\alpha)}(x)$ if $\lambda = -n - m$, $m \geq 0$.

Proof. Since $u \in \mathcal{B}_{\{0\}}(\mathbb{R}^n)$, (6.3.2) holds true. By the homogeneity of u we have (6.3.3) and the Lemma is therefore an immediate consequence of (6.3.4). $\square$

We now recall that both $u_+(0, x) - u_-(0, x)$ and $\frac{\partial u_\pm}{\partial t}(0, x)$ are supported at the origin. Moreover, by definition, $u_+(0, x) - u_-(0, x)$ is homogeneous of degree $(-n - +1)$ while $\frac{\partial u_\pm}{\partial t}(0, x)$ is homogeneous of degree $-n$. By Lemma 6.3.1 we obtain that

$$u_+(0, x) = u_-(0, x)$$

and

$$\frac{\partial u_\pm}{\partial t}(0, x) = c_\pm \delta(x),$$

where $c_\pm$ is a constant which can be explicitly computed and is given by (see e.g. [103])

$$c_\pm = \pm \frac{2\pi^{\frac{n+1}{2}} i}{\Gamma(\frac{n-1}{2})}.$$

$\square$

We are ready to prove the existence of a hyperfunction solution to the Cauchy problem for the wave equation.

Theorem 6.3.1 *Let φ_0 and φ_1 be arbitrary hyperfunctions on $\mathbb{R}^n$. Then there exists a (unique) hyperfunction $u(x)$ which solves (6.3.1).*

Proof. Uniqueness is a consequence of Holmgren's Theorem, so we will just prove the existence of a solution u. Following the computations of the previous pages we define two new hyperfunctions on $\mathbb{R}^{n+1}$:

$$K_1(t, x) = \frac{1}{2c_\pm}\{u_+(t, x) - u_-(t, x)\}$$

and

$$K_0(t, x) = \frac{\partial}{\partial t} K_1(t, x).$$

To begin with, we immediately notice that

(6.3.5) $$\begin{cases} QK_1(t, x) = 0 \\ K_1(0, x) = 0 \\ \dfrac{\partial K_1}{\partial t}(0, x) = \delta(x) \end{cases}$$

so that K_1 is a hyperfunction solution of the Cauchy problem with data $(0, \delta(x))$. On the other hand, (6.3.5) implies that

$$QK_0 = \frac{\partial}{\partial t}(QK_1) = 0,$$

$$K_0(0, x) = \delta(x)$$

and

$$\frac{\partial K_0}{\partial t}(0, x) = 0.$$

This last equality follows from the fact that

$$\frac{\partial K_0}{\partial t} = \frac{\partial^2 K_1}{\partial t^2} = \Delta K_1$$

and since Δ does not depend on t, we have

$$\frac{\partial K_0}{\partial T}(0, x) = (\Delta K_1)(0, x) = \Delta(K_1(0, x)) = 0.$$

We also note (see Chapter I) that

$$\text{supp}(K_0) \subseteq \text{supp}(K_1) \subseteq \{(t, x) \in \mathbb{R}^{n+1} : |x|^2 \leq t^2\}$$

and we can therefore define the following hyperfunctions:

$$u(t, x) := \int K_0(t, x - y)\varphi_0(y)dy + \int K_1(t, x - y)\varphi_1(y)dy.$$

Now we see that u is the required solution of (6.3.1), since:

$$Qu(t,x) = \int QK_0(t, x-y)\varphi_0(y)dy + \int QK_1(t, x-y)\varphi_1(y)dy = 0,$$

$$u(0,x) = \int \delta(x-y)\varphi_0(y)dy = \varphi_0(x)$$

$$\frac{\partial u}{\partial t}(0,y) = \int \frac{\partial K_0}{\partial t}(0, x-y)\varphi_0(y)dy + \frac{\partial K_1}{\partial t}(0, x-y)\varphi_1(y)dy =$$

$$= \int \delta(x-y)\varphi_1(y)dy = \varphi_1(x).$$

□

The next question we would like to ask about the wave operator is the structure of its microfunction solutions sheaf. We know that this sheaf is trivial outside the characteristic variety

$$V(Q) = \left\{(t, x; i(\tau, \xi)\infty) \in iS^* I\!\!R^{n+1}\infty : \tau^2 = \xi^2\right\},$$

but we would like to know what happens on $V(Q)$. The answer to this question is hidden in a concept which is well known in the theory of differential equations and which shows (for the first time in this book) an interesting interplay between contact geometry and differential equations. The notion we need is that of **bicharacteristic strip**:

Definition 6.3.2 *Let $P(x, D)$ be a linear differential operator and let $p(x, \xi)$ be its principal symbol. Any integral curve $(x(t), \xi(t))$ of the system*

$$\frac{dx_1}{\frac{\partial p}{\partial \xi_1}} = \ldots = \frac{dx_n}{\frac{\partial p}{\partial \xi_n}} = \frac{-d\xi_1}{\frac{\partial p}{\partial x_1}} = \ldots = \frac{-d\xi_n}{\frac{\partial p}{\partial x_n}}$$

which satisfies

$$p(x(t), \xi(t)) = 0$$

is said to be a bicharacteristic strip for the equation $P(x, D)u(x) = 0$. Its first projection $\{x(t)\}$ is called the bicharacteristic curve.

Our next theorem (again a consequence of Sato's Fundamental Principle) will show that the microfunction solutions of the wave equation only propagate along bicharacteristic strips. We need a preliminary computation:

Lemma 6.3.2 *Let $E(t, x) := K_1(t, x) \cdot H(t)$. Then:*

(i) $QE(t, x) = \delta(t, x)$;

(ii) $supp(E) \subseteq \{(t, x) : |x| \leq t\}$;

(iii)

$$S.S.(E) \subseteq \{(0,0;i(\tau,\xi)\infty) : (\tau,\xi) \neq (0,0)\} \cup$$

$$\left\{(t,x;i(\tau,\xi)\infty) : \tau^2 = \xi^2 \neq 0, t \geq 0, x = -\left(\frac{\xi}{\tau}\right)t\right\}.$$

Proof.   We first note that

$$\frac{\partial E}{\partial t}(t,x) = \frac{\partial K_1}{\partial t} \cdot H(t)$$

and therefore

$$\frac{\partial^2 E}{\partial t^2}(t,x) = \frac{\partial^2 K_1}{\partial t^2} \cdot H(t) + \delta(t,x).$$

Since, moreover, H does not depend on x, we have

$$\Delta E(t,x) = \Delta K_1 \cdot H(t),$$

and therefore (i) follows immediately. As for (ii), this is an immediate consequence of the fact that $\mathrm{supp}(K_1) \subseteq \{(t,x) : |x|^2 \leq t^2\}$ and $\mathrm{supp}(H) \subseteq \{(t,x) : t \geq 0\}$. Finally we need to prove (iii). To begin with, by (i) one has that

$$S.S(E) \cap \{t \leq 0\} \subseteq \{t = x = 0\}.$$

Moreover, for $t > 0$, $E(t,x) = K_1(t,x)$ and therefore, by taking the boundary value from Im $t > 0$, one deduces that

$$S.S.(E) \subseteq \left\{(x,t;i(\xi,\tau)\infty) : x^2 - t^2 = 0, (\xi,\tau) = 2(x,t)\right\},$$

and so $x = -\left(\frac{\xi}{\tau}\right)t$. However, for $t > 0$, $QE = 0$, and so, the first corollary to Sato's Fundamental Principle, $\tau^2 = \xi^2 \neq 0$ which yields our result.   □

We are now ready for the propagation of singularities theorem:

Theorem 6.3.2 *Let Ω be an open subset of $S^*\mathbb{R}^{n+1}$. Let u be a microfunction solution of*

$$Qu = 0,$$

and let B be a bicharacteristic strip for the wave equation. Then $\mathrm{supp}(u) \cap B$ is a union of connected subsets of $B \cap \Omega$.

Proof. We will show that if $u(t, x)$ is a microfunction solution of $Qu = 0$, defined in a neighborhood of

$$P_0(0, 0; i(\tau_0, \xi_0)\infty), \quad \tau_0^2 = \xi_0^2,$$

and if $u(t, x)$ vanishes at P_0, then $u(t, x)$ vanishes as well in

$$P_1 = (t, -\left(\frac{t}{\tau_0}\right)\xi_0; i(\tau_0, \xi_0)).$$

But to do so it will suffice to prove that

$$\tilde{u} := u \cdot H(t) = 0.$$

To show the vanishing of $\tilde{u}$ we note that

$$Q\tilde{u} = 0$$

in a neighborhood of

$$\{(t, x; i(\tau, \xi)\infty) : \tau = \tau_0, \xi = \xi_0\},$$

and therefore by (i) of Lemma 6.3.2 we get

$$\tilde{u}(t, x) = \int \delta(t - s, x - y)\tilde{u}(s, y)dsdy =$$

$$= \int [QE(t - s, x - y)]\,\tilde{u}(s, y)dsdy.$$

The result now follows via integration by parts. □

As we shall see shortly, this relevant role of bicharacteristic strips is much more general than what the content of Theorem 6.3.2 might lead to think. Before we get to its most interesting generalizations we may just mention that a result along these same lines holds for the so-called regularly hyperbolic operators (see [123], [206] for details).

6.4 Contact Transformations

As we have shown in section 6.3, a deep interplay links contact geometry with differential equations. This interplay can be exploited to obtain extremely powerful results. In this section we will review some basic notions from contact geometry, we will define the notion of contact transformations, and we will finally microlocalize this notion. The tools we will construct in this section are central to the classification results which will end our book.

For the sake of simplicity, we will work on a complex analytic manifold X but if X were taken to be real, one could repeat similar constructions with obvious modifications.

Let then L be a one-dimensional subbundle of the cotangent bundle T^*X, and let L^* denote its dual bundle and $L^\perp$ its orthogonal complement, i.e. the kernel of the map

$$TX \to$$

One can define a multilinear homomorphism of vector bundles

$$L^\perp \times L^\perp \times L \to \mathbb{C} \times X$$

by

$$(v_1, v_2, d\omega) \to \langle d\omega, v_1 \wedge v_2 \rangle,$$

and this provides an alternating bilinear homomorphism

(6.4.1) $$L^\perp \times L^\perp \to L^{\otimes -1}.$$

Definition 6.4.1 *We say that (X, L) is a contact manifold if the map (6.4.1) is non-degenerate.*

Remark 6.4.1 Definition 6.4.1 is equivalent to require that dim $X = 2n - 1$ (in particular, odd) and that for a nowhere vanishing section ω of L, the product

(6.4.2) $$\omega \wedge (d\omega)^{n-1}$$

never vanishes. Note that this last condition is independent of the choice of ω, which will be called a **fundamental 1-form** (or also a **canonical 1-form**) for X; under this definition we often write (X, ω_X) rather than (X, L).

There is of course a strict relationship between symplectic and contact geometry. In particular, if we write $\hat{X} = L^*\backslash X$, then for s a cross-section of $\hat{X}$, we define a 1-form θ on $\hat{X}$ by setting

$$s^*(\theta) = \omega.$$

Then $(d\theta)^n$ never vanishes and $\hat{X}$ is called the **symplectic manifold associated** to (X, L), with **canonical 1-form** θ.

Historically, the first examples of contact manifolds were born out of classical mechanics and are given by

$$X = \mathbb{P}^*Y$$

the projective cotangent bundle of an n-dimensional manifold Y; in this case $\hat{X} = T^*Y \backslash Y$. In this special case, Darboux theorem states that for a local coordinate system

$$(x_1, \ldots, x_n, p_1, \ldots, p_{n-1}),$$

a canonical 1-form can be written as

$$\omega = dx_n - (p_1 dx_1 + \cdots + p_{n-1} dx_{n-1}).$$

In this case, the associated symplectic manifold has a local coordinate system

$$(x_1, \ldots, x_n, \eta_1, \ldots, \eta_n)$$

with $p_j = -\frac{\eta_i}{\eta_n}$ and the symplectic structure is given by

$$\theta = \eta_1 dx_1 + \cdots + \eta_n dx_n = \eta_n \omega.$$

This example is not only historically relevant, but theoretically important, since it is well known that every contact manifold is locally isomorphic to a projective cotangent bundle.

Definition 6.4.2 *If (X, ω_X) and (Y, ω_Y) are two contact manifolds of the same dimension, we say that a map*

$$f : X \to Y$$

is a **contact transformation** *if $f^* \omega_Y$ is a fundamental 1-form for X.*

Many of the fundamental notions in classical mechanics are really notions from contact geometry. Let us quickly review them:

Definition 6.4.3 *Let f, g be functions on a symplectic manifold $\hat{X}$, $\dim \hat{X} = 2n$. Their* **Poisson bracket** *is defined by*

$$\{f, g\}(d\theta)^n := n df \wedge dg \wedge (d\theta)^{n-1}.$$

Remark 6.4.2 If $(x_1, \ldots, x_n, \eta_1, \ldots, \eta_n)$ are canonical coordinates on $\hat{X}$, then

$$\{f, g\} = \sum_{j=1}^{n} \left(\frac{\partial f}{\partial \eta_j} \frac{\partial g}{\partial x_j} - \frac{\partial f}{\partial x_j} \frac{\partial g}{\partial \eta_j} \right).$$

Remark 6.4.3 The Poisson bracket is sometimes referred to as the **Lagrangian bracket**.

Remark 6.4.4 To see why bicharacteristic strips are really concepts in contact geometry, just note that they are integral curves of the so called **Hamiltonian vector field**

$$H_f := \sum_{j=1}^{n} \left(\frac{\partial f}{\partial \eta_j} \frac{\partial}{\partial x_j} - \frac{\partial f}{\partial x_j} \frac{\partial}{\partial \eta_j} \right).$$

Bicharacteristic strips arise therefore when considering the integral curves of $H_{\sigma(P)}$.

Remark 6.4.5 By the definitions above it follows immediately that

$$H_{\{f,g\}} = [H_f, H_g]$$

where $[A, B] := AB - BA$ is the so called commutator or Lie bracket of A and B.

We have defined above the notion of contact transformation in a rather abstract way. There is, however, a canonical way to construct contact transformations, which seems to have been first utilized by Maslov [163], Egorov [45] and Hörmander [89].

Let X, Y be two complex manifolds of dimension n, and let Λ be the non-singular hypersurface of $X \times Y$ defined by $\Omega(x, y) = 0$, for Ω some holomorphic function on $X \times Y$.

Assume furthermore that, on Λ, it is

$$\det \begin{bmatrix} 0 & d_y \Omega \\ d_x \Omega & d_x d_y \Omega \end{bmatrix} \neq 0$$

(where the matrix is an $(n+1) \times (n+1)$ matrix).

We want to show how to use Ω to construct a contact transformation from $I\!P^* X$ to $I\!P^* Y$. To do so, one first consider

$$I\!P_\Lambda^*(X \times Y) := \{ (x, y; \xi, \eta) \in I\!P^*(X \times Y) : \Omega(x, y) = 0 \text{ and }$$

$$(\xi, \eta) = c \cdot \mathrm{grad}\Omega(x, y) \text{ for some } c \neq 0 \}.$$

Then, by the implicit function theorem, we know that the first projection

$$\pi_1 : I\!P_\Lambda^*(X \times Y) \to I\!P^* X$$

is a local isomorphism, and so is

$$\pi_2 : I\!P_\Lambda^*(X \times Y) \to I\!P^* Y.$$

Finally, π_1 and π_2 induce local isomorphisms

$$\pi_1 \circ \pi_2^{-1} : I\!P^* Y \to I\!P^* X$$

and

$$\pi_2 \circ \pi_1^{-1} : I\!\!P^* X \to I\!\!P^* Y$$

which are clearly contact transformations which will be said to have Ω as a **generating function**.

Remark 6.4.6 It can be shown that every contact transformation can be expressed by composition of two contact transformations with generating functions.

Remark 6.4.7 The most classical contact transformation with generating function is the so called Legendre transformation. The correspondence between (x, ξ) and (y, η) is given by

$$\begin{cases} x_j = -\dfrac{\eta_j}{\eta_n} & j < n \\[2mm] x_n = \dfrac{y \cdot \eta}{\eta_n} \\[2mm] \xi_j = y_j \eta_n & j < n \\[2mm] \xi_n = \eta_n \end{cases}$$

and

$$\begin{cases} y_j = -\dfrac{\xi_j}{\xi_n} & j < n \\[2mm] y_n = \dfrac{x \cdot \xi}{\xi_n} \\[2mm] \eta_j = -x_j \xi_n & j < n \\[2mm] \eta_n = \xi_n \end{cases}$$

and the generating function is

$$\Omega(x, y) = x_n - y_n + \sum_{j=1}^{n-1} x_j y_j.$$

Our next step will be to show that these contact transformations can be lifted from the manifolds on which they act, to the sheaves of differential (and microdifferential) operators on such manifolds. The lifting are the so called **quantized contact transformations**. The process of quantization can of course be done for any contact transformation but, in view of Remark 6.4.6, we will state our next result for the case of contact transformations with a generating function. We omit its proof which can be found in [185] and in [206].

Theorem 6.4.1 *Let X, Y be real analytic manifolds of dimension n, and let Ω satisfy the conditions stated before:*

$$\Omega \text{ is non singular;}$$

$$\det \begin{bmatrix} 0 & d_y\Omega \\ d_x\Omega & d_xd_y\Omega \end{bmatrix} \neq 0 \text{ on } \Lambda.$$

Then, for any microdifferential operator $P(x, D_x)$, there exists a microdifferential operator $Q(y, D_y)$ such that, for every microfunction u:

$$(6.4.3) \quad \int P(x, D_x)\delta(\Omega(x, y))u(y)dy = \int \delta(\Omega(x, y))Q(y, D_y)u(y)dy.$$

Moreover, (6.4.3) induces isomorphisms of sheaves

$$p^{-1}\mathcal{E}_X \cong q^{-1}\mathcal{E}_Y$$

(p being the map which, given Q, produces P and q being the map which, given P, produces Q, according to (6.4.3)),

$$p^{-1}\mathcal{E}_X(m) \cong q^{-1}\mathcal{E}_Y(m)$$

and finally

$$p^{-1}\mathcal{A}_{S \cdot X} \cong q^{-1}\mathcal{A}_{S \cdot Y}.$$

Definition 6.4.4 *The isomorphism p, q in Theorem 6.4.1 are called* **quantized contact transformations.**

Remark 6.4.8 If we apply this process to the Legendre transform, we obtain that the maps which relate $P(x, D_x)$ to $Q(y, D_y)$ are given by

$$\begin{cases} x_j = -\dfrac{\partial}{\partial y_j}\left(\dfrac{\partial}{\partial y_n}\right)^{-1} & j < n \\[2mm] x_n = y \cdot D_y \left(\dfrac{\partial}{\partial y_n}\right)^{-1} \\[2mm] \dfrac{\partial}{\partial x_j} = y_j \dfrac{\partial}{\partial y_n} & j < n \\[2mm] \dfrac{\partial}{\partial x_n} = \dfrac{\partial}{\partial y_n} \end{cases}$$

and its inverse

$$
\begin{cases}
y_j = \dfrac{\partial}{\partial x_j}\left(\dfrac{\partial}{\partial x_n}\right)^{-1} & j < n \\[2ex]
y_n = x \cdot D_x \left(\dfrac{\partial}{\partial x_n}\right)^{-1} \\[2ex]
\dfrac{\partial}{\partial y_j} = -x_j \dfrac{\partial}{\partial x_n} & j < n \\[2ex]
\dfrac{\partial}{\partial y_n} = \dfrac{\partial}{\partial x_n}
\end{cases}
$$

The reader should note here that since

$$
\Omega = x_n - y_n + \sum_{j=1}^{n-1} x_j y_j,
$$

the operators $\left(\frac{\partial}{\partial y_n}\right)^{-1}$ and $\left(\frac{\partial}{\partial x_n}\right)^{-1}$ are well defined microdifferential operators for $\eta_n \neq 0$ and $\xi_n \neq 0$.

One more important result is the fact that quantized contact transformations are isomorphisms at the level of microfunctions.

Theorem 6.4.2 *Assume the same hypotheses as in Theorem 6.4.1. Then the map*

$$
u(y) \rightarrow \int \delta(\Omega(x,y)) u(y) dy
$$

is an isomorphism between C_X and C_Y.

Proof. Consider

$$
K(z,y) := \int \delta(\Omega(x,z)) \delta(\Omega(x,y)) dx.
$$

It will be sufficient to show that K is a kernel for an invertible microdifferential operator on $S^* X$. This follows immediately from Sato's Fundamental Theorem as given in Chapter V (see [185], [206] for details). □

6.5 Structure of Systems of Differential Equations

In this last section we want to prove the fundamental theorem which classifies the structure of systems of microdifferential equations at generic points. This result is similar in spirit to all classification results from algebraic geometry as it will show how to reduce quite general systems to a small number of canonical forms.

It is in this context that we will be able to see the great power of the theory of microfunctions versus the theory of distributions or even microdistributions.

To begin with, we will deal with the simple case of systems of microdifferential equations with simple characteristics and one unknown: we will show how to microlocally reduce it to the so called partial de Rham system. Let us first recall a few very classical facts and definitions.

Definition 6.5.1 *An analytic subset Λ of a contact manifold X is said to be* **involutory** *if whenever $f_{|\Lambda} = g_{|\Lambda} = 0$, then $\{f, g\}_{|\Lambda} = 0$.*

The notion of involutory manifold is particularly relevant as Sato has proved that the characteristic variety of an arbitrary system of microdifferential equations is always involutory. Also note that the property of being involutory necessarily implies that if $\dim X = 2n - 1$, then $\dim \Lambda \geq n - 1$.

Definition 6.5.2 *An involutory submanifold λ of (X, ω) is* **regular** *if ω never vanishes on λ.*

On the other extreme, we say that an involutory submanifold λ of (X, ω) is **Lagrangian** if $\dim \lambda = n - 1$ (and in particular this implies that $\omega_{|\lambda} \equiv 0$).

Remark 6.5.1 Lagrangian manifolds are particularly important as they are the characteristic varieties of a fundamental class of systems, i.e. the **holonomic** or **maximally overdetermined** systems (see section 5.3).

Let us now describe the general strategy which we will use to deal with the structure (or classification) theorems for systems of microdifferential equations. Suppose (we use here the notations introduced in the previous Chapters) that we start with a system $\mathcal{M}$ of microdifferential equations on $\mathbb{R}^n$ of finite order and one unknown function. The first, classical, step consists in applying some contact transformation to $\mathbb{P}^*\mathbb{R}^n$ in such a way as to make the characteristic variety of $\mathcal{M}$ linear. This is nothing but the following famous Jacobi integration result:

Theorem 6.5.1 *Any involutory regular codimension d submanifold of a contact manifold can be expressed as*

$$\{(x, \xi) : \xi_1 = \ldots = \xi_d = 0\},$$

for some canonical coordinate system (x, ξ).

One usually says that Theorem 6.5.1 deals with the **geometrical optics**. The second step consists in applying an invertible quantized contact transformation to $\mathcal{M}$, so to transform it into a system $\mathcal{M}'$ with linear characteristic variety. Finally, one looks for suitable invertible microdifferential operators to transform $\mathcal{M}'$ into the partial de Rham system, say $\mathcal{N}$. This last part is usually referred to as the treatment of the **wave optics**.

As we already mentioned, the treatment of the systems is quite algebraic and it relies on a formal calculus which can be developed for microdifferential operators of finite order, in analogy with what is known for functions of several complex variables. We will need, in particular, the following version of the Weierstrass' division theorem, whose proof can be found in:

Theorem 6.5.2 *Let $P(x, D)$ be a microdifferential operator of order m such that, for some positive integer p,*

$$\frac{\sigma(P)(x, \xi)}{\xi_n^p}\Big|_{(0; i(1,0,\ldots,0,\xi_n)\infty)}$$

is a holomorphic function of ξ_n which never vanishes in some neighborhood of $\xi_n = 0$. Then, in a neighborhood U of $(0; i(1, 0, \ldots, 0)\infty)$, $P(x, D)$ can be uniquely decomposed as

$$P(x, D) = Q(x, D) \cdot R(x, D),$$

where Q is invertible in U and

$$R(x, D) = \frac{\partial^p}{\partial x_h^p} + \sum_{j=0}^{p-1} R_j(x, D')\frac{\partial^j}{\partial x_h^j},$$

with $D' = \left(\frac{\partial}{\partial x_1}, \ldots, \frac{\partial}{\partial x_{n-1}}\right)$, $\operatorname{ord}(R_j) \leq p - j$ and $\sigma(R_j)(0; i(1, 0, \ldots, 0)\infty) = 0$.

Our first result deals with the case of simple characteristics, and if we are looking at the case with only one unknown, we can use Theorem 6.5.1 to note that, without loss of generalities, on may assume $\sigma(P) = \xi_1$.

Theorem 6.5.3 *Let $P(x, D)$ be a microdifferential operator of the first order defined in a neighborhood of $(0; i(0, \ldots, 1)\infty)$, then the equation $P(x, D)u = 0$ is microlocally equivalent as a left $\mathcal{E}$-module, to the partial de Rham equation*

$$\frac{\partial u}{\partial x_1} = 0,$$

i.e., for $\mathcal{E}$ the sheaf of microdifferential operators,

(6.5.1)
$$\frac{\mathcal{E}}{\mathcal{E}P} \cong \frac{\mathcal{E}}{\mathcal{E}\frac{\partial}{\partial x_1}}.$$

Proof. Theorem 6.5.2 allows us to write

$$P = Q\left(\frac{\partial}{\partial x_1} - A(x, D')\right),$$

with $D' = (D_2, \dots, D_n)$ and Q invertible in U. It is therefore sufficient to prove the theorem for $\frac{\partial}{\partial x_1} - A(x, D')$. To conclude (6.5.1) we need to construct an invertible microdifferential operator $R(x, D')$ such that

(6.5.2)
$$\begin{cases} \dfrac{\partial}{\partial x_1} = R^{-1}PR \\ R = \displaystyle\sum_{k=0}^{+\infty} R_k(x, D') \end{cases}$$

with

(6.5.3)
$$R_0 = 1$$

and

(6.5.4)
$$\begin{cases} \dfrac{\partial R_k}{\partial x_1} = A \cdot R_{k-1} \\ R_{k|x_1=0} = 0 \quad \text{for} \quad k \geq 1. \end{cases}$$

But (6.5.4) can actually be considered as a set of defining equations for the R_k, and in fact one can use the well known lemma of Boutet de Monvel and Kree (see [185] as well as Chapter V) to show that the operator R defined by (6.5.3), (6.5.4) and the second formula in (6.5.2) is a well defined microdifferential operator (see [185], [206] for more details). On the other hand, (6.5.3) and the second formula in (6.5.4) imply that $\sigma(R)_{|x_1=0} = 1 \neq 0$ (note, in particular, that ord $(R_k) \leq 0$ for any k so is ord (R)). By Sato's Fundamental Theorem (in the form which was proved in Chapter V), R is therefore invertible. However, by (6.5.3) and (6.5.4) it also follows that

$$\left(\frac{\partial}{\partial x_1}_A\right) \circ R = R \circ \frac{\partial}{\partial x_1}$$

which concludes our proof.   □

Our next step will consist in showing how the general simple characteristic works (i.e. what happens when we still have a non-singular characteristic variety, whose codimension is, however, bigger than one). Before doing so, however, we want to briefly discuss what happens when the simplicity assumption is removed. As a matter of fact, such an assumption is quite restrictive, and it would be quite natural to try to remove it (or at least weaken it). In fact, as we will show in a second, this problem was at the heart of the motivations which stimulated

Sato and his coworkers to introduce infinite order differential operators. The next Example, Proposition and Theorem are taken from [206] and illustrate this argument.

Example 6.5.1 Consider the following linear (constant coefficients!) differential equation

$$(6.5.5) \qquad\qquad P_1(D)u := \frac{\partial^2 u}{\partial x_1^2} = 0$$

and

$$(6.5.6) \qquad\qquad P_2(D)v := \frac{\partial^2 v}{\partial x_2^2} - \frac{\partial v}{\partial x_2} = 0,$$

in the space of **distributions**. Even though the principal symbols (and therefore the characteristic varieties) coincide even up to multiplicities, the structure of their solutions are quite different. So, an immediate generalization of Theorem 6.5.3 will have to account for this phenomenon. In stark contrast with this example we have the following result.

Proposition 6.5.1 *Equations (6.5.5) and (6.5.6) are microlocally equivalent as left $\mathcal{E}^\infty$-modules, i.e.*

$$\frac{\mathcal{E}^\infty}{\mathcal{E}^\infty P_1} \cong \frac{\mathcal{E}^\infty}{\mathcal{E}^\infty P_2}.$$

In particular, the sheaves $\mathcal{B}^{P_1}$ and $\mathcal{B}^{P_2}$ of hyperfunction solutions to P_1 and P_2 are isomorphic (and so are the sheaves $\mathcal{C}^{P_1}$ and $\mathcal{C}^{P_2}$ of microfunction solutions).

Proof. To show such a result, it suffices to construct linear differential operators $A_1(x, D)$, $A_2(x, D)$, $A_3(x, D)$ and $A_4(x, D)$ such that

$$(6.5.7) \qquad\qquad \begin{cases} P_2 A_1 = A_2 P_1 \\ A_3 P_2 = P_1 A_4 \end{cases}$$

and

$$(6.5.8) \qquad\qquad \begin{cases} A_4 A_1 \equiv 1 \mod \mathcal{E}^\infty P_1 \\ A_1 A_4 \equiv 1 \mod \mathcal{E}^\infty P_2. \end{cases}$$

In fact, if (6.5.7) and (6.5.8) hold, then the equations $P_1(D)u = 0$ and $P_2(D)v = 0$ are equivalent by the correspondence

$$u = A_4 v \quad \text{and} \quad v = A_1 u.$$

To prove that (6.5.7) and (6.5.8) hold, it suffices to consider the following infinite order differential operators:

$$A_1(x, D) := \left(\cosh\left(x_1\sqrt{\frac{\partial}{\partial x_2}}\right)\right)\left(1 - x_1\frac{\partial}{\partial x_1}\right) + \frac{\sinh\left(x_1\sqrt{\frac{\partial}{\partial x_2}}\right)}{\sqrt{\frac{\partial}{\partial x_2}}}\frac{\partial}{\partial x_1};$$

$A_2(x, D)$:

$$= \left(\cosh\left(x_1\sqrt{\frac{\partial}{\partial x_2}}\right)\right)\left(1 - x_1\frac{\partial}{\partial x_1}\right) + \frac{\sinh\left(x_1\sqrt{\frac{\partial}{\partial x_2}}\right)}{\sqrt{\frac{\partial}{\partial x_2}}}\left(\frac{\partial}{\partial x_1} - 2x_1\frac{\partial}{\partial x_2}\right);$$

$A_3(x, D)$:

$$= \left(\cosh\left(x_1\sqrt{\frac{\partial}{\partial x_2}}\right)\right)\left(1 + x_1\frac{\partial}{\partial x_1}\right) + \frac{\sinh\left(x_1\sqrt{\frac{\partial}{\partial x_2}}\right)}{\sqrt{\frac{\partial}{\partial x_2}}}\left(-\frac{\partial}{\partial x_1} + x_1\frac{\partial}{\partial x_2}\right);$$

$A_4(x, D)$:

$$= \left(\cosh\left(x_1\sqrt{\frac{\partial}{\partial x_2}}\right)\right)\left(1 + x_1\frac{\partial}{\partial x_1}\right) + \frac{\sinh\left(x_1\sqrt{\frac{\partial}{\partial x_2}}\right)}{\sqrt{\frac{\partial}{\partial x_2}}}\left(-\frac{\partial}{\partial x_1} - x_1\frac{\partial}{\partial x_2}\right).$$

The reader should note that all of these operators are well defined infinite order differential operators (with variable coefficients), in view of the growth conditions which we have described in Chapter I; also, the square roots are purely formal, in view of the Taylor expansion of $\cosh t$ and $\sinh t$. The proof can now be concluded by a direct computation.   □

Remark 6.5.2 The proof of Proposition 6.5.1 shows the reason why (6.5.5) and (6.5.6) are not equivalent in the framework of distributions; the reason is that the operators $A_j(x, D)$, $j = 1, \ldots, 4$, which realize the isomorphism are of finite order. On the other hand, in contrast with the general situation which we will outline in our next result, equations (6.5.5) and (6.5.6) are equivalent even at the level of hyperfunctions (and not just microfunctions) because the operators A_j are differential operators (though of infinite order) rather than more general microdifferential operators.

Theorem 6.5.4 *Let $P(x, D)$ be a microdifferential operator of order m defined in a neighborhood of $(0; i(0, \ldots, 1)\infty)$ and such that $\sigma(P) = \xi_1^m$. Then, in a neighborhood of $(0; i(0, \ldots, 1)\infty)$, the equation $P(x, D)u = 0$ is microlocally equivalent, as a left $\mathcal{E}$-module, to the differential equation*

$$\frac{\partial^m u}{\partial x_1^m} = 0.$$

If, moreover, $P(x, D)$ is actually a differential equation, then the equivalence holds as left $\mathcal{D}$-modules, $\mathcal{D}$ being the sheaf of linear differential operators of finite order. In the first case, only microfunction solutions are equivalent, while in the second case, also hyperfunction solutions are equivalent.

Proof. The proof follows exactly the same lines as for the Theorem 6.5.3, but the computations are necessarily more complex since it becomes necessary to involve infinite order differential operators. $\qquad\square$

We are now ready to go back to the case of simple characteristic to prove the analogue of Theorem 6.5.3 for higher codimension.

Theorem 6.5.5 *Let $\mathcal{M}$ be an $\mathcal{E}$-module defined in a neighborhood of $(x_0, i\xi_0\infty)$ be such that:*

(6.5.9) there is a left ideal $\mathcal{I}$ such that $\mathcal{M} = \mathcal{M}/\mathcal{I}$ (i.e. we are looking at a system with one unknown function);

(6.5.10) let $J := \cup_m\{\sigma_m(P) : P \in \mathcal{I} \cap \mathcal{E}(m)\}$ and let $V(J)$ be its zero set: then $V(J)$ is a non-singular codimension d manifold in a neighborhood of $(x_0, i\xi_0\infty)$ and $\omega_{|V(J)} \neq 0$ (for $\omega = \sum \xi_j dx_j$ the canonical form);

(6.5.11) $V(J)$ is real;

(6.5.12) the totality of ξ-homogeneous analytic functions which vanish on $V(J)$ is J.

Then, via a quantized contact transformation, we can transform $\mathcal{M}$ into the system

$$\mathcal{N} := \frac{\mathcal{E}}{\mathcal{E}\frac{\partial}{\partial x_1} + \cdots + \mathcal{E}\frac{\partial}{\partial x_n}}.$$

Proof. By the comment we made after Definition 6.5.1 we know that $V(J)$ is involutory and therefore (6.5.10) and (6.5.11) imply that, by Theorem 6.5.1, $V(J)$ can be written as

$$V(J) = \{(x, \xi) : \xi_1 = \ldots = \xi_d = 0\},$$

and by (6.5.12) we can choose $P_1, \ldots, P_d \in \mathcal{I}$ such that $\sigma(P_j) = \xi_j$ for $j = 1, \ldots, d$. In particular, by just following the proof of Theorem 6.5.3, we can assume P_1 to be $\frac{\partial}{\partial x_1}$. We now show, by induction, that we can choose P_j to be $\frac{\partial}{\partial x_j}$ for $j = 1, \ldots, d$. Assume that, for some $k < d$, we have

$$P_1 = \frac{\partial}{\partial x_1}, \quad P_2 = \frac{\partial}{\partial x_2}, \ldots, P_k = \frac{\partial}{\partial x_k}.$$

Now, since

$$\frac{\partial u}{\partial x_j} = 0 \text{ for } j = 1, \ldots, k$$

we may assume

$$P_j = P_j(x, \frac{\partial}{\partial x_{k+1}}, \ldots, \frac{\partial}{\partial x_n}) \text{ for } j = k+1, \ldots, n.$$

Weierstrass' division theorem (in its version for microdifferential operators) shows that there exists a microdifferential operator $Q_{k+1}(x, \frac{\partial}{\partial x_{k+1}}, \ldots, \frac{\partial}{\partial x_n})$ of order at most zero, and an invertible microdifferential operator $R(x, \frac{\partial}{\partial x_{k+1}}, \ldots, \frac{\partial}{\partial x_n})$ such that

$$(6.5.9) \qquad P_{k+1}(x, \frac{\partial}{\partial x_j}k+1, \ldots, \frac{\partial}{\partial x_j}n) = R(\frac{\partial}{\partial x_{k+1}}) + Q_{k+1}.$$

From (6.5.13), and using the Späth theorem in its natural version for microdifferential operators (see [123]), we can actually deduce that the generators $P_1, \ldots, P_{k+1}$ can be chosen so that

$$(6.5.10) \qquad \begin{cases} P_j(x, D) = \frac{\partial}{\partial x_j} & j = 1, \ldots, k \\ P_{k+1}(x, D) = R\left(\frac{\partial}{\partial x_{k+1}} + Q_{k+1}(x, \frac{\partial}{\partial x_{k+1}}, \ldots, \frac{\partial}{\partial x_n})\right). \end{cases}$$

Note that, for $j = 1, \ldots, k$, the Lie bracket

$$\left[\frac{\partial}{\partial x_j}, P_{k+1}\right] := \frac{\partial}{\partial x_j} P_{k+1} - P_{k+1} \frac{\partial}{\partial x_j}$$

belongs to $\mathcal{I}$ and so, by (6.5.14), also $[\frac{\partial}{\partial x_j}, Q_{k+1}] \in \mathcal{I}$ for $j = 1, \ldots, k$. Moreover, again by (6.5.14), one can even show that $[\frac{\partial}{\partial x_j}, Q_{k+1}] = 0$ for $j = 1, \ldots, k$, i.e. Q_{k+1} only depends on $(x_{k+1}, \ldots, x_n, \frac{\partial}{\partial x_{k+1}}, \ldots, \frac{\partial}{\partial x_n})$. Following the same steps as in Theorem 6.5.3 we can find an invertible microdifferential operator $R(x_{k+1}, \ldots, x_n, \frac{\partial}{\partial x_{k+1}}, \ldots, \frac{\partial}{\partial x_n})$ such that

$$R^{-1}P_{k+1} = \frac{\partial}{\partial x_k} R^{-1}.$$

Since R commutes with $\frac{\partial}{\partial x_j}$ for $j = 1, \ldots, k$, if we replace a generator u of $\mathcal{M}$ by Ru, we can now choose generators $P_1, \ldots, P_d$ for $\mathcal{I}$ such that

$$P_1 = \frac{\partial}{\partial x_1}, \quad P_2 = \frac{\partial}{\partial x_2}, \ldots, P_{k+1} = \frac{\partial}{\partial x_{k+1}},$$

so that we have shown, inductively, that $P_1 = \frac{\partial}{\partial x_1}, \ldots, P_d = \frac{\partial}{\partial x_d}$. We now proceed to prove that $\frac{\partial}{\partial x_1}, \ldots, \frac{\partial}{\partial x_d}$ generate $\mathcal{I}$, which would conclude our proof. If this were not so, we could find $R(x, \frac{\partial}{\partial x_{d+1}}, \ldots, \frac{\partial}{\partial x_n}) \in \mathcal{I}$, $R \neq 0$. Then, it would be $\sigma(R)_{|V(J)} = 0$, but since $\sigma(R)$ does not depend on $\xi_1, \ldots, \xi_d$, this would imply $\sigma(R) \equiv 0$, which would contradict the fact that $R \not\equiv 0$. $\qquad\square$

This last result can be extended to the case of non-simple characteristic by allowing the use of infinite order differential operators exactly in the same way which Theorem 6.5.4 extended the result of Theorem 6.5.3. Since the treatment of higher codimension non-simple characteristics is a little more complicated, we will not give the precise statement here, but refer the reader to [206], Theorem 5.3.7.

We can now use Theorem 6.5.5 to emphasize the role of bicharacteristics in the study of some systems of differential equations.

Definition 6.5.3 *Let V be an involutory submanifold of $S^*\mathbb{R}^n$ satisfying (6.5.10) and (6.5.11), and suppose*

$$V = \{(x, i\xi\infty) \in S^*M : f_1(x, \xi) = \mathcal{D}d = f_d(x, \xi) = 0\}.$$

The **bicharacteristic manifold** $B = B_{(x_0, i\xi_0\infty)}$ *associated to V and passing* **through** $(x_0, i\xi_0\infty) \in V$ *is the dimension d integral manifold through $(x_0, i\xi_0\infty)$ of the d Hamiltonian operators*

$$H_j := \sum_{t=1}^{n} \left(\frac{\partial f_j}{\partial \xi_t} \frac{\partial}{\partial x_t} - \frac{\partial f_j}{\partial x_t} \frac{\partial}{\partial \xi_t} \right), \quad j = 1, \ldots, d.$$

The next result shows the great power of our previous analysis of systems of microdifferential equations.

Theorem 6.5.6 *Let $\mathcal{M}$ be an $\mathcal{E}$-module as in Theorem 6.5.5. Then (in a neighborhood of $(x_0, i\xi_0\infty)$) the microfunction solution sheaf $Hom_\mathcal{E}(\mathcal{M}, \mathcal{C})$ is supported in V and is locally constant along each bicharacteristic manifold. Moreover, $Hom_\mathcal{E}(\mathcal{M}, \mathcal{C})$ is a flabby sheaf in the direction transversal to bicharacteristic manifolds, and, for $j \neq 0$, $Ext_\mathcal{E}^j(\mathcal{M}, \mathcal{C}) = 0$.*

Proof. The nature of this statement is clearly invariant under quantized contact transformations and so, in view of Theorem 6.5.5, we may assume that $\mathcal{M}$ is the partial de Rham system

$$\frac{\partial u}{\partial x_1} = \ldots = \frac{\partial u}{\partial x_d} = 0,$$

for which the result is clearly true.   □

Remark 6.5.3 When we say that a sheaf S is flabby in the transversal direction to the bicharacteristic manifolds, we mean that there is a manifold U_0, a flabby sheaf $\tilde{S}$ on U_0 and a smooth morphism $\varphi : U \cap V \to U_0$ such that the bicharacteristic manifolds in $U \cap V$ are the fibers of φ and $S_{|U\cap V} \cong \varphi^{-1}\tilde{S}$.

Remark 6.5.4 A completely analogous result holds for the case of non-simple, codimension d, characteristics. The reader may see Theorem 2.1.8 in [206].

Remark 6.5.5 Theorem 6.5.6 really shows two fundamental features. On one hand the flabbiness of the sheaf of microfunction solutions is nothing but an elegant restatement of the propagation of singularities along bicharacteristic manifolds. On the other hand, the vanishing of the higher Ext functors shows, in particular, that the system $\mathcal{M}$ is locally solvable, given suitable algebraic compatibility conditions.

Not all systems of course, have characteristic varieties which satisfy the conditions of Theorem 6.5.5. There are at least (and, in a sense, only) two more important cases which we need to discuss (but we refer the reader to [185], [206] for more details and proofs).

Theorem 6.5.7 *Let $\mathcal{M} = \mathcal{E}/\mathcal{I}$ be a system of microdifferential equations in one unknown with simple characteristics. Assume, moreover, that its characteristic variety V satisfies the following conditions:*

(i) $V \cap \overline{V}$ is a non-singular involutory manifold;

(ii) $V \cap \overline{V}$ intersect transversally;

(iii) $\omega_{|V\cap\overline{V}} \neq 0$.

Then the system $\mathcal{M}$ is microlocally equivalent to the **partial Cauchy-Riemann** *system*

$$\mathcal{N} := \frac{\partial u}{\partial \bar{z}_j} := \frac{1}{2}\left(\frac{\partial}{\partial x_{2j-1}} + i\frac{\partial}{\partial x_{2j}}\right) u = 0, \quad j = 1,\ldots,d,$$

where d is the codimension of V.

Systems of this sort enjoy the same properties one would expect from the (partial) Cauchy-Riemann. One has, in particular, strong propagation of regularity results. To state them (their proofs being essentially obvious after Theorem 6.5.6) we need an extra definition.

Definition 6.5.4 *Let $\mathcal{M}$ be a system of finite order microdifferential operators whose codimension d characteristic variety V satisfies conditions (i), (ii) and (iii) of Theorem 6.5.7. The 2d-dimensional bicharacteristic manifold of $V \cap \overline{V}$ through $(x_0, i\xi_0\infty)$ is called* **virtual bicharacteristic manifold** *of $\mathcal{M}$.*

Remark 6.5.6 Since the notion of bicharacteristic manifold is obviously invariant under contact transformations, then the notion of virtual bicharacteristic manifold is invariant under real contact transformations.

Remark 6.5.7 Let $\mathcal{N}$ be the partial Cauchy-Riemann system of Theorem 6.5.7. Its virtual bicharacteristic manifold through $(x_0, i\xi_0\infty)$ is given by

$$\{(x_0, i\xi\infty) : x_j = (x_0)_j \text{ for } j = 2d+1, \ldots, n; \xi = \xi_0\}.$$

Finally, we can state

Theorem 6.5.8 *Let $\mathcal{M} = \mathcal{E}/\mathcal{I}$ be as in Theorem 6.5.7 and let U be any open set in the virtual bicharacteristic manifold of $\mathcal{M}$. Then every microfunction solution of $\mathcal{M}$ which vanishes in U also vanishes everywhere in the virtual bicharacteristic manifold.*

Finally, the deep work of H. Lewy on linear partial differential equations with no solutions, has stimulated the study of one more important case, i.e. the so called Lewy-Mizohata type systems.

Definition 6.5.5 *Let V be an involutory submanifold of $S^*\mathbb{R}^n$ which, in the neighborhood of a point $(x_0, i\xi_0\infty)$ is written as*

$$\{(x, i\xi\infty) : p_1(x, i\xi) = \ldots = p_d(x, i\xi) = 0\}.$$

Then the **generalized Levi form** *of V is the hermitian matrix whose coefficients are the Poisson brackets*

$$\{p_j(x, \xi), \bar{p}_k(x, \xi)\}_{1 \leq j,k \leq d}.$$

Remark 6.5.8 Note that the signature of the generalized Levi form is independent of the choice of the defining functions p_j, and is also invariant under a real contact transformation.

The structure Theorem, in this case, runs as follows [185]:

Theorem 6.5.9 *Let $\mathcal{M} = \mathcal{E}/\mathcal{I}$ be an $\mathcal{E}$-module defined in a neighborhood of $(x_0, i\xi_0\infty)$ and which satisfies (6.5.10) and (6.5.12). If the generalized Levi form of $V(J)$ has p positive eigenvalues and $d - p$ negative eigenvalues at $(x_0, i\xi_0\infty)$, then $\mathcal{M}$ is microlocally equivalent to the $(p, 1 - p)$-***Lewy-Mizohata*** system*

$$\mathcal{N}_p := \begin{cases} \left(\dfrac{\partial}{\partial x_j} - ix_j\dfrac{\partial}{\partial x_n}\right)u = 0 & j = 1,\dots,p \\[3mm] \left(\dfrac{\partial}{\partial x_j} + ix_j\dfrac{\partial}{\partial x_n}\right)u = 0 & j = p+1,\dots,d. \end{cases}$$

The great interest of these results can only be appreciated in view of the following structure theorem for general systems [185]:

Theorem 6.5.10 *Let $\mathcal{M}$ be an admissible and regular system of microdifferential equations such that $V \cap \overline{V}$ is regular, $T_x(V) \cap T_x(\overline{V}) = T_x(V \cap \overline{V})$ for any $x \in V$, and such that its generalized Levi form is of constant signature (p,q). Then $\mathcal{M}$ is microlocally isomorphic to a direct summand of the direct sum of a finite number of copies of the system $\mathcal{N}$ which, in a suitable neighborhood, has the following form:*

$$\mathcal{N} := \begin{cases} \dfrac{\partial}{\partial x_j}ju = 0 & j = 1, \mathcal{D}d, r \\[3mm] \left(\dfrac{\partial}{\partial x_{r+2k-1}} + i\dfrac{\partial}{\partial x_{r+2k}}\right)u = 0 & k = 1,\dots,s \\[3mm] \left(\dfrac{\partial}{\partial x_{r+2s+l}} + ix_{r+2s+l}\dfrac{\partial}{\partial x_n}\right)u = 0 & l = 1,\dots,q \\[3mm] \left(\dfrac{\partial}{\partial x_{r+2s+l}} - ix_{r+2s+l}\dfrac{\partial}{\partial x_n}\right)u = 0 & l = q+1,\dots,p+q \end{cases}$$

where $r = 2\ \mathrm{codim}\ (V) - \mathrm{codim}\ (V \cap \overline{V})$ and $s = \mathrm{codim}\ (V \cap \overline{V}) - \mathrm{codim}\ (V) - (p+q)$.

6.6   Historical Notes

Most of the results described in this last Chapter are so recent that it is quite difficult to provide a meaningful historical appendix. As we have seen in section 6.1, Sato's Fundamental Theorem is, from a philosophical point of view, a far reaching generalization (at least in some sense) of Ehrenpreis' Fundamental Principle. As it stands now, and in the framework of hyperfunctions, it is also a very elegant result which, in a way, had been anticipated by some deep results of, e.g., Mizohata [166], [167] and Hörmander [80]. As for the techniques employed in its proof, they really consist in little more than a microlocalization of John's work on elliptic differential equations, [100], [101]. In section 6.2, on the other hand, we dealt with the study of the Cauchy problem and of the phenomenon of propagation of singularities along bicharacteristics. The first works of Sato in this direction, [200], had appeared before the birth of microfunctions, so the notations (and also some basic concepts) were quite different. An interesting comment on the psychological status at those times is given in [185]. In section 6.2 we have deliberately avoided the treatment of those operators for which it is natural to study the Cauchy problem, namely the hyperbolic operators (whose microlocal analysis is essentially due to Kawai [135]) and, more generally, the micro-hyperbolic operators, introduced by Kashiwara and Kawai in [118] and [119]. The fact that bicharacteristic strips should be the main carriers of singularities has, on the other hand, slowly developed from the works of F. John [101], to reach its fully maturity with the creation of microfunctions. Section 4 has dealt with contact geometry which in its modern form was probably originated by Jacobi. The evolution of different geometries, according to the different needs of mechanics is beautifully described in Arnold's treaty on Classical Mechanics. This section, however, also contains the very important process of quantization of a contact transformation. It is usually accepted that the first to push such an idea was Maslov [163], essentially a physicist, whose work was made famous by Egorov in [45]. Section 5, finally, has dealt with the so called structure theorems. The material we have described is taken from [185], [206]. There is an interesting difference between the treatment given in [185] (where operators on $I\!R^n$, or on real manifolds, are defined as restrictions of operators defined on C^n, or on complex manifolds) and the one given in [206] (which we have mostly followed, and where microdifferential operators are defined directly on $S^* I\!R^n$). As for the partial de Rham system, the simplest case (i.e. a single equation) was originally dealt with by Kawai. Of course, its corollary on the propagation of singularities along the characteristics was of great interest even in the C^∞ case (where a complete solution had to wait until the late sixties). The case of the partial Cauchy-Riemann system was (again for $d = 1$) treated by Kawai.

The treatment of the Lewy-Mizohata is obviously influenced by the striking discovery of Lewy [148] in 1957, as well as by the later works of Mizohata [165], [166], and [167], who first understood the role of equations such as

$$\left(\frac{\partial}{\partial x} \pm ix\frac{\partial}{\partial y}\right) u = 0.$$

Finally, one should point out that according to Hitotumatu, a conjecture on the general structure theorem was formulated by Bers in 1956 during the symposium on analytic functions at Princeton. Bers' conjecture apparently considered de Rham and Cauchy-Riemann systems, but failed to take into account the Lewy-Mizohata system which would have become well known only one year later.

Bibliography

[1] W.W. Adams, C.A. Berenstein, P. Loustaunau, I. Sabadini and D.C. Struppa, Regular functions of several quaternionic variables and the Cauchy-Fueter complex, to appear in J. Geom. Anal., 1997.

[2] W.W. Adams, P. Loustaunau and D.C. Struppa, Projective dimension of certain modules over polynomial rings with applications to the theory of regular functions, preprint, Ann. Inst. Fourier 47 (1997), 623-640.

[3] W.W. Adams, P. Loustaunau and D.C. Struppa, Applications of commutative and computational algebra to partial differential equations, Proceedings International Conference on Computational and Symbolic Mathematics, 1996.

[4] K.G. Andersson, Propagation of analyticity of solutions of partial differential equations with constant coefficients, Ark. Mat. 8 (1971), 277-302.

[5] A. Andreotti and H. Grauert, Theoremes de finitude pour la cohomologie des espaces complexes, Bull. Soc. Math. France 90 (1962), 193-259.

[6] A. Andreotti and M. Nacinovich, Complexes of Partial Differential Operators, *Ann. Scuola Norm. Sup. Pisa Cl. Sci.* (1976), 553-621.

[7] M. Artin, Algebra, Prentice Hall, 1991.

[8] C. Banica and O. Stanasila, *Algebraic Methods in the Global Theory of Complex Spaces*, John Wiley and Sons, 1976.

[9] A.A. Beilinson, J. Bernstein and P. Deligne, Faisceaux Pervers, *Analysis and topology on singular spaces*, *I*, 5-171, Astérisque 100, Soc. Math. France, Paris, 1982.

[10] A.A. Beilinson and J. Bernstein, Localisation of g-modules, C.R. Acad. Sci. Paris Ser. I 292 (1981), 15-18.

[11] C.A. Berenstein, T. Kawai and D.C. Struppa, Interpolating varieties and the Fabry-Ehrenpreis-Kawai gap theorem, Adv. Math. 122 (1996), no. 2, 280-310.

[12] C.A. Berenstein, T. Kawai, Y. Takei and D.C. Struppa, Exponential representation of a holomorphic solution of a system of differential equations associated with the theta-zerovalue, *Structure of Solutions of Differential Equations*, M. Morimoto and T. Kawai Eds., World Scientific Publ., River Edge, NJ (1996), 89-102.

[13] C.A. Berenstein and D.C. Struppa, On the Fabry-Ehrenpreis-Kawai gap theorem, Publ. Res. Inst. Math. Sci. 23 (1987), no. 3, 565-574.

[14] C.A. Berenstein and D.C. Struppa, Dirichlet series and convolution equations, Publ. Res. Inst. Math. Sci. 24 (1988), no. 5, 783-810.

[15] I.N. Bernstein, Modules over a ring of differential operators. An investigation of the fundamental solutions of equations with constant coefficients, Funkcional. Anal. i Prilozen, 5 (1971), no. 2, 1-16.

[16] I.N. Bernstein, The analytic continuation of generalized functions with respect to a parameter, Funct. Anal. and its Appl. 6 (1972), 272-285.

[17] G. Björck, Beurling distributions and linear partial differential equations, *Symposia Mathematica, Vol. VII*, Academic Press, London (1971), 367-379.

[18] J.E. Björk, Rings of differential operators, *North-Holland Mathematical Library, 21*, North-Holland Publishing Co., Amsterdam-NY, 1979.

[19] J.E. Björk, Analytic $\mathcal{D}$-Modules and Applications, *Mathematics and its Applications*, 247, Kluwer Academic Publishers Group, Dordrecht, 1993.

[20] S. Bochner, Partial differential equations and analytic continuation, Proc. Nat. Acad. Sci., 38 (1952), 227-230.

[21] S. Bochner and W.T. Martin, *Several Complex Variables*, Princeton Mathematical Series, vol. 10. Princeton University Press, Princeton, NJ, 1948.

[22] N.N. Bogoljubov and D.V. Shirkov, *Introduction to the theory of quantized fields*. Revised and enlarged by the authors. Translated from the Russian by G.M. Volkoff. Interscience Monographs in Physics and Astronomy, Vol. III Interscience Publishers, Inc., New York, Interscience Publishers Ltd., London, 1959.

[23] J.M. Bony, Équivalence des diverses notions de spectre singulier analytique, Séminaire Goulaouic-Schwartz (1976-1977), *Équations aux dérivées partielles et analyse fonctionnette*, Exp. No. 3, 12, Centre Math., École Polytech., Palaiseau, 1977.

[24] A. Borel et al., Sheaf theoretic intersection cohomology, *Intersection Cohomology*, 47-182, Progr. Math., 50, Birkhäuser Boston, Boston, MA, 1984.

[25] A. Borel et al., *Algebraic D-Modules*, Academic Press, 1994.

[26] L. Boutet de Monvel, Operateurs pseudo-differentiels analytiques et operateurs d'ordre infini, Ann. Inst. Fourier 22 (1972), 229-268.

[27] L. Boutet de Monvel and P. Kree, Pseudo-differential operators and Gevrey classes, Ann. Inst. Fourier 27 (1967), 295-323.

[28] H.J. Bremermann, R. Oehme and J.G. Taylor, Proof of dispersion relations in quantized field theory, Phys. Rev. (2) 109 (1958), 2187-2190.

[29] J. Bros, H. Epstein, V. Glaser and R. Stora, Quelques aspects globaux des problèmes d'edge-of-the-wedge, *Hyperfunctions and theoretical physics*, 185-218, Lecture Notes in Math, Vol. 449, Springer, Berlin, 1975.

[30] J. Bros and D. Iagolnitzer, Causality and local analyticity: mathematical study, Ann. Inst. Henri Poincare 18 (1973), 147-184.

[31] F.E. Browder, On the "edge of the wedge" theorem, Canad. J. Math 15 (1963), 125-131.

[32] O.I. Brylinski and M. Kashiwara, Kazdan-Lusztig conjecture and holonomic systems, Inv. Math. 64 (1981), 387-410.

[33] P. Berthelot, D-modules coherents, I, Operateurs differentiels de niveau fini, Institut de Recherche Mathematique de Rennes, prepublication, 1993.

[34] T. Carleman, *L'integrale de Fourier et questions qui s'y rattachent*, Institut Mittag-Leffler, Uppsala, 1944.

[35] H. Cartan and S. Eilenberg, *Homological Algebra*, Princeton University Press, Princeton, NJ, 1956.

[36] A. Cerezo, J. Chazarain and A. Piriou, Introduction aux hyperfonctions, in Springer LNM 449 (1975), 1-53.

[37] C.C. Chea, Microfunctions for sheaves of holomorphic functions with growth conditions, Ph.D. dissertation, University of Maryland, 1994.

[38] P. Deligne, *Equations differentielles a points singuliers reguliers*, Springer LNM 163, 1970.

[39] J. Dieudonnè and L. Schwartz, La dualitè dans les espaces (F) et (LF), Ann. Inst. Fourier 1 (1949), 61-101.

[40] P.A.M. Dirac, *The Principles of Quantum Mechanics*, Oxford, 1930.

[41] J.J. Duistermaat, *Fourier Integral Operators*, Courant Institute of Mathematical Sciences, New York, 1973.

[42] J.J. Duistermaat and L. Hormander, Fourier Integral Operators II, Acta Math. 128 (1972), no. 3-4, 183-269.

[43] F.J. Dyson, Connection between local commutativity and regularity of Wightman functions, Phys. Rev. (2) 110 (1958), 579-581.

[44] R.J. Eden, P.V. Landshoff, D.I. Olive and J.C. Polkinghorne,*The Analytic S-Matrix*, Cambridge University Press, 1966.

[45] Yu. V. Egorov, On canonical transformations of pseudodifferential operators, Uspehi Mat. Nauk 24 (1969), 235-236.

[46] Yu. V. Egorov, Conditions for the solvability of pseudo-differential operators, Dokl. Akad. Nauk SSSR 187 (1969), 1232-1234.

[47] L. Ehrenpreis, Sheaves and differential equations, Proc. A.M.S. 7 (1956), 1131-1138.

[48] L. Ehrenpreis, Some applications of the theory of distributions to several complex variables, Seminar on Analytic Functions, Princeton, 1957, 65-79.

[49] L. Ehrenpreis, Theory of infinite derivatives, Amer. J. Math. 81 (1959), 799-845.

[50] L. Ehrenpreis, Solutions of some problems of division IV, Amer. J. Math. 82 (1960), 522-588.

[51] L. Ehrenpreis, A fundamental principle for systems of linear differential equations with constant coefficients and some of its applications, Proc. Intern. Symp. on Linear Spaces, Jerusalem 1961, 161-174.

[52] L. Ehrenpreis, *Fourier Analysis in Several Complex Variables*, Wiley Interscience, New York, 1970.

[53] H. Epstein, Generalization of the "edge of the wedge" theorem, J. Math. Phys. 1 (1960), 524-531.

[54] A. Fabiano, G. Gentili and D.C. Struppa, Sheaves of quaternionic hyperfunctions and microfunctions, Compl. Var. Theory and Appl. 24 (1994), no. 3-4, 161-184.

[55] L. Fantappiè, Le funzionali analitiche e le loro singolarita', Rend. Acc. Lincei (6) 1 (1925), 502-508.

[56] L. Fantappiè, Nuovi Fondamenti della teoria dei funzionali analitici, Mem. Acc. d'Italia 12 (1941).

[57] L. Fantappiè, *Teoria de los funcionales analiticos y sus aplicaciones*, Barcelona, 1943.

[58] L. Fantappiè, Su un'espressione generale dei funzionali lineari mediante le funzioni "para-analitiche" di piu' variabili, Rend. Sem. Mat. Padova 22 (1953), 1-10.

[59] L. Fantappiè, *Opere Scelte*, U.M.I., Rozzano, 1973.

[60] L. Fantappiè, L'indicatrice proiettiva dei funzionali lineari e i prodotti funzionali lineari, Ann. di Mat. 22 (1943), 181.

[61] Fourier, *Theorie Analytique de la Chaleur*, Paris, 1822.

[62] O. Gabber, The integrability of the characteristic variety, Am. J. Math., 103 (1981), no. 3, 445-468.

[63] S.I. Gelfand and Yu. I. Manin, *Methods of Homological Algebra*, Springer Verlag, 1996.

[64] R. Godement, *Topologie Algebrique et Theorie des Faisceaux*, Hermann, Paris, 1958.

[65] M. Goresky and R. MacPherson, Intersection homology theory, II, Inv. Math. 72 (1983), no. 1, 77-129.

[66] H. Grauert, On Levi's problem and the embedding of real analytic manifolds, Ann. Math. 68 (1958), 460-472.

[67] P. Griffiths and J. Harris, *Principles of Algebraic Geometry*, John Wiley & Sons, 1978.

[68] A. Grothendieck, Éléments de géométrie algébrique. II. Étude cohomologique des faisceaux cohérents. I. Inst. Hautes Études Sci. Publ. Math. No. 11 (1961), 167.

[69] A. Grothendieck, Sur certain espaces de fonctions holomorphes, I, J. Reine Angew. Math. 192 (1953), 34-64.

[70] A. Grothendieck, *Local Cohomology*, Springer Lecture Notes in Mathematics 41, Berlin-Heidelberg-New York, 1967 (original notes appeared at Harvard University, 1961).

[71] R. Gunning, *Introduction to Holomorphic Functions of Several Variables*, Wadsworth & Brooks/Cole, 1990.

[72] H. Hahn, Uber eine Verallgemeinerung der Fourierschen Integral- formel, Acta Math. 49 (1926), 301-353.

[73] R. Hartshorne, *Residues and Duality*, Springer Lecture Notes in Mathematics 20, Springer Verlag, Berlin Heidelberg New York, 1966.

[74] R. Hartshorne, *Algebraic Geometry*, GTM 52, Springer Verlag, 1977.

[75] R. Harvey, Hyperfunctions and partial differential equations, Proc. Nat. Acad. Sci. USA 55 (1966), 1042-1046.

[76] O. Heaviside, On operators in physical mathematics, Proc. Roy. Soc. London 52 (1893), 504-529.

[77] O. Heaviside, *Electromagnetic Theory*, vol. I, II, London 1897, 1899.

[78] O. Heaviside, *Electromagnetic Theory*, vol. III, London, 1912.

[79] J. Hilgevoord, *Dispersion relations and causal description: An introduction to dispersion relations in field thoery*, Thesis, Univ. of Amsterdam. Series in Physics North-Holland Publishing Co., Amsterdam, 1960.

[80] L. Hörmander, Hypoelliptic differential operators, Ann. Inst. Fourier 11 (1961), 477-492.

[81] L. Hörmander, Pseudo-differential operators, Comm. Pure Appl. Math. 18 (1965), 501-517.

[82] L. Hörmander, L^2 estimates and existence theorems for the $\bar{\partial}$ operator, Acta Math. 113 (1965), 89-152.

[83] L. Hörmander, *An Introduction to Complex Analysis in Several Variables*, Van Nostrand, Princeton, 1966.

[84] L. Hörmander, Fourier Integral Operators I, Acta Math. 127 (1971), no. 1-2, 79-183.

[85] L. Hörmander, Uniqueness theorems and wave front sets for solutions of linear differential equations with analytic coefficients, Comm. Pure Appl. Math. 24 (1971), 671-704.

[86] L. Hörmander, A remark on Holmgren's uniqueness theorem, J. Diff. Geom. 6 (1971), 129-134.

[87] L. Hörmander, On the existence and the regularity of solutions of linear pseudo-differential equations, Enseign. Math. 17 (1971), 99-163.

[88] L. Hörmander, Spectral analysis of singularitites, "Seminar on Singularities of Solutions of Linear Partial Differential Equations" (L. Hormander ed.), Ann. of Math. Studies, Princeton, NJ, 1979, 3-50.

[89] L. Hörmander, *The Analysis of Linear Partial Differential Operators I, II, III, IV*, Springer Grundlehren 256, 257, 274, 275, Berlin 1983, 1983, 1985, 1985.

[90] R. Hotta, *Introduction to D-modules*, Madras, India, 1987.

[91] P. Hilton and U. Stammbach, *A Course in Homological Algebra*, Springer, 1970.

[92] D. Iagolnitzer, *Introduction to S-Matrix Theory*, Paris, 1973.

[93] D. Iagolnitzer, Macrocausality, physical-region analyticity and independence property in S-matrix theory, *Hyperfunctions and theoretical physics*, 102-120, Lecture Notes in Math., Vol. 449, Springer, Berlin, 1975.

[94] D. Iagolnitzer, Appendix: Microlocal essential support of a distribution and decomposition theorems - an introduction, *Hyperfunctions and theoretical physics*, 121-132, Lecture Notes in Math., Vol. 449, Springer, Berlin, 1975.

[95] D. Iagolnitzer and H. Stapp, Macroscopic causality and physical region analyticity in S-matrix theory, Comm. Math. Phys. 14 (1969).

[96] L. Illusie, Crystalline cohomologie, *Motives*, 43-70, Proc. Sympos. Pure Math., 55, Part 1, Amer. Math. Soc., Providence, RI, 1994.

[97] S. Iyanaga, Three personal reminiscences, "Algebraic Analysis: Papers Dedicated to Professor Mikio Sato on the Occasion of His Sixtieth Birthday", M.Kashiwara and T.Kawai eds., pp.9-12, Academic Press, Tokyo, 1988.

[98] U. Jannsen et al., *Motives*, Proc. Symposia in Pure Math., Vol. 55, Part 1, A.M.S., 1994.

[99] R. Jost, *The Generalized Theory of Quantized Fields*, AMS, Providence, Rhode Island, 1956.

[100] F. John, The fundamental solution for linear elliptic differential equations wiht analytic coefficients, Comm. Pure Appl. Math. 3 (1950), 273-304.

[101] F. John, *Plane Waves and Spherical Means Applied to Partial Differential Equations*, Interscience Publishers, New York, 1955.

0BIBLIOGRAPHY

[102] A. Kaneko, Fundamental principle and extension of solutions of linear differential equations with constant coefficients, in Springer LNM 187 (1971), 122-134.

[103] A. Kaneko, *Introduction to Hyperfunctions*, Kluwer, 1988 (English translation of original Japanese version published by University of Tokyo Pres, 1980).

[104] M. Kashiwara, Duality no ippanron to derived category (General theory of dualities and derived category), Sugaku Shinkokai Seminar Report, 1969 (in Japanese).

[105] M. Kashiwara, Algebraic foundation of the theory of hyperfunctions, Surikaiseki-kenkyusho Kokyuroku 108 (1969), 58-71 (in Japanese).

[106] M. Kashiwara, On the structure of hyperfunctions (after M. Sato), Sugaku no Ayumi 15 (1970), 9-72 (in Japanese).

[107] M. Kashiwara, On the flabbiness of the sheaf C and the Radon transformation, Surikaiseki-kenkyusho Kokyuroku 114 (1971), 1-4 (in Japanese).

[108] M. Kashiwara, Algebraic study of systems of partial differential equations, Master's thesis, Univ. of Tokyo, 1971 (English translation available by G.Kato).

[109] M. Kashiwara, Chokansuron no kojinteki tenbo (Personal view on hyperfunction theory), Sugaku no ayumi, 16-1 (1971), 108-112 (in Japanese).

[110] M. Kashiwara, On the maximally overdetermined systems of linear differential equations I, Publ. RIMS Kyoto Univ. 10 (1975), 563-579.

[111] M. Kashiwara, B-functions and holonomic systems. Rationality of roots of B-functions, Invent. Math. 38 (1976-1977), no. 1, 33-53.

[112] M. Kashiwara, B-functions and holonomic systems II, Inv. Math. 38 (1976), no. 2, 121-134.

[113] M. Kashiwara, On the holonomic systems of linear differential equations, II, Inv. Math. 49 (1978), 121-135.

[114] M. Kashiwara, Faisceaux constructibles et systèmes holonomes d'équations aux dérivées partielles linéaires à points singuliers réguliers, Séminaire Goulaouic-Schwartz, 1979-1980, expose 19, 7, École Polytech., Palaiseau, 1980.

[115] M. Kashiwara, *Systems of Microdifferential Equations*, Birkhauser, 1983.

[116] M. Kashiwara, The Riemann-Hilbert problem for holonomic systems, Publ. RIMS Kyoto, 20 (1984), 319-365.

[117] M. Kashiwara and T. Kawai, Pseudo-differential operators in the theory of hyperfunctions, Proc. Japan Acad. 46 (1970), 1130-1134.

[118] M. Kashiwara and T. Kawai, Micro-hyperbolic pseudo-differential operators, I, J. Math. Soc. Japan 27 (1975), no. 3, 359-404.

[119] M. Kashiwara and T. Kawai, Micro-hyperbolic pseudo-differential operators, in Springer LNM 449 (1975), 70-82.

[120] M. Kashiwara and T. Kawai, On holonomic systems of microdifferential equations, III, Publ. RIMS Kyoto Univ. 17 (1981), 813-979.

[121] M. Kashiwara and T. Kawai, Microlocal Analysis, Publ. RIMS Kyoto Univ. 19 (1983), no. 3, 1003-1032.

[122] M. Kashiwara and T. Kawai, Introduction to "Algebraic Analysis: Papers Dedicated to Professor Mikio Sato on the Occasion of His Sixtieth Birthday", M. Kashiwara and T. Kawai eds., Academic Press, Tokyo (1988), 1-8.

[123] M. Kashiwara, T. Kawai and T. Kimura, *Foundations of Algebraic Analysis*. Translated from the Japanese by Goro Kato. Princeton Mathematical Series, 37, Princeton University Press, Princeton, NJ, 1986.

[124] M. Kashiwara and P. Schapira, Micro-hyperbolic systems, Acta Math., no. 1-2, (1979).

[125] M. Kashiwara and P. Schapira, Microlocal study of sheaves, Asterisque 128 (1985).

[126] M. Kashiwara and P. Schapira, *Sheaves on Manifolds*, Springer Verlag, 1994.

[127] K. Kataoka, On the theory of Radon transformations of hyperfunctions, J. Fac. Sci. Univ. Tokyo 28 (1981), no. 2, 331-413.

[128] N. Katz, An overview of Deligne's work on Hilbert's 21st problem, *Mathematical developments arising from Hilbert problems*, 537-557, Amer. Math. Soc., Providence, RI, 1976.

[129] N. Katz and T. Oda, On the differentiation of the de Rham cohomology classes with respect to parameters, J. Kyoto Univ. 8 (1969), 119-213.

[130] T. Kawai, On the theory of Fourier hyperfunctions and its applications to partial differential equations with constant coefficients, J. Fac. Sci. Univ. Tokyo 17 (1970), 467-517.

[131] T. Kawai, On the global existence of real analytic solutions of linear differential equations, in Springer LNM 287 (1971), 99-121.

[132] T. Kawai, Construction of local elementary solutions for linear partial differential operators with real analytic coefficients I, the case with real principal symbols, Publ. RIMS 7 (1971), 363-397.

[133] T. Kawai, Construction of local elementary solutions for linear partial differential operators with real analytic coefficients II, the case with complex principal symbols, Publ. RIMS 7 (1971), 399-426.

[134] T. Kawai, A survey of the theory of linear (pseudo-) differential equations from the view point of phase functions - existence, regularity, effect of boundary conditions, transformations of operators, etc. Reports of the Symposium on the Theory of Hyperfunctions and Differential Equations, RIMS Kyoto Univ. (1971), 84-92 (in Japanese).

[135] T. Kawai, Pseudo-differential operators acting on the sheaf of microfunctions, in Springer LNM 449 (1975), 54-69.

[136] T. Kawai and H.P. Stapp, Microlocal study of the S-matrix singularity structure, *International Symposium on Mathematical Problems in Theoretical Physics*, 38-48, Lecture Notes in Physics, 39, Springer, Berlin, 1975.

[137] T. Kawai and D.C. Struppa, On the existence of holomorphic solutions of systems of linear differential equations of infinite order and with constant coefficients, Intern. J. Math. 1 (1990), 63-82,

[138] G. Köthe, Die Randverteilungen analytischer Funktionen, Math. Z. 57 (1952), 13-33.

[139] G. Köthe, Dualitat in der Funktionentheorie, J. Reine Angew. Math. 181 (1953), 30-49.

[140] H. Komatsu, Resoulution by hyperfunctions of sheaves of solutions of differential equations with constant coefficients, Math. Ann. 176 (1968), 77-86.

[141] H. Komatsu (ed.), Proceedings of the Katata Conference on "Hyperfunctions and Pseudo-Differential Equations", Springer Lecture Notes in Mathematics 287, Berlin-Heidelberg-New York, 1971.

[142] H. Komatsu, An introduction to the theory of hyperfunctions, in Springer LNM 287 (1971), 3-40.

[143] H. Komatsu, Hyperfunctions and linear partial differential equations, in Springer LNM 287 (1971), 180-191.

[144] H. Komatsu, Relative cohomology of sheaves of solutions of differential equations, in Springer LNM 287 (1971), 192-263.

[145] H. Komatsu and T. Kawai, Boundary values of hyperfunction solutions of linear partial differential equations, Publ. RIMS 7 (1971), 95-104.

[146] A. Kolm and B. Nagel, A generalized edge of the wedge theorem, Comm. Math. Phys. 8 (1968), 185-203.

[147] D.T. Le and Z. Mebkhout, Introduction to linear differential systems, *Singularities Part 2*, 31-63, Proc. Sympos. Pure Math, 40, Amer. Math. Soc., Providence, RI, 1983.

[148] H. Lewy, An example of a smooth linear partial differential equation without solution, Ann. of Math. 66 (1957), 155-158.

[149] O. Liess, Intersection properties of weak analytically uniform classes of functions, Ark. for Mat. 14 (1976), no. 1, 93-111.

[150] S. Lubkin, *Cohomology of Completions*, North-Holland Mathematics Studies, 42. Notas de Matemática, 71, North-Holland Publishing Co., Amsterdam-NY, 1980.

[151] J. Lutzen, *The Prehistory of the Theory of Distributions*, Springer Verlag, Berlin-Heidelberg-New York, 1972.

[152] B. Malgrange, Existence et approximation des solutions des equations aux derivees partielles et des equations de convolution, Ann. Inst. Fourier 6 (1955-56), 271-355.

[153] B. Malgrange, Faisceaux sur des varietes analytiques reelles, Bull. Soc. Math. France 85 (1957), 231-237.

[154] B. Malgrange, Sur les systemes differentiels a coefficients constants, Sem. Leray 8, 8a (1961-62), College de France.

[155] B. Malgrange, Systèmes differéntiels à coefficients constants, *Séminaire Bourbaki, Vol. 8* Exp. No. 246, 79-89, Soc. Math. France, Paris, 1995.

[156] A. Martineau, Les hyperfonctions de M.Sato, Séminaire Bourbaki 214 (1960), 13.

[157] A. Martineau, Indicatrices des fonctions analytiques et inversion de la transformation de Fourier-Borel par la transformation de Laplace, C. R. Acad. Sc. Paris 255 (1962), 1845-1847.

[158] A. Martineau, Sur les fonctionnelles analytiques et la transformation de Fourier-Borel, J. An. Math. 11 (1963), 1-164.

[159] A. Martineau, Distributions et valeurs au bord des fonctions holomorphes, Proc. Intern. Summer Course on the theory of Distributions, Lisbon, 1964, 195-326.

[160] A. Martineau, Sur la topologie des espaces de fonctions holomorphes, Math. Ann. 163 (1966), 62-88.

[161] A. Martineau, Theoremes sur le prolongement analytique du type "Edge of the Wedge Theorem", Sem. Bourbaki, 20-ieme annee, No. 340, 1967/68.

[162] A. Martineau, Le "edge of the wedge theorem" en theorie des hyperfonctions de Sato, Proc. Intern. Conf. on Functional Analysis, Tokyo, 1969, Univ. Tokyo Press, 1970, 95-106.

[163] V.P. Maslov, *Theorie des Perturbations et Methodes Asymptotiques*, Dunod, Paris, 1972 (French translation of the Russian original published in Moscow, 1965).

[164] Z. Mebkhaut, *Théoremes de dualité pour les $\mathcal{D}_X$-modules cohérents*, Note aux C.R. Acad. Sci. Sér A-B 287 (1977), 785-787.

[165] S. Mizohata, Analyticity of solutions of hyperbolic systems with analytic coefficients, Comm. Pure Appl. Math. 14 (1961), 547-559.

[166] S. Mizohata, Some remarks on the Cauchy problem, J. Math. Kyoto Univ. 1 (1961), 109-127.

[167] S. Mizohata, Solutions nulles et solutions non analytiques, J. Math. Kyoto Univ. 1 (1961), 271-302.

[168] M. Morimoto, Une remarque sur un theoreme de "edge of the wedge" de A. Martineau, Proc. Japan Acad. 45 (1969), 446-448.

[169] M. Morimoto, Edge of the wedge theorem and hyperfunctions, in Springer LNM 287 (1971), 41-81.

[170] M. Morimoto, Support et support singulier de l'hyperfonction, Proc. Japan Acad. 47 (1971), 648-652.

[171] L. Nirenberg, Pseudo-differential operators, 1970 *Global Analysis*, 149-167, Amer. Math. Soc., Providence, RI, 1970.

[172] M. Noumi, Constructible C-kagun to holonomic $\mathcal{D}$-kugun (Constructible C-modules and holonomic $\mathcal{D}$-modules), Sugaku 36 (1984), no. 2, 125-136 (in Japanese).

[173] T. Oda, Introduction to algebraic analysis on complex manifolds, *Algebraic varieties and analytic varieties*, 29-48, Adv. Stud. Pure Math., 1, North-Holland, Amsterdam-NY, 1983.

[174] Y. Ohyama, $\mathcal{D}$-kagun nyumon, I (Introduction to the theory of $\mathcal{D}$-modules, I), RIMS Kokyuroku 67 (1988), 1-98.

[175] P. Painleve, Sur les lignes singulieres des fonctions analytiques, Ann. Fac. Sci. Univ. Toulouse 2 (1938), 26.

[176] V.P. Palamodov, On the general form of the solution of a homogeneous differential equation with constant coefficients, Dokl. Akad. Nauk USSR 137 (1961), 774-777.

[177] V.P. Palamodov, The general theorems on the system of linear equations with constant coefficients, 1963 *Outlines Joint Sympos. Partial Differential Equations*, 206-213, Acad. Sci. USSR Siberian Branch, Moscow, 1963.

[178] V.P. Palamodov, Linear differential operators with constant coefficients. Translated from the Russian by A.A. Brown. Die Grundlehren der mathematischen Wissenschaften, Band 168 Springer-Verlag, New York-Berlin, 1970.

[179] V.P. Palamodov, Systems of linear differential equations, Progress in Math. 10 (1971).

[180] V.P. Palamodov, From hyperfunctions to analytic functionals, Sov. Math. Dokl. 18 (1977), 975-979.

[181] F. Pellegrino, La theorie des fonctionnelles analytiques et ses applications, in Problemes concrets d'analyse fonctionnelle, by P. Levy, Paris, 1951.

[182] I.G. Petrowsky, Sur l'analyticite des solutions des systemes d'equations differentielles, Mat. Sb. 5 (1938), 1-74.

[183] I.G. Petrowsky, On the diffusion of waves and the lacunas for hyperbolic equations, Rec. Math. N.S. 17 (1945), 289-370.

[184] F. Pham (ed.), *Hyperfunctions and Theoretical Physics*, Springer LNM 449 (1975).

[185] F. Pham, *Microanalyticite de la matrice S*, in Springer LNM 449 (1975), 83-101.

[186] S. Pincherle, Memoire sur le Calcul Fonctionnel distributif, Math. Ann.
49 (1897), 325-382.

[187] S. Pincherle and U. Amaldi, *Le operazioni distributive e le loro applicazioni all'analisi*, Zanichelli, Bologna, 1901.

[188] Poisson, Memoire sur la theorie du magnetism, Mem. Acad. Roy. Sci. Paris 5 (1821-22), 247-338.

[189] D.G. Quillen, Formal properties of over-determined systems of linear partial differential equations, Ph.D. thesis Harvard University, 1964.

[190] G.C. Rota, M. Sato: Impressions, in "Algebraic Analysis: Papers Dedicated to Professor Mikio Sato on the Occasion of His Sixtieth Birthday", M. Kashiwara and T. Kawai eds., pp.13-16, Academic Press, Tokyo, 1988.

[191] W. Rudin, Lectures on the edge-of-the-wedge theorem, Conference Board of teh Mathematical Sciences Regional Conference Series in Mathematics, No. 6, AMS, Providence, Rhode Island, 1971.

[192] W. Rudin, *Real and Complex Analysis*, McGraw-Hill, NY-Toronto, Ont.-London, 1966.

[193] I. Sabadini and D.C. Struppa, Topologies on quaternionic hyperfunctions and duality theorems, Compl. Var. Theory and Appl. 30 (1996), no. 1, 19-34.

[194] M. Saito, Induced $\mathcal{D}$-modules and differential complexes, Bull. Soc. Math. France 117 (1989), no. 3, 361-387.

[195] M. Sato, Theory of hyperfunctions, Sugaku, 10 (1958), 1-27 (in Japanese).

[196] M. Sato, On a generalization of the concept of function, Proc. Jap. Acad. 34 (1958), 126-130 and 604-608.

[197] M. Sato, Theory of hyperfunctions, J. Fac. Sci. Univ. Tokyo Sect.I, 8 (1959), 139-193 and 387-436.

[198] M. Sato, Hyperfunctions and partial differential equations, Proc. Intern. Conf. on Functional Analysis and Related Topics, 1969, Univ. Tokyo Press, Tokyo, 1970, 91-94.

[199] M. Sato, Structure of hyperfunctions, Reports of the Katata Symp. on algebraic geometry and hyperfunctions 1969, 4.1-30 (notes by T. Kawai, in Japanese).

[200] M. Sato, Structure of hyperfunctions, Sugaku no Aymi 15 (1970), 9-27 (notes by M. Kashiwara, in Japanese).

[201] M. Sato, Chokansu to so C ni tsuite (On hyperfunctions and the sheaf C), RIMS Kokyuroku 126 (1971), in Japanese.

[202] M. Sato, Recent developments in hyperfunction theory and its applications to physics, Springer LNP 39 (1975), 13-39.

[203] M. Sato, Lectures delivered at Kyoto University in 1984-1985, RIMS Lecture Notes 5 (1989), in Japanese.

[204] M. Sato, Private Communication, 1992.

[205] M. Sato and M. Kashiwara, Structure of hyperfunctions, Sugakuno-Ayumi 15 (1970), 9-71 (in Japanese).

[206] M. Sato, T. Kawai and M. Kashiwara, Microfunctions and pseudo- differential equations, in Hyperfunctions and Pseudo-Differential Equations, in Springer LNM 287 (1971), 265-529.

[207] P. Schapira, *Theorie des Hyperfonctions*, Springer LNM 126 (1970).

[208] P. Schapira, *An Introduction to the Theory of Microdifferential Systems in the Complex Domain*, Springer, 1985.

[209] L. Schwartz, Étude des Sommes d'exponentielles réelles, Actualités Sci. Ind., no. 959 Hermann et Cie., Paris, 1943.

[210] L. Schwartz, Théorie des distributions, Tome I, Actualités Sci. Ind., no. 1091 = Publ. Inst. Math. Univ. Strasbourg 9, Hermann and Cie., Paris, 1950.

[211] J.S. e Silva, As funcoes analiticas e a Analise functional, Port. Math. 9 (1950), 1-130.

[212] J.S. e Silva, Su certe classi di spazi localmente convessi importanti per le applicazioni, Rend. Mat. Roma 14 (1955), 388-410.

[213] B. Stenstrom, *Rings of Quotients*, Springer, 1972.

[214] D.C. Struppa, Cenni sulla nascita della teoria delle iperfunzioni, Dip. Mat. Univ. Milano 9/S, 1983.

[215] D.C. Struppa, Luigi Fantappie' e la teoria dei funzionali analitici, in Proceedings of "La matematica italiana tra le due guerre mondiali", Milano-Gargnano, October 1986, Pitagora, Bologna, 1987, 393-429.

[216] D.C. Struppa, The first eighty years of Hartogs' theorem, Sem. Geom. Dip. Mat. Univ. Bologna, 1987, 127-211.

[217] D.C. Struppa, An extension to Fantappiè's theory of analytic functions, *Geometry and complex variables*, 329-356, Lecture Notes in Pure and Appl. Math., 132, Dekker, New York, 1991.

[218] D.C. Struppa and C. Turrini, Alcuni aspetti della teoria delle iperfunzioni, Dip. Mat. Univ. Milano 13/S, 1984.

[219] D.C. Struppa and C. Turrini, Hyperfunctions and boundary values of holomorphic functions, Nieuw Archief voor Wiskunde, 4 (1986), no. 2, 91-118.

[220] D.C. Struppa and C. Turrini, Pincherle's contribution to the Italian school of analytic functionals, Proc. "Giornate di Storia della Matematica", M. Galuzzi ed., Cetraro 1988, Edit El, Rende, 1990.

[221] H.G. Tillmann, Distributionen als Randverteilungen Analytischer Funktionen II, Math. Z. 76 (1961), 5-21.

[222] H.G. Tillmann, Darstellung der Schwartzchen Distributionen durch Analytische Funktionen, Math. Z. 77 (1961), 106-124.

[223] F. Treves, *Topological Vector Spaces, Distributions and Kernels*, Academic Press, New York, 1967.

[224] F. Treves, *Introduction to Pseudodifferential and Fourier Integral Operators, vol. I, Pseudodifferential Operators*, Plenum Press, New York, 1980.

[225] J.L. Verdier, *Categories derivees (etat 0)*, SGA 4, Lecture Notes in Mathematics 569, Springer Verlag, 1977.

[226] V.S. Vladimirov, On Bogoljubov's "edge of the wedge" theorem, Izv. Akad. Nauk SSSR, Ser. Mat. 26 (1962), 825-838 (in Russian).

[227] V.S. Vladimirov, *Methods of the Theory of Functions of Many Complex Variables*, Nauka, Moscow, 1964. English transl. MIT Press, Cambridge, Mass., 1966.

[228] R.O. Wells Jr., *Differential Analysis on Complex Manifolds*, Prentice Hall, 1973.

[229] N. Wiener, The Fourier integral and certain of its applications. Reprint of the 1933 edition. With a foreword by Jean-Pierre Kahane. *Cambridge Mathematical Library*, Cambridge University Press, Cambridge-NY, 1988.

[230] K. Yosida, Sato, a Perfectionist, in "Algebraic Analysis: Papers Dedicated to Professor Mikio Sato on the Occasion of His Sixtieth Birthday", M. Kashiwara and T. Kawai eds., Academic Press, Tokyo, (1988), 1-8.

Index

Milton Keynes UK
Ingram Content Group UK Ltd.
UKHW020021071024
449327UK00032B/2871